Advances in Delays and Dynamics

Volume 13

Delay systems are largely encountered in modeling propagation and transportation phenomena, population dynamics and representing interactions between interconnected dynamics through material, energy and communication flows. Thought as an open library on delays and dynamics, this series is devoted to publish basic and advanced textbooks, explorative research monographs as well as proceedings volumes focusing on delays from modeling to analysis, optimization, control with a particular emphasis on applications spanning biology, ecology, economy and engineering. Topics covering interactions between delays and modeling (from engineering to biology and economic sciences), control strategies (including also control structure and robustness issues), optimization and computation (including also numerical approaches and related algorithms) by creating links and bridges between fields and areas in a delay setting are particularly encouraged.

Indexed by zbMATH.

Wen Kang · Emilia Fridman

Delay-Robust Control of Distributed Parameter Systems

Springer

Wen Kang
Beijing Institute of Technology
Beijing, China

Emilia Fridman
School of Electrical and Computer Engineering
Tel Aviv University
Tel Aviv, Israel

ISSN 2197-117X ISSN 2197-1161 (electronic)
Advances in Delays and Dynamics
ISBN 978-981-96-6343-9 ISBN 978-981-96-6344-6 (eBook)
https://doi.org/10.1007/978-981-96-6344-6

This Springer imprint is published by the registered company Springer Nature Singapore Pte Ltd.
The registered company address is: 152 Beach Road, #21-01/04 Gateway East, Singapore 189721, Singapore

This work is dedicated to

Yu-ping with tolerance, patience, encouragement, support, and wonderful frame of mind

—Wen Kang

Eugenii, Boris, Paz, and Libby

—Emilia Fridman

Preface

Processes that exhibit significant spatial variations are ubiquitous among various fields, such as engineering, fluid mechanics, biological mathematics, and so on. For these processes, the distinguishing attribute of the control problem is that it involves the regulation of distributed variables using spatially distributed or boundary control actuators and measurement sensors. In general, these processes are described by infinite-dimensional systems-partial differential equations (PDEs). In contrast to ordinary differential equations (ODEs), sometimes no general methodology is applicable for PDEs, which motivates our work. However, it is exciting that it provides an endless stream of opportunities for learning and creativity for researchers.

Control of PDEs has been under development since at least the 1960s. There are two main PDE control settings. One is the distributed control: actuation and sensing are located inside the domain of the PDE system or are distributed everywhere in the domain, and the other is the boundary control: actuation and sensing are only located at the boundary. The present book focuses on the boundary or distributed control of PDEs. There exist a large number of different classes of PDEs involving time and space derivatives, such as reaction-diffusion equation, Korteweg-de Vries (KdV) equation, Korteweg-de Vries-Burgers (KdVB) equation, and Kuramoto-Sivashinsky equations (KSE). The control design and analysis of PDE systems become more challenging due to higher-order spatial derivatives. Note that the PDE systems mentioned above are usually regarded as real-valued systems. However, some types of PDEs, such as Schrödinger equation, are commonly studied as complex-valued systems, which expands our research scope. It should be noticed that most existing results were devoted to one-dimensional single PDE systems, while the present book introduces some significant results of cascaded ODE-PDE systems, cascaded PDE-PDE systems, and high-dimensional PDEs, which are more complicated and challenging. Owing to the complexity of PDE systems, developing boundary or distributed control design for various PDEs definitely belongs to a remarkable and challenging topic.

The present book studies delay-robust control of PDE systems. Time delays may appear in many real-world processes. Sometimes, even a small enough delay may lead to the instability of the systems. Thus, delay is crucial for control of PDEs in theory and practice. Note that the existing Lyapunov-Krasovskii functionals for parabolic PDEs are not applicable to the nonlinear complex PDE systems, such as KdV/KdVB systems and so on. New Lyapunov-Krasovskii functionals are suggested to derive the stability conditions, and an upper bound on the delay that preserves the stability that can be found from linear matrix inequalities.

The present book studies also sampled-data control of PDE systems via a time-delay approach. In practical applications, communication network media brings great advantages, such as low cost, reduced weight, simple installation/maintenance, and long-distance control. It is important to provide a stability and performance certificate that takes into account network imperfections, such as variable sampling intervals, communication delays, and control constraints. Furthermore, event-triggered control for PDE systems is considered to reduce the workload of the communication network. To avoid the adverse effects of undesired disturbances, disturbance rejection approaches-adaptive disturbance rejection control (ADRC) and disturbance observer-based control (DOBC) are developed for PDE systems. Via a combination of time-delay approaches to network-based control with Lyapunov-Krasovskii functionals, sufficient stability conditions are established for various PDE systems.

The book presents the reader some classical results to recent developments on Lyapunov-based analysis and design with applications to the hot topics of sampled-data delay-robust control for PDE systems. It should be of interest for researchers working in the field, for graduate students in engineering and applied mathematics, and for practicing engineers. It may also be used as a textbook for a graduate course on Control Theory. The sufficient background to follow most of the material are the undergraduate courses in mathematics and an introduction to control.

Although most of the results are available in a wide variety of journals, the authors felt that a monograph is required. The purpose of this book is to act as a single source from which researchers and students could obtain core material on the topic.

The book is organized into five chapters. Chapter 1 provides a literature review on boundary/distributed control of PDEs along with basic inequalities and introduction to Lyapunov methods for time-delay ODE systems. Chapter 2 is devoted to constrained control of PDE systems: delayed nonlinear Schrödinger equation in Sect. 2.1 and delayed ODE-heat system in Sect. 2.2. Chapter 3 presents sampled data and delayed control of PDE systems: KSE in Sects. 3.1–3.2, parabolic PDEs with spatially scheduled actuators and sensors in Sect. 3.3, cascaded heat-heat system in Sect. 3.4 and KdVB equation in Sect. 3.5. Chapter 4 addresses distributed event-triggered control of PDEs. Chapter 5 discusses disturbance rejection approaches of PDE systems.

We would like to strongly encourage readers to send us suggestions, comments, and corrections by e-mail (kangwen@amss.ac.cn, emilia@tauex.tau.ac.il).

We wish to acknowledge our great debt to many colleagues and students who helped us in writing this book. We extend our thanks to Bao-Zhu Guo at Chinese Academy of Science and Jun-Min Wang at Beijing Institute of Technology for their many productive discussions on research ideas. Particularly, discussions on well-posedness with Bao-Zhu Guo were very useful. Of the former students and colleagues with whom we have worked on problems directly pertinent to this book, it is a pleasure to acknowledge: Anatoly Blighovsky, Nezer Bar Am, Anton Selivanov,

Rami Katz, Yury Orlov, Sabine Mondie, Alexander Seuret. We are deeply grateful to the students Jing Zhang, Er-Xin Pang, Yu-Qian Tang, Qi Wang, and Ming Qu for their meticulous reviews and invaluable suggestions, which greatly improved the content of this book and enhanced the figures. Heartfelt thanks to Yuping Li, Eugenii, and Boris Shustin for moral and emotional support. We would also like to extend our sincere appreciation to Jing Dou from Springer. Our collaboration, which began in 2024, has been immensely productive and has played a crucial role in bringing this book to fruition. In addition, we acknowledge Wiley, Elsevier, and IEEE for granting us permission to reuse materials from our previously published works in this book.

We are grateful to Beijing Institute of Technology and Tel Aviv University for an environment that allowed us to write this book, and to the National Natural Science Foundation of China (NSFC) under Grant 62273046 and Grant 12361141817, Israel Science Foundation (ISF) ISF-NSFC under Grant 3054/23, ISF grant 446/24, Chana and Heinrich Manderman Chair on System Control at Tel Aviv University, Guangdong Basic and Applied Basic Research under Grant 2022A1515010039 and Beijing Nova Program for supporting our research on "Delay-robust control of Distributed Parameter system".

Beijing, China
Tel Aviv, Israel
March 2025

Wen Kang
Emilia Fridman

Contents

About the Authors

Wen Kang received the B.S. degree in Mathematics from Wuhan University, Wuhan, China, in 2009, and the Ph.D. degree in Mathematics from Academy of Mathematics and Systems Science, Chinese Academy of Sciences, Beijing, China, in 2014. She was a Lecturer with the Harbin Institute of Technology, Harbin, China, from 2014 to 2017. From 2015 to 2017, she was a Postdoctoral Researcher with the School of Electrical Engineering, Tel Aviv University, Tel Aviv-Yafo, Israel. She was an Associate Professor in University of Science and Technology Beijing, China. In 2022, she jointed Beijing Institute of Technology as a Research Professor. Her research interests include distributed parameter systems, time-delay systems, multi-agent systems, and ADRC. She serves as Associate Editor for *IMA Journal of Mathematical Control and Information*, *Information and Control*, and *IFAC MICNON*.

Emilia Fridman received the M.Sc. and Ph.D. in Mathematics in Russia. Since 1993 she has been at Tel Aviv University, where she is currently Professor in the Department of Electrical Engineering—Systems. She has held numerous visiting positions in Europe, China, and Australia. Her research interests include time-delay systems, networked control systems, distributed parameter systems, robust control and extremum seeking. She has published more than 200 journal articles and 2 monographs. She serves/served as Senior Editor in *Systems and Control Letters* and as Associate Editor in *Automatica*, *SIAM Journal on Control and Optimization* and

IMA Journal of Mathematical Control and Information. She is IEEE Fellow and was a member of the IFAC Council. In 2014, she was ranked as a Highly Cited Researcher by Thomson ISI. Since 2018, she has been the incumbent for Chana and Heinrich Manderman Chair on System Control at Tel Aviv University. In 2021, she was recipient of IFAC Delay Systems Life Time Achievement Award and of Kadar Award for outstanding research in Tel Aviv University. She is currently IEEE CSS Distinguished Lecturer. In 2023, her monograph *Introduction to Time-Delay Systems: Analysis and Control* (Birkhauser, 2014) was the winner of IFAC Harold Chestnut Control Engineering Textbook Prize.

Acronyms

$\forall$	For all
$\in$	Belong to
i	$i = \sqrt{-1}$ denotes the imaginary unit
$\Re$ (resp. $\Im$)	The real (resp. imaginary) part of a complex number
$\mathbb{R}$	The set of real numbers
$\mathbb{C}$	The set of complex numbers
$\mathbb{R}^+$	The set of non-negative real numbers
$\mathbb{R}^n$	n-dimensional Euclidean space
$\mathbb{R}^{n\times m}$	The set of $n \times m$ real matrices
$\Omega \subset \mathbb{R}^2$	The spatial domain
$\partial\Omega$	The boundary of Ω
$\mathbb{Z}^+$	The set of non-negative integers
$\mathbb{N}$	The set of positive integers
$\lfloor x \rfloor$	The greatest integer less than or equal to x
$\lceil x \rceil$	The smallest integer more than or equal to x
C^2	The space of functions that are twice continuously differentiable
$E\{\cdot\}$	The mathematical expectation
$\lvert \cdot \rvert$	Euclidean norm of a vector
$\lvert\lvert \cdot \rvert\rvert_2$	L_2 norm of a vector
$\lvert\lvert \cdot \rvert\rvert_\infty$	Essential supremum of a vector and the induced infinity norm of a matrix
z_x	The partial derivative with respect to a variable x $z_x = \frac{\partial z}{\partial x}$
$l_N^2(\mathbb{N})$	For $N \in \mathbb{Z}^+$, $l_N^2(\mathbb{N})$ denotes the space satisfying $l^2(\mathbb{N}) = \mathbb{R}^N \times l_N^2(\mathbb{N})$
$L^2(0, L)$	The Hilbert space of square integrable scalar functions $f(x)$ on $(0, L)$ with the corresponding norm $\lVert f \rVert_{L^2(0,L)} = \left[\int_0^L \lvert f(x)\rvert^2 dx\right]^{\frac{1}{2}}$
$\mathcal{L}^2(0, L)$	The complex space of square integrable functions with the norm $\lVert f \rVert_{\mathcal{L}^2(0,L)} = \left[\int_0^L \lvert f(x)\rvert^2 dx\right]^{\frac{1}{2}} = \left[\int_0^L f(x)\overline{f(x)} dx\right]^{\frac{1}{2}}$

$L^2(\Omega)$	The space of measurable squared-integrable functions over Ω with the corresponding norm $\|f\|_{L^2(\Omega)} = \left[\int_\Omega |f(x)|^2 dx\right]^{\frac{1}{2}}$
$(L^2(\Omega))^2$	The Hilbert space of square intergrable vector-valued functions $f = (f_1, f_2)^T$
$L^\infty(0, L)$	$L^\infty(0, L)$ denotes the space of essentially bounded function $f(x)$ on $(0, L)$ with the corresponding norm $\|f\|_{L^\infty(0, L)} = \underset{x\in(0, L)}{esssup} |f(x)|$
$L_n^2(0, L)$	$L_n^2(0, L) = L^2((0, L); R^n)$ stands for the Hilbert space of square interable vector functions $f(x)$ on $(0, L)$ with the norm $\|f\|_{L_n^2(0, L)} = \left[\int_0^L |f(x)|^2 dx\right]^{\frac{1}{2}}$
$C^k(\Omega)$	$C^k(\Omega)$ is the space of functions k times continuously differentiable on ($k \geq 1$ is an integer)
$C^\infty(\Omega)$	$C^\infty(\Omega) = \bigcap_k C^k(\Omega)$
$C_c(\Omega)$	$C_c(\Omega)$ is the space of continuous functions on Ω with compact support in Ω, i.e., which vanish outside some compact set $K \subset \Omega$
$C_c^\infty(\Omega)$	$C_c^\infty(\Omega) = C_c(\Omega) \cap C^\infty(\Omega)$
$H^k(0, L)$	$H^k(0, L) = \{f : f^{(\alpha)} \in L^2(0, L),\ \forall\, 0 \leq \|\alpha\| \leq k\}$ with the norm $\|f\|_{H^k(0, L)} = \{\sum_{0 \leq \|\alpha\| \leq k} \|f^{(\alpha)}\|^2_{L^2(0, L)}\}^{\frac{1}{2}}$
$H_0^k(0, L)$	$H_0^k(0, L) = \{f \in H^k(0, L) \| f(0) = f'(0) = ... = f^{(k-1)}(0) = 0, f(L) = f'(L) = ... = f^{(k-1)}(L) = 0\}$
$H_1^n(0, L)$	$H_1^n(0, L) \triangleq \{f \in L_n^2(0, L),\ f' \in L_n^2(0, L)\}$ with the norm $\|f\|^2_{H_1^n(0,L)} = \|f\|^2_{L_n^2(0,L)} + \|f'\|^2_{L_n^2(0,L)}$
$H^k(\Omega)$	$H^k(\Omega) = \{f : f^{(\alpha)} \in L^2(\Omega),\ \forall\, 0 \leq \|\alpha\| \leq k\}$ with the norm $\|f\|_{H^k(\Omega)} = \{\sum_{0 \leq \|\alpha\| \leq k} \|f^{(\alpha)}\|^2_{L^2(\Omega)}\}^{\frac{1}{2}}$
$H_0^k(\Omega)$	The closure of $C_c^\infty(\Omega)$ in the space $H^k(\Omega)$ with the norm $\|f\|_{H_0^k(\Omega)} = \{\sum_{\|\alpha\| = k} \|f^{(\alpha)}\|^2_{L^2(\Omega)}\}^{\frac{1}{2}}$
$W^{k,p}(\Omega)$	The Sobolev space $\{f : f^{(\alpha)} \in L^p(\Omega), \forall\, 0 \leq \|\alpha\| \leq k\}$ with the norm $\|f\|_{W^{k,p}(\Omega)} = \{\sum_{0 \leq \|\alpha\| \leq k} \|f^{(\alpha)}\|^p_{L^p(\Omega)}\}^{\frac{1}{p}}$
$C([a, b]; \mathcal{H})$	The Hilbert space of the continuous $\mathcal{H}$-valued functions $f : [a, b] \to \mathcal{H}$ with the induced norm $\|f\|_{C([a, b]; \mathcal{H})} = \max_{s\in[a, b]} \|f(s)\|_{\mathcal{H}}$
supp$g$	The support of a function g
conv(supp$g$)	The convex hull of supp$g$
I	The identity matrix with appropriate dimension
$\mathbf{0}$	The zero matrix with appropriate dimensions
$*$	The symbol $*$ represents symmetric elements
P^T	Transpose of matrix P
P^{-1}	Inverse of matrix P
$P > 0$	The symmetric matrix P is positive definite
$P \geq 0$	The symmetric matrix P is positive semi-definite
$P < 0$	The symmetric matrix P is negative definite
$P \leq 0$	The symmetric matrix P is negative semi-definite

$\lambda_{\max}(P)$	Maximum eigenvalue of matrix P
$\lambda_{\min}(P)$	Minimum eigenvalue of matrix P
$\mathrm{diag}\{\cdots\}$	Block-diagonal matrix
$\mathrm{sat}(P, \bar{p})$	$\mathrm{sign}(P)\min(\|P\|, \bar{p})$
$\mathrm{col}\{x, y\}$	Vector $[x^T y^T]^T$
ISS	Input-to-state stability
LMI	Linear matrix inequality

Chapter 1
Introduction

Partial differential equations (PDEs) are commonly used to model various phenomena involved in space and time in different fields, such as engineering science, quantum mechanics, financial mathematics and so on. This chapter introduces several PDE systems and cutting-edge control techniques for PDEs. Furthermore, the time-delay approach to sampled-data control is introduced, offering an effective way to address the stabilization problems of PDE systems.

Time-delay naturally appears in many control systems, and it is frequently a source of instability. However, for some systems, the presence of delay can have a stabilizing effect. Therefore, stability and control of time-delay systems is of theoretical and practical importance.

1.1 Stabilization of PDE Systems

1.1.1 PDE Systems

1.1.1.1 Heat Equation

Heat equations play an important role in modeling practical phenomena including heat transfer processes [1], traffic flow [2], phase transition [3], chemical reactions [4] and dynamics of biological groups [5]. A classical heat equation is governed by

$$z_t(x,t) = z_{xx}(x,t) + \mu z(x,t),$$

where $z(x,t)$ represents the temperature that depends on the spatial variable $x \in [0, L]$ and time $t > 0$, whereas $\mu > 0$ stands for the reaction rate. The topic of stabilization of systems described by a single PDE system has emerged during the last 60 years. For boundary control, a variety of references employed the backstepping

W. Kang and E. Fridman, *Delay-Robust Control of Distributed Parameter Systems*, Advances in Delays and Dynamics 13, https://doi.org/10.1007/978-981-96-6344-6_1

method to stabilize the heat equation (see e.g. [6, 7]). In [6], a feedback control scheme was proposed to stabilize the heat equation with long input delay. On the basis of modal decomposition approach, a constructive method was proposed to suggest the finite-dimensional observer-based controller of 1D heat equation in [8]. Reference [9] suggested a state feedback law for stabilizing an unstable heat equation in a general multi-dimensional domain via the modal decomposition and dynamics compensation approach.

1.1.1.2 Korteweg-De Vries/Korteweg-De Vries Burgers Equation

In fluid mechanics, the Korteweg-de Vries (KdV) equation

$$z_t(x,t) = -\nu z_{xxx}(x,t) - z(x,t)z_x(x,t),$$

where $z(x,t)$ is the state of KdV and $\nu > 0$ represents the dispersion coefficient, is a mathematical model of waves on shallow water surfaces in a rectangular channel, equation in which the effects of dispersion, dissipation and nonlinearity are taken into account. When adding a diffusion term $\varepsilon z_{xx}(x,t)$, with $\varepsilon > 0$ the KdV equation becomes Korteweg-de Vries Burgers (KdVB) equation as follows:

$$z_t(x,t) = -\nu z_{xxx}(x,t) + \varepsilon z_{xx}(x,t) - z(x,t)z_x(x,t).$$

The study of KdV/ KdVB systems has been an active research topic because of its potential applications (see e.g. [10–14]). In the field of automatic control, a backstepping approach has been applied in [11] for the feedback stabilization of KdV equation, and Lyapunov-based arguments have been employed to ensure the stability of the original system under the proposed control law. On the other hand, the survey paper [12] gives a detailed overview of boundary controllability and internal stabilization approaches and results for the KdV equation. One can find in [10] two different approaches (from a Lyapunov functional or from an observability inequality) employed to exponentially stabilize the nonlinear KdV equation via delayed boundary damping terms. Some researchers have contributed to the study of KdV/KdVB equations with periodic boundary conditions in various applications, such as generation of electrical short pulses (see e.g. [15]) and traffic flow models (see e.g. [16]).

1.1.1.3 Kuramoto-Sivashinsky Equation

Kuramoto-Sivashinsky equation (KSE)

$$z_t(x,t) = -\nu z_{xxxx}(x,t) - z_{xx}(x,t) - z(x,t)z_x(x,t),$$

where $z(x, t)$ is the state of KSE and ν is a positive constant, describes a variety of physical and chemical phenomena including magnetized plasmas, flame front propagation, viscous flow problems and chemical reaction-diffusion processes (see e.g. [17, 18]). Boundary control of 1D KSE was studied in [19, 20]. The local rapid stabilization problem for a controlled KSE on a bounded interval was considered in [19]. In [20], a Neumann feedback law was designed to guarantee L^2-global exponential stability and H^2-global asymptotic stability for small values of the anti-diffusion parameter.

Distributed control of KSE was studied in [18, 21–23]. In [21, 22], a finite-dimensional controller was designed on the basis of a finite-dimensional system that captures the dominant (slow) dynamics of the infinite-dimensional system. In [23], the problem of global exponential stabilization of the KSE subject to periodic boundary conditions was considered. In [18], a distributed finite-dimensional feedback controller based on either point or averaged measurements of the state was proposed. In [24, 25], boundary control of KSE via modal decomposition was studied.

1.1.1.4 Schrödinger Equation

Schrödinger equation is central to all applications of quantum mechanics. A classical nonlinear Schrödinger equation is governed by

$$\mathrm{i}z_t(x, t) = -z_{xx}(x, t) - \mu|z(x, t)|^2 z(x, t), \tag{1.1}$$

where $\mathrm{i} = \sqrt{-1}$, z is a complex-valued function of time and space, $\mu \in \mathbb{R}$ is the Landau coefficient. In optics, the nonlinear Schrödinger equation is a model of wave propagation in fiber optics. Its state represents a wave and the nonlinear Schrödinger equation describes the propagation of the wave through a nonlinear medium. For water waves, the nonlinear Schrödinger equation describes the evolution of the envelope of modulated wave groups. Time delay in nonlinear Schrödinger equation may appear due to the Raman effect of nonlinear fiber optics (see e.g. [26–28]). Raman self-frequency shift can down-shift the central frequency and delay the pulse arrival time.

Boundary control of linearized Schrödinger equation has been studied in [29–31]. Recent results for the Schrödinger equation include the multiplier technique (see e.g. [31]), the backstepping method (see e.g. [30]), sliding mode control (see e.g. [29]) and extension of backstepping to Heat-Schrödinger equation cascade [32] and ODE-Schrödinger equation cascade [33]. Stabilization of linear Schrödinger equation with time delay by boundary or internal feedbacks was dealt with in [34].

1.1.1.5 Cascaded ODE-PDE/PDE-PDE System

In the last few years, coupled systems have attracted considerable attention in research communities. Stabilization of the cascade of PDE systems was dealt with in [4, 35]. Controller design for ODE-PDE cascade systems has been extensively studied for many types of coupling such as ODE-Reaction diffusion equation (see e.g. [36–38]), ODE-Wave equation (see e.g. [39]), and ODE-Schrödinger equation (see e.g. [33]). In order to stabilize the cascaded PDE-ODE systems, the backstepping method has been applied in [33, 36–39]. The idea is to use a Volterra integral transformation to transform the original system to a target system [7].

Recently a cascaded system consisting of heat equation and PDE(partial differential equation)/ODE(ordinary differential equation) system has drawn considerable attention, because it can describe more physical and chemical phenomena. An important example of a nonisothermal reaction taking place in a tubular reactor was discussed in [4]. Various approaches have been proposed for stabilizing unstable coupled systems, including the Lyapunov method [40], the modal decomposition approach [41], and the backstepping method (see e.g. [36, 42–47]). In [41], a finite-dimensional observer-based controller was introduced to stabilize the ODE-heat system with a long interconnection delay. In [36], predictor-like feedback control laws were suggested to stabilize the coupled ODE-heat system. In [45], an output-feedback boundary control law was proposed for stabilizing a heat PDE sandwiched between two ODEs. In [44, 46], the discontinuous boundary feedback controller was designed to stabilize an unstable coupled ODE-heat system in presence of boundary disturbance by combining the sliding mode control approach and the backstepping method. The above references focused on the study of stabilization of a cascaded of ODE-heat system.

For the case of stabilization of cascaded PDE-PDE system, very few works focus on this topic. Though a coupled heat-heat system with the same reaction coefficients was studied in [43], the case of the coupled heat-heat system with different reaction coefficients has not been studied yet, since the backstepping method is not applicable to this case. Reference [48] studied heat exchangers for actuation of transcription. The heat exchangers were made of materials compatible with living cells, and they can facilitate volumetric heat patterning in artificial tissues. The cascaded model was governed by two subsystems: one subsystem describes the heat transfer from the perfusion channel to the hydrogel volume, while the other subsystem describes the heat transfer within the hydrogel volume. The heat transfer through different mediums leads to the different reaction coefficients.

1.1.2 Constrained Control

In control systems, handling constraints is a critical aspect for the reason that the performance of systems is usually limited by the presence of physical constraints on system components. For example, the actuators and sensors have the maximum and minimum allowable signals, which can be regarded as input and output constraints. Besides, for safety or other operational reasons, a vehicle's speed remains within safe limits or a temperature stays within a specified range, which are the examples of state constraints. While one could potentially ignore these constraints and hope for the best, such an approach is generally not advisable, as it could lead to suboptimal or unsafe performance.

In many cases, the constraints on the control input should be taken into account for practical application of control laws. Stabilization of finite-dimensional systems in the presence of actuator saturation has been extensively studied (see e.g. [49–52] and the references therein). However, there are only a few works on constrained control for distributed parameter systems. In [53], the internal feedbacks with input constraints of quasi-linear heat equation were designed and the domains of attractions were found via the Galerkin's method. In [54, 55], global stabilization by distributed state-feedback saturated control of 1D KdV and wave equations was studied. Global stabilization of linear or semilinear system in the Hilbert space by using constrained control was presented in [56, 57]. There have been some important results about PDEs subject to distributed control constraints (see e.g. [53–55, 58]). Backstepping-based boundary control of PDEs in the presence of actuator saturation was studied in [42].

1.1.3 Sampled-Data and the Switching Control

Sampled-data control refers to a control system approach in which continuous-time signals, such as system inputs and outputs, are discretized at either fixed or variable time intervals and processed using discrete-time control algorithms. This method is widely used in digital control systems, where analog signals are sampled at a predetermined rate (fixed sampling intervals) or based on some adaptive criterion (variable sampling intervals), and the controller operates based on these discrete samples rather than continuous signals.

Sampled-data control of PDEs became recently an active research area (see e.g. [13, 59–61]) for practical application of finite-dimensional controllers for PDEs, where LMI conditions for the exponential/regional stability of the closed-loop systems were derived in the framework of time-delay approach by employing appropriate Lyapunov functionals. The sampled-data control was proposed for stabilization of abundant systems governed by 1D heat equation [62, 63], 1D KdVB equation [13], 1D KSE [64] and 2D KSE [65]. It should be noticed that the mentioned above results were confined to one single parabolic equation. Sampled-data control of coupled PDE-PDE system was first addressed in [66].

Distributed sampled-data control of PDEs under the point or spatially averaged measurements was suggested in [59, 60, 67], where LMI conditions for the exponential stability and L_2-gain analysis of the closed-loop systems were derived in the framework of time-delay approach to sampled-data control by employing appropriate Lyapunov functionals. However, the above results were confined to diffusion equations and to globally Lipschitz nonlinearities, where stabilization is global. Distributed sampled-data control of various classes of PDEs are of great interest. Most of the existing results deal with 1D PDEs system. Sampled-data observers and controllers for ND and 2-D heat equations with globally Lipschitz nonlinearities have been suggested in [67, 68]. Sampled-data control of various classes of high dimensional PDEs is an interesting and challenging problem. Reference [69] addressed the sampled-data boundary stabilization of 2D semilinear parabolic stochastic PDEs with globally Lipschitz nonlinearities.

Extension from 1D ([61, 70]) to 2D is far from being straightforward. Thus, in the case of heat equation, sampled data extension under the point measurements seems to be not possible (see [67]). This is due to the fact that stability analysis of the closed-loop system is based on the bound of the L^2-norm of the difference between the state and its point value. However, according to Friedrich's inequality [67, 71], this bound depends on the L^2-norm of the second-order spatial derivatives of the state. Differently from the heat equation, stability analysis for KSE in H^2 allows to compensate such terms.

Stabilization of unstable systems by switching is a classical, challenging and well-studied problem for ODEs [72]. Thus, for linear switched systems with a stable convex combination, stabilization can be achieved by state or output-dependent switching (see e.g. [73, 74] and the references therein). In recent years, substantial efforts have been taken to develop switched control of PDEs (see e.g. [75, 76]). In [75], optimal and switching policies of spatially scheduled actuators were suggested that were based on finite horizon Linear Quadratic Regulator problem. In [76] the following problem was formulated: assuming that one can control a system using two or more actuators, does there exist a control strategy such that at all times, only one actuator is active? The positive answer for the controllability of some classes of PDEs along with the corresponding switching laws was given in [76]. The key idea of switching control design for PDEs is to schedule the position of the actuator and sensor in order to achieve the control goal.

The switching control laws can be implemented either by stationary actuators or by one moving actuator that can move to the active subdomain in the negligible time [75]. In [77–79], moving actuators and sensors guiding laws were suggested for PDEs. Intermittent control of reaction-diffusion equation by time-dependent switching between all working pairs of collocated mobile actuators and sensors and the rest (all not working) has been studied in [80]. Furthermore, reference [81] has dealt with the stabilization problem of linear reaction-diffusion equation with time-varying delay via the projection modification algorithm using collocated mobile actuators and sensors. In [82], switching control of closed quantum systems governed by Schrödinger equation has been proposed for the degenerate case via multiple Lyapunov function technique. Note that the mentioned above methods for switching control or for control by mobile actuators and sensors may be inefficient for the unstable open-loop systems, which motivates our study [83].

In [64, 84], for KSE, a switching law reconfigured the control actuators following fault detection. It should be noticed that stabilization by switching of open-loop unstable PDE with several actuators, where the system is not stabilizable by using only one actuator, has not been achieved yet. Thus, the design of a switching controller for open-loop unstable parabolic PDEs is a challenging topic.

1.1.4 Event-Triggered Control

The previous section focuses on the study of sampled-data control, which belongs to time-triggered control, while this section focuses on the study of event-triggered control. An event-triggered control system is a sampled-data system in which the sampling time instants are determined by events generated by the real-time system state. In general, event-triggered control includes two main components: a feedback control law which stabilizes the system and an event-triggered mechanism which contains a triggering condition that determines the time instants at which the control needs to be updated [85]. Event-triggered control has become popular and promising due to not only its efficient way of using communication and computational resources by updating the control value aperiodically (only when needed) but also due to its rigorous way of implementing continuous-time controllers into digital platforms. By taking the advantage of the inter-sample behavior, event-triggered sampling may realize improved control performance over periodic sampling [86].

In recent years, event-triggered control systems have been extensively studied (see e.g. [87–91]), bringing an important alternative to periodic sampling of control laws. There are many important results on event-triggering mechanisms (see e.g. [90–92]). The event-triggering mechanism (ETM) is used to minimize the number of signals transmitted over a communication network [90, 93], and three main types of ETMs have been proposed to reduce the frequency of updates: the continuous event-triggering mechanism [90], the periodic event-triggering mechanism [92], and the event-triggering mechanism with a dwell time [91, 94]. Two main time-delay approaches to event-triggered control have been suggested: the discrete-time (that includes as a particular case periodic) ETM [95, 96] and the continuous time ETM with a dwell time (the switched system approach [94]. It is worth pointing out that most works focus on event-triggered control of finite-dimensional systems. However, to the best of our knowledge, there are few papers studying this technique in the infinite-dimensional systems framework (see e.g. [85, 87, 88, 97]).

In [13], distributed control of KdVB system has been suggested under point or averaged localised measurements in space but the proof rely strongly on the presence of a diffusion term that is missing in the KdV equation. Such distributed control was introduced for heat equation under point [60] and under averaged [59] measurements. In the latter papers, sampled-data control via time-delay approach and Lyapunov-Krasovskii functionals were studied, and the results of [59, 60] were extended to event-triggered control in [88]. However, since the Lyapunov-Krasovskii functionals for sampled-data control depend on the state-derivative (see Chap. 7 of [98]), this method cannot be applied to sampled-data control of KdVB equation. So reference [13] considered the constant input delay case.

To the best of our knowledge, the first result on event-triggered control of KdV equation has been suggested in [99]. The goal of event-triggering mechanism is to update the control input only at meaningful instants. Its drawback, well-known in hybrid systems problematics, could be the exhibition of a Zeno behaviour. This can be summed up as the law bringing an infinite number of updates in a finite time interval.

We aim to address the event-triggered control design for KdV system under in domain measurements averaged in space, and for the record, reference [12] gathers the results for distributed continuous-in-time controller to stabilize the KdV equation exponentially. Our concern here is then mainly to prove that distributed event-triggered control can still bring, under appropriate assumptions and choice of triggering mechanism, the expected exponential stability.

1.1.5 Anti-Disturbance Control

Copying with disturbances and uncertainties is the eternal theme in control theory due to the ubiquitousness of the uncertainties and disturbances in most of the industrial control systems, which most often cause negative effects on performance and even stability of control systems [102]. Many control approaches have been developed since 1970s to cope with disturbances or uncertainties through disturbance attenuation and disturbance rejection. To avoid the adverse effects of the undesired disturbances, researchers have studied some effective control strategies, including H_∞ control (see e.g. [103]), sliding mode control (SMC) (see e.g. [104]), active disturbance rejection control (ADRC) (see e.g. [105, 106]), disturbance observer-based control (DOBC) (see e.g. [107, 108]) and so on. The ideology of disturbance rejection could be found in DOBC and ADRC, where the disturbances or uncertainties (or both of them) are estimated in real time and are eliminated in feedback loop so that these disturbance rejection control approaches have good anti-disturbance performance.

As an efficient tool, ADRC can deal with the "total disturbance" containing the internal uncertainty dynamics and the external unknown disturbances (see [105]). By constructing an extended state observer (ESO), the total disturbance can be compensated in the feedback loop. In the framework of PDEs, fruitful theoretical results on ADRC have been developed over the past decade (see e.g. [106, 107, 109]). Researchers have investigated different types of estimators/compensators to attenuate the disturbance. In [110], a constant high-gain estimator has been adopted to handle boundary disturbance for beam equation. With the advantages of relaxing the restriction of disturbance and reducing the peak value, a time-varying high-gain estimator has been applied for wave equation in [111]. In [109], a novel dynamic first-order compensator has been proposed for heat equation, and ADRC has been designed with a time-varying/constant high-gain estimator. The above existing works focus on the boundary control of PDEs, which motives us to study the distributed control of PDEs by applying the ADRC technique.

In the late of 1980s, DOBC method was established for linear frequency-domain systems in [107]. The basic idea of DOBC approach is to construct a disturbance observer (DO) to reject the disturbance actively in the feed-foreword channel. Due to the perfect anti-disturbance performance, DOBC approach has been introduced to various plants described by ODEs (see e.g. [112]). As for PDEs, DOBC approach has been considered for a stochastic heat equation [108], a flexible string [113], a Timoshenko beam system [114] and so on. DOBC was studied for distributed stabilization of KdVB equation in our work [115], where the comparison of ADRC and DOBC for KdVB system was also present.

1.2 Useful Inequalities

For easy reference, we collect here some useful inequalities in functional analysis that will be used throughout this work.

Lemma 1.1 (1D Poincaré inequality [116]) *Let $f \in H^1(0, L)$ be a scalar function with $\int_0^L f(x)dx = 0$. Then*

$$\int_0^L f^2(x)dx \leq \frac{L^2}{\pi^2}\int_0^L \left[\frac{df}{dx}(x)\right]^2 dx.$$

Lemma 1.2 (1D Wirtinger inequality [117]) *Let $f \in H^1(0, L)$ and either $f(0) = 0$ or $f(L) = 0$, then*

$$\int_0^L f^2(x)dx \leq \frac{4L^2}{\pi^2}\int_0^L \left[\frac{df(x)}{dx}\right]^2 dx.$$

If $f \in H^1(0, L)$ and $f(0) = f(L) = 0$, then

$$\int_0^L f^2(x)dx \leq \frac{L^2}{\pi^2}\int_0^L \left[\frac{df(x)}{dx}\right]^2 dx.$$

Lemma 1.3 *Let $\Omega = (0, L_1) \times (0, L_2)$. Assume $f : \Omega \to \mathbb{R}$ and $f \in H^1(\Omega)$.*
(i) (2D Poincaré's inequality) If $\int_\Omega f(x)dx = 0$, then according to [116]

$$\|f\|^2_{L^2(\Omega)} \leq \frac{L_1^2 + L_2^2}{\pi^2}\|\nabla f\|^2_{L^2(\Omega)}. \tag{1.2}$$

(ii) (2D Wirtinger's inequality [68]) If $f|_{\partial\Omega} = 0$, then the following inequality holds:

$$\|f\|^2_{L^2(\Omega)} \leq \frac{L_1^2 + L_2^2}{\pi^2}\|\nabla f\|^2_{L^2(\Omega)}. \tag{1.3}$$

The following lemma gives a classical Friedrich's inequality (Theorem 18.1 of [118]) with tight bounds on the coefficients of terms $\left\|\frac{\partial f}{\partial x_1}\right\|^2_{L^2(\Omega)}$, $\left\|\frac{\partial f}{\partial x_2}\right\|^2_{L^2(\Omega)}$ and $\left\|\frac{\partial^2 f}{\partial x_1 \partial x_2}\right\|^2_{L^2(\Omega)}$. The inequality (1.4) bounds the L^2-norm of a function by the reciprocally convex combination of the L^2-norm of its derivatives.

Lemma 1.4 (See (2) of [67]) *Let* $\Omega = (0, l)^2$, $f \in H^2(\Omega)$ *with* $f(0, 0) = 0$. *Then the following inequality holds:*

$$\begin{aligned}\|f\|^2_{L^2(\Omega)} \le \frac{1}{\alpha_1}\left(\frac{2l}{\pi}\right)^2\left\|\frac{\partial f}{\partial x_1}\right\|^2_{L^2(\Omega)} + \frac{1}{\alpha_2}\left(\frac{2l}{\pi}\right)^2\left\|\frac{\partial f}{\partial x_2}\right\|^2_{L^2(\Omega)} \\ + \frac{1}{\alpha_3}\left(\frac{2l}{\pi}\right)^4\left\|\frac{\partial^2 f}{\partial x_1 \partial x_2}\right\|^2_{L^2(\Omega)}\end{aligned} \tag{1.4}$$

where $\alpha_1, \alpha_2, \alpha_3$ *are positive constants satisfying*

$$\alpha_1 + \alpha_2 + \alpha_3 = 1.$$

Lemma 1.5 ([67]) *Let* $\Omega = (0, l)^2$, $f \in H^2(\Omega)$ *with* $f(0, 0) = 0$, $\eta > 0$. *Then*

$$\eta\|f\|^2 \le \beta_1\left(\frac{2l}{\pi}\right)^2\left\|\frac{\partial f}{\partial x_1}\right\|^2 + \beta_2\left(\frac{2l}{\pi}\right)^2\left\|\frac{\partial f}{\partial x_2}\right\|^2 + \beta_3\left(\frac{2l}{\pi}\right)^4\left\|\frac{\partial^2 f}{\partial x_1 \partial x_2}\right\|^2 \tag{1.5}$$

for any $\beta_1, \beta_2, \beta_3$ *satisfying*

$$\operatorname{diag}\{\beta_1, \beta_2, \beta_3\} \ge \eta\begin{bmatrix} 1 & 1 & 1 \\ 1 & 1 & 1 \\ 1 & 1 & 1 \end{bmatrix}. \tag{1.6}$$

Cauchy-Schwartz's inequality leads to the next lemma:

Lemma 1.6 (Jensen's Inequality [98]) *For* $a < b$, *let* $c : [a, b] \to [0, \infty)$ *and* $f : [a, b] \to \mathbb{R}$ *be such that the integration concerned is well defined. Then* $\left[\int_a^b c(x) f(x)dx\right]^2 \le \int_a^b c(x)dx \int_a^b c(x) f^2(x)dx.$

Lemma 1.7 (Sobolev's embedding and inequality) *The embedding* $H^1(0, L) \subset C([0, L])$ *is compact and for any* $f \in H^1(0, L)$ *with either* $f(0) = 0$ *or* $f(L) = 0$, *it holds* $\|f\|_{L^\infty(0,L)} \le \sqrt{L}\|f'\|_{L^2(0,L)}$.

Lemma 1.8 (1D Sobolev's inequality [13]) *Let* $z(x) \in H^1(0, 1)$, *then for all* $\Gamma > 0$

$$\max_{x\in[0,1]} |z(x)|^2 \le (1 + \Gamma)\int_0^1 z^2(x)dx + \frac{1}{\Gamma}\int_0^1 z_x^2(x)dx$$

Proof Since $z(\cdot) \in H^1(0, 1)$ implies $z(\cdot) \in C[0, 1]$ (cf. [119, 120]), by mean value theorem, there exists $c \in (0, 1)$ such that

$$z(c) = \int_0^1 z(x)dx$$

Then, by integration by parts and further application of Jensen's and Young's inequalities, for all $x_1 \in [0, 1]$ we have

$$\begin{aligned} z^2(x_1) &= z^2(c) + 2\int_c^{x_1} z(x)z_x(x)dx \\ &= \left[\int_0^1 z(x)dx\right]^2 + 2\int_c^{x_1} z(x)z_x(x)dx \\ &\leq \int_0^1 z^2(x)dx + \Gamma\int_0^1 z^2(x)dx + \frac{1}{\Gamma}\int_0^1 z_x^2(x)dx \\ &\leq (1+\Gamma)\int_0^1 z^2(x)dx + \frac{1}{\Gamma}\int_0^1 z_x^2(x)dx \quad \forall\Gamma > 0. \end{aligned} \tag{1.7}$$

Lemma 1.9 (2D Sobolev's inequality [67]) *Let* $\Omega = (0, 1)^2$ *and* $w = w(x_1, x_2) \in H^2(\Omega) \cap H_0^1(\Omega)$, *where* $(x_1, x_2) \in \Omega$. *Then*

$$\|w\|^2_{C^0(\bar{\Omega})} \leq \frac{1}{2}(1+\Gamma)\left[\|w_{x_1}\|^2_{L^2(\Omega)} + \|w_{x_2}\|^2_{L^2(\Omega)}\right] + \frac{1}{\Gamma}\|w_{x_1x_2}\|^2_{L^2(\Omega)}, \ \forall\Gamma > 0. \tag{1.8}$$

Proof Due to $w \in H_0^1(\Omega)$, application of 1D Sobolev's inequality to w in x_2 yields

$$\max_{x_2\in[0,1]} w^2(x_1, x_2) \leq \int_0^1 w_{x_2}^2(x_1, \xi_2)d\xi_2. \tag{1.9}$$

Further application of Lemma 1.8 to w_{x_2} in x_1 leads to

$$\max_{x_1\in[0,1]} w_{x_2}^2(x_1, \xi_2) \leq (1+\Gamma)\int_0^1 w_{x_2}^2(\xi_1, \xi_2)d\xi_1 + \frac{1}{\Gamma}\int_0^1 w_{x_1x_2}^2(\xi_1, \xi_2)d\xi_1, \forall\Gamma > 0. \tag{1.10}$$

Substitution of (1.10) into (1.9) yields

$$\begin{aligned} &\|w\|^2_{C^0(\bar{\Omega})} = \max_{(x_1,x_2)\in\bar{\Omega}} w^2(x_1, x_2) = \max_{x_1\in[0,1]}\left[\max_{x_2\in[0,1]} w^2(x_1, x_2)\right] \\ &\leq \max_{x_1\in[0,1]}\left[\int_0^1 w_{x_2}^2(x_1, \xi_2)d\xi_2\right] \leq \int_0^1 \max_{x_1\in[0,1]} w_{x_2}^2(x_1, \xi_2)d\xi_2 \\ &\leq (1+\Gamma)\int_0^1\int_0^1 w_{x_2}^2(\xi_1, \xi_2)d\xi_1 d\xi_2 + \frac{1}{\Gamma}\int_0^1\int_0^1 w_{x_1x_2}^2(\xi_1, \xi_2)d\xi_1 d\xi_2. \end{aligned} \tag{1.11}$$

Following the same procedure, we can obtain

$$\|w\|^2_{C^0(\bar{\Omega})} \le (1+\Gamma)\int_0^1\int_0^1 w^2_{x_1}(\xi_1,\xi_2)d\xi_1 d\xi_2 + \frac{1}{\Gamma}\int_0^1\int_0^1 w^2_{x_1x_2}(\xi_1,\xi_2)d\xi_1 d\xi_2. \tag{1.12}$$

From (1.11) and (1.12) it follows that (1.8) holds. □

Remark 1.1 In Lemma 1.9, we give a new 2-D Sobolev inequality with constants depending on a free parameter $\Gamma > 0$. Lemma 1.9 is very useful and plays an important role in the stability analysis leading to LMIs (with Γ as a decision parameter) that guarantee stability and give a bound on the domain of attraction.

1.3 Time-Delay Systems

Time-Delay Systems (TDS) are also called systems with aftereffect or dead-time, hereditary systems, equations with deviating argument or differential-difference equations. They belong to the class of *functional differential equations* which are infinite-dimensional, as opposed to ordinary differential equations (ODEs). The simplest example of such a system is

$$\dot{x}(t) = -x(t-h), \;\; x(t) \in \mathbb{R},$$

where $h > 0$ is the time-delay.

Time-delays appear in many engineering systems—aircraft, chemical control systems, in lasers models, in internet, biology, medicine [121, 122]. Delays are strongly involved in challenging areas of communication and information technologies: stability of networked control systems or high-speed communication networks [123].

Time-delay is, in many cases, a source of instability. However, for some systems, the presence of delay can have a stabilizing effect. In the well-known example

$$\ddot{y}(t) + y(t) - y(t-h) = 0,$$

the system is unstable for $h = 0$, but is asymptotically stable for $h = 1$. The approximation $\dot{y}(t) \simeq [y(t) - y(t-h)]\,h^{-1}$ explains the damping effect. The stability analysis and robust control of time-delay systems are, therefore, of theoretical and practical importance.

As in systems without delay, an efficient method for stability analysis of TDSs is the Lyapunov method. For TDSs, there exist two main Lyapunov methods: the *Krasovskii* method of Lyapunov *functionals* [124] and the *Razumikhin* method of Lyapunov *functions* [125]. The two Lyapunov methods for linear TDSs result in Linear Matrix Inequalities (LMIs) conditions. The LMI approach to analysis and design of TDSs provides constructive finite-dimensional conditions, in spite of significant model uncertainties [126].

Modern control systems usually employ digital technology for controller implementation, i.e. sampled-data control. Consider a sampled-data control system

$$\dot{x}(t) = Ax(t) + BKx(t_k),\ t \in [t_k, t_{k+1}),\quad k = 0, 1, \ldots, \tag{1.13}$$

where $x(t) \in \mathbb{R}^n$, A, B, K are constant matrices and $\lim_{k\to\infty} t_k = \infty$. This system can be represented as a continuous system with time-varying delay $\tau(t) = t - t_k$ [127, 128]:

$$\dot{x}(t) = Ax(t) + BKx(t - \tau(t)),\ t \in [t_k, t_{k+1}), \tag{1.14}$$

where the delay is piecewise-linear (sawtooth) with $\dot{\tau} = 1$ for $t \neq t_k$. Modeling of continuous-time systems with *digital control* in the form of *continuous-time systems with time-varying delay* and the extension of Krasovskii method to TDSs without any constraints on the delay derivative [129] and to discontinuous delays [130] have allowed the development of the *time-delay approach to sampled-data and to network-based control.*

Bernoulli, Euler and Concordet were (among) the first to study equations with delay (the 18-th century). Systematical study started at the 1940s by A. Myshkis and R. Bellman. Since 1960 there have appeared more than 50 monographs on the subject (see e.g. [121, 122, 131–133] to name a few). The beginning of the 21st century can be characterized as the "time-delay boom" leading to numerous important results.

1.4 Time-Delay Approach to Sampled-Data Control

Consider the linear time-invariant (LTI) system

$$\dot{x}(t) = Ax(t) + Bu(t),\quad x(t) \in \mathbb{R}^n, u(t) \in \mathbb{R}^{n_u}, \tag{1.15}$$

where $x(t)$ is the state and $u(t)$ is the control input. The control signal is assumed to be generated by the Zero-Order Hold (ZOH) function $u(t) = u_d(t_k),\ t_k \leq t < t_{k+1}$ with a sequence of hold times

$$0 = t_0 < t_1 < \ldots < t_k < \ldots,\quad \lim_{k\to\infty} t_k = \infty,$$

where u_d is a discrete-time control signal. The sampling interval can be either constant $t_{k+1} - t_k \equiv h$ or variable with $t_{k+1} - t_k \leq h$. In the context of networked control systems (NCSs), the sampling interval may be variable and uncertain due to packet dropouts. We consider a state-feedback control law with the following form:

$$u(t) = Kx(t_k),\ t_k \leq t < t_{k+1}.$$

Then the closed-loop system of (1.15) becomes

$$\dot{x}(t) = Ax(t) + BKx(t_k),\ t \in [t_k, t_{k+1}),\ k \in \mathbb{Z}^+. \tag{1.16}$$

The classical approach to sampled-data control is discretization, which loses information on the inter-sampling behavior. To avoid the latter problem, the time-delay approach (see [98]) was introduced, where the closed-loop system (1.16) was represented as a continuous system with a time-varying delay as follows:

$$\dot{x}(t) = Ax(t) + BKx(t - \tau(t)), \ \tau(t) = t - t_k, \ t \in [t_k, t_{k+1}), \ k \in \mathbb{Z}^+. \quad (1.17)$$

Here the time-varying delay $\tau(t)$ is piecewise linear with derivative $\dot{\tau}(t) = 1$ for $t \neq t_k$. The time-delay approach has become popular in recent years. The stability of (1.17) can be established by using Lyapunov-Krasovskii technique ([98, 134]).

1.5 Lyapunov Method for Time-Delay Systems and Halanay's Inequality

Consider the following TDS:

$$\dot{x}(t) = f(t, x_t), \quad t \geq t_0, \quad (1.18)$$

where $x_t(\theta) \triangleq x(t+\theta), \ \theta \in [-h, 0], \ f : \mathbb{R} \times C[-h, 0] \to \mathbb{R}^n$ is continuous in both arguments and is locally Lipschitz continuous in the second argument. We assume that $f(t, 0) = 0$, which guarantees that (1.18) possesses a trivial solution $x(t) \equiv 0$.

Let $C[-h, 0]$ be the Banach space of continuous functions mapping the interval $[-h, 0]$ into $\mathbb{R}^n$ with the norm $\|x_t\|_C = \max_{\theta\in[-h,0]} |x(t+\theta)|$.

Definition 1.1 The trivial solution of (1.18) is:

(1) uniformly (in t_0) stable if $\forall t_0 \in \mathbb{R}$ and $\forall \epsilon > 0$, there exists a $\delta = \delta(\epsilon) > 0$ such that $\|x_{t_0}\|_C < \delta(\epsilon)$ implies $|x(t)| < \epsilon$ for all $t \geq t_0$;

(2) uniformly asymptotically stable if it is uniformly stable and there exists a $\delta_a > 0$ such that for any $\eta > 0$ there exists a $T(\delta_a, \eta)$ such that $\|x_{t_0}\|_C < \delta_a$ implies $|x(t)| < \eta$ for all $t \geq t_0 + T(\delta_a, \eta)$ and $t_0 \in \mathbb{R}$;

(3) globally uniformly asymptotically stable if δ_a can be an arbitrarily large, finite number.

1.5.1 Lyapunov–Krasovskii Approach

Let $V : \mathbb{R} \times C[-h, 0] \to \mathbb{R}^n$ be a continuous functional, and let $x_\tau(t, \phi)$ be the solution of (1.18) at time $\tau \geq t$ with the initial condition $x_\tau = \phi$. We define the right upper derivative $\dot{V}(t, \phi)$ along (1.18) as follows:

$$\dot{V}(t,\phi) = \lim \sup_{\Delta t \to 0^+} \frac{1}{\Delta t}\left[V(t+\Delta t, x_{t+\Delta t}(t,\phi)) - V(t,\phi)\right].$$

Intuitively, a non-positive $\dot{V}(t, x_t)$ indicates that x_t does not grow with t, meaning that the system under consideration is stable.

Theorem 1.1 (Lyapunov-Krasovskii Theorem [131]) *Suppose* $f : \mathbb{R} \times C[-h, 0] \to \mathbb{R}^n$ *maps* $\mathbb{R}\times$ *(bounded sets in* $C[-h, 0]$ *) into bounded sets of* $\mathbb{R}^n$ *and that* $u, v, w : \mathbb{R}_+ \to \mathbb{R}_+$ *are continuous nondecreasing functions,* $u(s)$ *and* $v(s)$ *are positive for* $s > 0$*, and* $u(0) = v(0) = 0$*. The trivial solution of (1.18) is uniformly stable if there exists a continuous functional* $V : \mathbb{R} \times C[-h, 0] \to \mathbb{R}^+$*, which is positive-definite, i.e.*

$$u(|\phi(0)|) \le V(t,\phi) \le v(\|\phi\|_C), \tag{1.19}$$

and such that its derivative along (1.18) is non-positive in the sense that

$$\dot{V}(t,\phi) \le -w(|\phi(0)|). \tag{1.20}$$

If $w(s) > 0$ *for* $s > 0$*, then the trivial solution is uniformly asymptotically stable. If in addition* $\lim_{s\to\infty} u(s) = \infty$*, then it is globally uniformly asymptotically stable.*

In some cases functionals $V(t, x_t, \dot{x}_t)$ that depend on the state-derivatives are useful (see [122], p. 337 and [135]). Denote by $W[-h, 0]$ the Banach space of absolutely continuous functions $\phi : [-h, 0] \to \mathbb{R}^n$ with $\dot{\phi} \in L^2(-h, 0)$ (the space of square integrable functions) with the norm

$$\|\phi\|_W = \max_{s\in[-h,0]} |\phi(s)| + [\int_{-h}^{0} |\dot{\phi}(s)|^2 ds]^{\frac{1}{2}}.$$

Theorem 1.1 is extended then to continuous functionals

$$V : \mathbb{R} \times W[-h, 0] \times L^2(-h, 0) \to \mathbb{R}_+,$$

where inequalities (1.19) and (1.20) are modified as follows:

$$u(|x(t)|) \le V(t, x_t, \dot{x}_t) \le v(\|x_t\|_W) \tag{1.21}$$

and

$$\dot{V}(t, x_t, \dot{x}_t) \le -w(|x(t)|). \tag{1.22}$$

Note that the functionals $V(t, x_t, \dot{x}_t)$ can be applied to solutions of RDEs with the initial functions x_{t_0} from $W[-h, 0]$. However, for RDE the stability results corresponding to continuous initial functions and to absolutely continuous initial functions are equivalent [132].

1.5.2 Lyapunov–Razumikhin Approach

To give a precise formulation of Razumikhin method, we consider a differentiable function $V : \mathbb{R} \times \mathbb{R}^n \to \mathbb{R}_+$ and define the derivative of V along the solution $x(t)$ of (1.18) as

$$\dot{V}(t, x(t)) = \frac{d}{dt}V(t, x(t)) = \frac{\partial V(t, x(t))}{\partial t} + \frac{\partial V(t, x(t))}{\partial x} f(t, x_t).$$

Theorem 1.2 (Lyapunov–Razumikhin Theorem [131]) *Suppose* $f : \mathbb{R} \times C[-h, 0] \to \mathbb{R}^n$ *maps* $R\times$ *(bounded sets in* $C[-h, 0]$*) into bounded sets of* $\mathbb{R}^n$ *and that* $u, v, w : \mathbb{R}_+ \to \mathbb{R}_+$ *are continuous non-decreasing functions* $u(s)$ *and* $v(s)$ *are positive for* $s > 0$ *and* $u(0) = v(0) = 0$*,* v *is strictly increasing. The trivial solution of (1.18) is uniformly stable if there exists a differentiable function* $V : \mathbb{R} \times \mathbb{R}^n \to \mathbb{R}^+$*, which is positive-definite, i.e.*

$$u(|x|) \le V(t, x) \le v(|x|),$$

and such that the derivative of V *along the solution* $x(t)$ *of (1.18) satisfies*

$$\dot{V}(t, x(t)) \le -w(|x(t)|) \quad \text{if } V(t+\theta, x(t+\theta)) \le V(t, x(t)) \quad \forall \theta \in [-h, 0].$$

If, in addition, $w(s) > 0$ *for* $s > 0$ *and there exists a continuous non-decreasing function* $\rho(s) > s$ *for* $s > 0$ *such that condition is strengthened to*

$$\dot{V}(t, x(t)) \le -w(|x(t)|) \quad \text{if } V(t+\theta, x(t+\theta)) \le \rho(V(t, x(t))) \quad \forall \theta \in [-h, 0],$$

then the trivial solution of (1.18) is uniformly asymptotically stable. If, in addition, $\lim_{s\to\infty} u(s) = \infty$*, then it is globally uniformly asymptotically stable.*

1.5.3 Stability of Linear Systems with Discrete Delays

Consider a simple linear TDS

$$\dot{x}(t) = Ax(t) + A_1 x(t - \tau(t)), \quad t \ge t_0, \tag{1.23}$$

where $x(t) \in \mathbb{R}^n$, $\tau(t) \in [0, h]$ is a bounded time-varying delay. Here A and A_1 are constant $n \times n$-matrices. In this section, delay-independent and delay-dependent conditions will be presented. For simplicity only we consider a linear system with a single discrete delay. The results can be easily extended to a finite number of discrete delays.

1.5.3.1 Delay-Independent Conditions

The choice of Lyapunov-Krasovskii functional (that we will call also Lyapunov functional) is crucial for deriving stability criteria. Special forms of the functional lead to *delay-independent* and *delay-dependent* conditions. In this section we consider the delay-independent (i.e. h-independent) conditions for systems with time-varying delays.

A simple Lyapunov functional for (1.23) has the form:

$$V(t, x_t) = x^T(t)Px(t) + \int_{t-\tau(t)}^{t} x^T(s)Qx(s)ds, \tag{1.24}$$

where $P > 0$ and $Q > 0$ are $n \times n$ matrices. Note that V of (1.24) depends on t because of $\tau(t)$. Note that in the case of constant delay $\tau \equiv h$, the functional in (1.24) is time-independent (i.e. $V(x_t)$). We further assume that the delay τ is a differentiable function with $\dot{\tau} \le d < 1$ (this is the the case of *slowly-varying delays*). It is clear that V satisfies the positivity condition $\beta|x(t)|^2 \le V(t, x_t)$ (for some $\beta > 0$).

Halanay's inequality [136] that extends the Razumikhin method to the exponential stability can also lead to delay-independent conditions for systems with fast-varying delays.

1.5.3.2 Delay-Dependent Conditions

The first delay-dependent (both, Krasovskii and Razumikhin-based) conditions were derived by using the relation

$$x(t - \tau(t)) = x(t) - \int_{t-\tau(t)}^{t} \dot{x}(s)ds \tag{1.25}$$

via different model transformations and by bounding the cross terms [137–139]. The widely used 1st Model Transformation, where (1.25) is substituted into (1.23) with $\dot{x}(s)$ substituted by the right-hand side of (1.23), has the form

$$\dot{x}(t) = [A + A_1]x(t) - A_1 \int_{t-\tau(t)}^{t} [Ax(s)+A_1x(s-\tau(s))]ds. \tag{1.26}$$

Note that this transformation is valid for $t - \tau(t) \ge t_0$. The latter system is not equivalent to the original one possessing some additional dynamics [140, 141]. The stability of the transformed system (1.26) guarantees the stability of the original one, but not vise versa.

The first delay-dependent conditions treated only the *slowly*-varying delays with $\dot{\tau} \le d < 1$, whereas the fast-varying delay (without any constraints on the delay derivative) was analyzed via Lyapunov-Razumikhin functions.

For the first time, systems with *fast varying delays* were analyzed by using Krasovskii method in [129], via the *descriptor model transformation* introduced in [142]:

$$\dot{x}(t) = y(t),\ 0 = -y(t) + (A + A_1)x(t) - A_1 \int_{t-\tau(t)}^{t} y(s)ds. \tag{1.27}$$

The descriptor system (1.27) is equivalent to (1.23) in the sense of stability. In the descriptor approach, $\dot{x}(t)$ is not substituted by the right-hand side of the differential equation. Instead, it is considered as an additional state variable of the resulting descriptor system (1.27). Therefore, the novelty of the descriptor approach is not in $V = x^T(t)Px(t) + \ldots\ (P > 0)$, but in $\dot{V}$, where $\frac{d}{dt}\left[x^T(t)Px(t)\right]$ is found as

$$\begin{aligned}&\tfrac{d}{dt}\left[x^T(t)Px(t)\right] = 2x^T(t)P\dot{x}(t) + 2[x^T(t)P_2^T + \dot{x}^T(t)P_3^T]\\&\times[-\dot{x}(t) + (A + A_1)x(t) - A_1 \int_{t-\tau(t)}^{t} \dot{x}(s)ds],\end{aligned} \tag{1.28}$$

and where $P_2 \in \mathbb{R}^{n\times n}$ and $P_3 \in \mathbb{R}^{n\times n}$ are "slack variables". This leads to $\dot{V} \le -\epsilon(|x(t)|^2 + |\dot{x}(t)|^2),\ \epsilon > 0$.

The advantages of the descriptor method are:

- less conservative conditions (even without delay) for *uncertain* systems,
- "unifying" LMIs for the discrete-time and for the continuous-time systems, having almost the same form and the same advantages [143],
- simple conditions in terms of LMIs can be derived for *neutral type systems* (these are systems with the delayed highest-order state derivative), where the LMIs imply the stability of the difference operator [144],
- efficient design is obtained for systems with state, input and output delays by choosing $P_3 = \epsilon P_2$ with a tuning scalar parameter ϵ [145],
- simple delay-dependent conditions can be derived for diffusion partial differential equations [146].

Most of the recent Krasovskii-based results do not use model transformations and cross terms bounding. They are based on the application of Jensen's inequality [131] and Park's inequality [147].

The corresponding Lyapunov functional has the form

$$\begin{aligned}&V(t, x_t, \dot{x}_t) = x^T(t)Px(t) + \int_{t-h}^{t} x^T(s)Sx(s)ds\\&+h\int_{-h}^{0}\int_{t+\theta}^{t} \dot{x}^T(s)R\dot{x}(s)dsd\theta + \int_{t-\tau(t)}^{t} x^T(s)Qx(s)ds,\end{aligned} \tag{1.29}$$

where $P > 0$, $R \ge 0$, $S \ge 0$, $Q \ge 0$. This functional depends on the state derivative. Moreover, this functional with $Q = 0$ leads to delay-dependent conditions for systems with fast-varying delays, whereas for $R = S = 0$ it leads to delay-independent conditions (for systems with slowly-varying delays) and coincides with (1.24). The above V with $S = 0$ was introduced in [129], whereas the S-dependent term was added in [148].

Proposition 1.1 *Given $h \geq 0$, $d \in [0, 1)$. If there exist $n \times n$ matrices $P > 0$, $S > 0$, $Q > 0$, $R > 0$ and S_{12} such that the following LMIs are feasible:*

$$\begin{bmatrix} A^TP + PA + S + Q - R & S_{12} & PA_1 + R - S_{12} & hA^TR \\ * & -S - R & R - S_{12}^T & 0 \\ * & * & \Phi_{33} & hA_1^TR \\ * & * & * & -R \end{bmatrix} < 0, \tag{1.30}$$

$$\begin{bmatrix} R & S_{12} \\ * & R \end{bmatrix} \geq 0, \tag{1.31}$$

where $\Phi_{33} = -(1-d)Q - 2R + S_{12} + S_{12}^T$, then the system (1.23) is uniformly asymptotically stable for all delays $\tau(t) \in [0, h]$ such that $\dot{\tau}(t) \leq d$. Moreover, if the above conditions hold with $Q = 0$ (or, equivalently, with $d = 1$), then the system is uniformly asymptotically stable for all fast varying delays $\tau \in [0, h]$.

Proof Differentiating V given by (1.29), we find

$$\begin{aligned} \tfrac{d}{dt}V &\leq 2x^T(t)P\dot{x}(t) + h^2\dot{x}^T(t)R\dot{x}(t) - h\int_{t-h}^t \dot{x}^T(s)R\dot{x}(s)ds + x^T(t)[S+Q]x(t) \\ &-x^T(t-h)Sx(t-h) - (1-d)x^T(t-\tau(t))Qx(t-\tau(t)). \end{aligned} \tag{1.32}$$

Let the LMI (1.31) be feasible. Then by Jensen's inequality [131] and Park's inequality [147]

$$\begin{aligned} -h\int_{t-h}^t \dot{x}^T(s)R\dot{x}(s)ds &\leq -\begin{bmatrix} e_1 \\ e_2 \end{bmatrix}^T \begin{bmatrix} \frac{h}{\tau(t)}R & 0 \\ * & \frac{h}{h-\tau(t)}R \end{bmatrix} \begin{bmatrix} e_1 \\ e_2 \end{bmatrix} \\ &\leq -\begin{bmatrix} e_1 \\ e_2 \end{bmatrix}^T \begin{bmatrix} R & S_{12} \\ * & R \end{bmatrix} \begin{bmatrix} e_1 \\ e_2 \end{bmatrix}, \end{aligned} \tag{1.33}$$

where $e_1 = x(t) - x(t-\tau(t))$ and $e_2 = x(t-\tau(t)) - x(t-h)$.
Denote $\eta(t) = col\{x(t), x(t-h), x(t-\tau(t))\}$. Then employing (1.32), (1.33) and substituting for $\dot{x}(t)$ the right-hand side of (1.23), we arrive at

$$\tfrac{d}{dt}V \leq \eta^T(t)\Phi\eta(t) + \eta^T(t)[hRA \;\; 0 \;\; hRA_1]^TR^{-1}[hRA \;\; 0 \;\; hRA_1]\eta(t), \tag{1.34}$$

where

$$\Phi = \begin{bmatrix} A^TP + PA + S + Q - R & S_{12} & PA_1 + R - S_{12} \\ * & -S - R & R - S_{12}^T \\ * & * & -(1-d)Q - 2R + S_{12} + S_{12}^T \end{bmatrix}.$$

Applying the Schur complement to the last term in (1.34), we find that $\frac{d}{dt}V \leq -\epsilon|x(t)|^2$ for some $\epsilon > 0$ if (1.30) is feasible. □

1.5.3.3 Time-Dependent Lyapunov Functionals for Sampled-Data systems

Consider the sampled-data system given by (1.13), i.e.

$$\dot{x}(t) = Ax(t) + A_1 x(t_k), \quad x(t) \in \mathbb{R}_+^n, \ k = 0, 1, \ldots, \tag{1.35}$$

where $A_1 = BK$. It is assumed that the sampling intervals may be variable and uncertain (e.g. due to packet dropouts in networked control systems). However, it is assumed that $t_{k+1} - t_k \leq h$, where $h > 0$ is a known upper bound.

Till [149] the conventional time-independent Lyapunov functionals $V(x_t, \dot{x}_t)$ for systems with fast-varying delays were applied to sampled-data systems [130]. These functionals did not take advantage of the sawtooth evolution of the delays induced by sampled-and-hold. The latter drawback was removed in [149], where time-dependent Lyapunov functionals (inspired by [150]) were introduced for sampled-data systems.

Lemma 1.10 ([135]) *Let there exist positive numbers α, β and a functional $V : \mathbb{R} \times W[-h, 0] \times L^2[-h, 0] \to \mathbb{R}_+$ such that*

$$\alpha|\phi(0)|^2 \leq V(t, \phi, \dot{\phi}) \leq \beta\|\phi\|_W^2.$$

Consider the function $\bar{V}(t) = V(t, x_t, \dot{x}_t)$, which is continuous from the right for $x(t)$ satisfying (1.35), absolutely continuous for $t \neq t_k$ and which satisfies

$$\lim_{t \to t_k^-} \bar{V}(t) \geq \bar{V}(t_k). \tag{1.36}$$

If along (1.35) $\frac{d}{dt}\bar{V}(t) \leq -\epsilon|x(t)|^2, \quad t \neq t_k$ for some scalar $\epsilon > 0$, then (1.35) is asymptotically stable.

Consider the following simple Lyapunov functional:

$$V_s(t, x(t), \dot{x}_t) = \bar{V}(t) = x^T(t)Px(t) + V_U(t, \dot{x}_t), \ P > 0, \tag{1.37}$$

where

$$\begin{array}{l} V_U(t, \dot{x}_t) = (h - \tau(t)) \int_{t-\tau(t)}^t \dot{x}^T(s)U\dot{x}(s)ds, \\ \tau(t) = t - t_k, \quad U > 0. \end{array} \tag{1.38}$$

The discontinuous term V_U does not increase along the jumps since $V_U \geq 0$ and V_U vanishes after the jumps because $x(t)_{|t=t_k} = x(t - \tau(t))_{|t=t_k}$, i.e. (1.36) holds.

Since $\frac{d}{dt}x(t - \tau(t)) = (1 - \dot{\tau}(t))\dot{x}(t - \tau(t)) = 0$, we find

$$\frac{d}{dt}V_U(t, \dot{x}_t) = -\int_{t-\tau(t)}^t \dot{x}^T(s)U\dot{x}(s)ds + (h - \tau(t))\dot{x}^T(t)U\dot{x}(t)$$

and thus

$$\frac{d}{dt}\bar{V}(t) \le 2\dot{x}^T(t)Px(t) - \int_{t-\tau(t)}^{t} \dot{x}^T(s)U\dot{x}(s)ds + (h-\tau(t))\dot{x}^T(t)U\dot{x}(t). \quad (1.39)$$

Denoting

$$v_1 = \frac{1}{\tau(t)}\int_{t-\tau(t)}^{t} \dot{x}(s)ds,$$

we understand by $v_{1|\tau(t)=0}$ the following: $\lim_{\tau(t)\to 0} v_1 = \dot{x}(t)$. We apply further Jensen's inequality

$$\int_{t-\tau(t)}^{t} \dot{x}^T(s)U\dot{x}(s)ds \ge \tau(t)v_1^T U v_1,$$

and the descriptor method, where

$$0 = 2[x^T(t)P_2^T + \dot{x}^T(t)P_3^T][(A+A_1)x(t) - \tau(t)A_1 v_1 - \dot{x}(t)],$$

with some $n \times n$ matrices $P_2, P_3,$ is added to (1.39). Setting $\eta_1(t) = col\{x(t), \dot{x}(t), v_1\}$, we obtain that $\frac{d}{dt}\bar{V}(t) \le \eta_1^T(t)\Psi_s\eta_1(t) < -\epsilon|x(t)|^2$ for some $\epsilon > 0$ if

$$\Psi_s = \begin{bmatrix} \Phi_{11} & P - P_2^T + (A+A_1)^T P_3 & -\tau(t)P_2^T A_1 \\ * & -P_3 - P_3^T + (h-\tau(t))U & -\tau(t)P_3^T A_1 \\ * & * & -\tau(t)U \end{bmatrix} < 0,$$

where $\Phi_{11} = P_2^T(A + A_1) + (A + A_1)^T P_2$. The latter matrix inequality for $\tau(t) \to 0$ and $\tau(t) \to h$ leads to two LMIs:

$$\begin{bmatrix} \Phi_{11} & P - P_2^T + (A+A_1)^T P_3 \\ * & -P_3 - P_3^T + hU \end{bmatrix} < 0, \quad \begin{bmatrix} \Phi_{11} & P - P_2^T + (A+A_1)^T P_3 & -hP_2^T A_1 \\ * & -P_3 - P_3^T & -hP_3^T A_1 \\ * & * & -hU \end{bmatrix} < 0. \quad (1.40)$$

We arrived at the following:

Proposition 1.2 *Let there exist $n \times n$ matrices $P > 0$, $U > 0$, P_2 and P_3 such that the LMIs (1.40) are feasible. Then (1.35) is asymptotically stable for all variable sampling instants $t_{k+1} - t_k \le h$.*

The conditions of Proposition 1.2 cannot be applied to (1.35) with A_1 from uncertainty polytope, since in the matrix Ψ_s the matrix A_1 is multiplied by $\tau(t)$. Moreover, additional terms in the Lyapunov functional may further improve the results. See [149, 151] for various time-dependent construction of V. A different discontinuous in time Lyapunov functional was suggested in [152] which is based on **Wirtinger's inequality** [153]:

Let $z(t) : (a, b) \to \mathbb{R}^n$ be absolutely continuous with $\dot{z} \in L^2[a, b]$ and $z(a) = 0$. Then for any $n \times n$ matrix $W > 0$ Wirtinger's inequality holds:

$$\int_a^b z^T(\xi)Wz(\xi)d\xi \le \frac{4(b-a)^2}{\pi^2}\int_a^b \dot{z}^T(\xi)W\dot{z}(\xi)d\xi. \tag{1.41}$$

Consider the following Lyapunov functional [154]:

$$V(t, x_t, \dot{x}_t) = x^T(t)Px(t) + V_W(t, x_t, \dot{x}_t), \quad P > 0,$$

where the Wirtinger-based term is given by

$$V_W(t, x_t, \dot{x}_t) = h^2\int_{t_k}^t \dot{x}^T(s)W\dot{x}(s)ds - \frac{\pi^2}{4}\int_{t_k}^t [x(s)-x(t_k)]^T W[x(s)-x(t_k)]ds,$$
$$W > 0, \quad t_k \le t < t_{k+1}, \quad k = 0, 1, 2, \ldots$$

Since $[x(s) - x(t_k)]_{|s=t_k} = 0$, by Wirtinger's inequality (1.41) $V_W \ge 0$. Moreover, V_W vanishes at $t = t_k$, i.e. the condition (1.36) holds.

Setting $v(t) = x(t_k) - x(t)$ and differentiating V_W, we have

$$\frac{d}{dt}V_W = h^2\dot{x}^T(t)W\dot{x}(t) - \frac{\pi^2}{4}v^T(t)Wv(t).$$

Then we arrive at the following stability condition (that recovers result of [153, 155] derived via the small-gain theorem):

$$\begin{bmatrix} P(A+A_1)+(A+A_1)^TP & -\frac{\pi^2}{4}WPA_1 & h(A+A_1)^TW \\ * & -W & hA_1^TW \\ * & * & -W \end{bmatrix} < 0, \quad P > 0,\ W > 0.$$

1.5.4 Halanay's Inequality

Delay-independent (as well as delay-dependent) stability conditions for systems with fast-varying delays can be derived via Razumikhin method. The exponential decay rate can also be found by using Razumikhin method (see, e.g. [122]). Instead of Razumikhin method one can use Halanay's inequality [98, 135, 136].

Lemma 1.11 (Halanay's Inequality [136]) *Let* $V : [t_0 - h, +\infty) \to \mathbb{R}_+$ *be bounded on* $[t_0 - h, t_0]$ *and continuous on* $[t_0, \infty)$. *Assume that for some positive constants* $\delta_1 < \delta_0$ *the following inequality holds:*

$$D_-V(t) \le -2\delta_0 V(t) + 2\delta_1 \sup_{-h\le\theta\le 0} V(t+\theta),\ t \ge t_0, \tag{1.42}$$

where

$$D_-V(t) = \liminf_{s\to 0^-}\frac{V(t+s)-V(t)}{s}.$$

Then

$$V(t) \le e^{-2\delta(t-t_0)} \sup_{-h\le\theta\le 0} V(t_0+\theta),\ t \ge t_0, \tag{1.43}$$

where $\delta > 0$ is a unique positive solution of the equation $\delta = \delta_0 - \delta_1 e^{2\delta h}$.

Lemma 1.12 (A generalized Halanay's inequality [156, 157]) *Given $\Delta_f > 0$, let $V : [t_0 - h, \infty) \to \mathbb{R}^+$ be a locally absolutely continuous function, and $f : [t_0, \infty) \to \mathbb{C}$ be a bounded continuous function satisfying $|f(t)| \le \Delta_f$. If there exist $0 < \delta_1 < \delta_0$ and $\gamma > 0$ such that for all $t \ge t_0$ the following inequality holds*

$$\frac{d}{dt}V(t) + 2\delta_0 V(t) - 2\delta_1 \sup_{-h\le\theta\le 0} V(t+\theta) - \gamma|f(t)|^2 \le 0,$$

then we have

$$V(t) \le e^{-2\delta(t-t_0)} \sup_{-h\le\theta\le 0} V(t_0+\theta) + \frac{\gamma}{2\sigma}\Delta_f^2,\ t \ge t_0, \tag{1.44}$$

where $\sigma = \delta_0 - \delta_1 > 0$, and δ is a unique solution of

$$\delta = \delta_0 - \delta_1 e^{2\delta h}. \tag{1.45}$$

1.6 Monograph Overview

Besides the Introduction and some inequalities in this chapter, this monograph is organized as follows:

Chapter 2 is devoted to constrained control of PDE systems. In Sect. 2.1, we study regional boundary stabilization of nonlinear Schrödinger equation with state delay and bounded internal disturbance. The boundary constrained control law is designed by using the backstepping method. We establish regional ISS of the delayed nonlinear target system via a Lyapunov function and a generalized Halanay's inequality [157]. We find an estimate on the set of initial conditions starting from which the state trajectories of the system are exponentially attracted to a bounded set. In Sect. 2.2, we introduce stabilizing backstepping-based boundary controllers for coupled heat-ODE systems with time-varying state delays in the presence of actuator saturation. We first extend the backstepping method to the latter class of delayed systems. For simplicity only, our conditions are based on delay-independent stability condition for finite-dimensional system with delay. Less conservative delay-dependent conditions can be derived by employing Lyapunov-Krasovskii functionals similar to [142, 158].

Chapter 3 presents sampled-data control of PDE systems. In Sect. 3.1, we introduce distributed sampled-data control of 1D nonlinear KSE with the Dirichlet or periodic boundary conditions. The sensors provide either point or averaged discrete-time measurements of the state. The controllers enter KSE through distributed in space shape functions and the control signals are generated by zero-order hold devices. In Sect. 3.2, we design a sampled-data controller for 2D KSE under averaged/point measurements. Section 3.3 addresses a switched sampled-data control design for stabilization of KSE under the Dirichlet/periodic boundary conditions with spatially scheduled actuators. Section 3.4 is devoted to sampled-data control design of a PDE-PDE cascade system. These PDEs are governed by heat equations with different reaction coefficients.

Chapter 4 addresses distributed event-triggered control of PDEs. This chapter addresses the event-triggered control design for KdV system under in domain measurements averaged in space, and for the record, [12] gathers the results for distributed continuous-in-time controller to stabilize the KdV equation exponentially. Our concern here is then mainly to prove that distributed event-triggered control can still bring, under appropriate assumptions and choice of triggering mechanism, the expected exponential stability.

Chapter 5 discusses disturbance rejection approaches for PDE systems. In this chapter, disturbance rejection approaches are suggested to stabilize KdVB equation under the averaged measurements. Here two approaches—ADRC and DOBC, are introduced to reject the external unknown disturbance actively. Continuous-time and event-triggered anti-disturbance controllers are further presented for distributed stabilization of KdVB system. An event-triggering mechanism is utilized and the Zeno behaviour is avoided. Sufficient stability conditions are established via Lyapunov functional method. The effectiveness of proposed approaches is verified by simulation results.

References

1. Boskovic, D.M., Krstic, M., Liu, W.J.: Boundary control of an unstable heat equation via measurement of domain-averaged temperature. IEEE Trans. Autom. Control. **46**, 2022–2028 (2001)
2. Martin, T.: Traffic Flow Dynamics. Springer, Berlin, Heidelberg (2013)
3. Liu, W.J.: Boundary feedback stabilization of an unstable heat equation. SIAM J. Control. Optim. **42**, 1033–1043 (2003)
4. Orlov, Y., Dochain, D.: Discontinuous feedback stabilization of minimum-phase semilinear infinite-dimensional systems with application to chemical tubular reactor. IEEE Trans. Autom. Control. **47**, 1293–1304 (2002)
5. Dwivedy, S., Eberhard, P.: Dynamic analysis of flexible manipulators, a literature review. Mech. Mach. Theory **41**, 749–777 (2006)
6. Krstic, M.: Control of an unstable reaction-diffusion PDE with long input delay. Syst. & Control. Lett. **58**, 773–782 (2009)
7. Krstic, M., Smyshlyaev, A.: Boundary Control of PDEs: A Course on Backstepping Designs. SIAM, Philadelphia, PA (2008)
8. Katz, R., Fridman, E.: Constructive method for finite-dimensional observer-based control of 1-D parabolic PDEs. Automatica **122**, 109285 (2020)
9. Feng, H., Lang, P.H., Liu, J.H.: Boundary stabilization and observation of a weak unstable heat equation in a general multi-dimensional domain. Automatica **138**, 110152 (2022)

10. Baudouin, L., Crépeau, E., Valein, J.: Two approaches for the stabilization of nonlinear KdV equation with boundary time-delay feedback. IEEE Trans. Autom. Control. **64**, 1403–1414 (2019)
11. Cerpa, E., Coron, J.M.: Rapid stabilization for a Korteweg-de Vries equation from the left Dirichlet boundary condition. IEEE Trans. Autom. Control. **58**(7), 1688–1695 (2013)
12. Cerpa, E.: Control of a Korteweg-de Vries equation: a tutorial. Math. Control Relat. Fields **4**, 45–99 (2014)
13. Kang, W., Fridman, E.: Distributed stabilization of Korteweg-de Vries-Burgers equation in the presence of input delay. Automatica **100**, 260–273 (2019)
14. Marx, S., Cerpa, E.: Output feedback stabilization of the Korteweg-de Vries equation. Automatica **87**, 210–217 (2018)
15. Essimbi, B.Z., Jger, D.: Generation of electrical short pulses using a Schottky transmission line periodically loaded with tunnelling diodes. Phys. Scr. **81**(3), 035801 (2010)
16. Zhai, C., Wu, W.: A continuous traffic flow model considering predictive headway variation and preceding vehicle's taillight effect. Phys. A Stat. Mech. Appl. **584**, 126364 (2021)
17. Kuramoto, Y., Tsuzuki, T.: On the formation of dissipative structures in reaction-diffusion systems. Prog. Theor. Phys. **54**, 687–699 (1975)
18. Lunasin, E., Titi, E.S.: Finite determining parameters feedback control for distributed nonlinear dissipative systems—a computational study. Evol. Equ. Control. Theory **6**, 535–557 (2017)
19. Coron, J.M., Lü, Q.: Fredholm transform and local rapid stabilization for a Kuramoto-Sivashinsky equation. J. Differ. Equ. **259**, 3683–3729 (2015)
20. Liu, W.J., Krstic, M.: Stability enhancement by boundary control in the Kuramoto-Sivashinsky equation. Nonlinear Anal. Ser. A Theory Methods **43**, 485–507 (2001)
21. Armaou, A., Christofides, P.D.: Feedback control of the Kuramoto-Sivashinsky equation. Phys. D **137**, 49–61 (2000)
22. Armaou, A., Christofides, P.D.: Wave suppression by nonlinear finite-dimensional control. Chem. Eng. Sci. **55**, 2627–2640 (2000)
23. Christofides, P.D., Armaou, A.: Global stabilization of the Kuramoto-Sivashinsky equation via distributed output feedback control. Syst. & Control. Lett. **39**, 283–294 (2000)
24. Katz, R., Fridman, E.: Regional stabilization of the nonlinear 1D Kuramoto-Sivashinsky equation via modal decomposition. IEEE Control. Syst. Lett. **6**, 1814–1819 (2022)
25. Katz, R., Fridman, E.: Finite-dimensional boundary control of the linear Kuramoto-Sivashinsky equation under point measurement with guaranteed L^2-gain. IEEE Trans. Autom. Control. **67**, 5570–5577 (2022)
26. Agrawal, G.P.: Nonlinear Fiber Optics. Academic press (2007)
27. Chen, C.J., Menyuk, C.R., Islam, M.N., Stolen, R.H.: Numerical study of the Raman effect and its impact on soliton-dragging logic gates. Opt. Lett. **21**(16), 1647–1649 (1991)
28. Zhao, Z.H., Ge, W.G.: Traveling wave solutions for Schrödinger equation with distributed delay. Appl. Math. Model. **35**(2), 675–687 (2011)
29. Guo, B.Z., Liu, J.J.: Sliding mode control and active disturbance rejection control to the stabilization of one-dimensional Schrödinger equation subject to boundary control matched disturbance. Int. J. Robust Nonlinear Control **24**(16), 2194–2219 (2014)
30. Krstic, M., Guo, B.Z., Smyshlyaev, A.: Boundary controllers and observers for the linearized schrödinger equation. SIAM J. Control Optim. **49**(4), 1479–1497 (2011)
31. Machtyngier, E., Zuazua, E.: Stabilization of the Schrödinger equation. Port. Math. **51**(2), 243–256 (1994)
32. Wang, J.M., Ren, B.B., Krstic, M.: Stabilization and Gevrey regularity of a Schrödinger equation in boundary feedback with a heat equation. IEEE Trans. Autom. Control **57**(1), 179–185 (2012)
33. Ren, B.B., Wang, J.M., Krstic, M.: Stabilization of an ODE-Schrödinger cascade. Syst. Control Lett. **62**(6), 503–510 (2013)
34. Nicaise, S., Rebiai, S.: Stabilization of the Schrödinger equation with a delay term in boundary feedback or internal feedback. Port. Math. **68**(1), 19–39 (2011)

35. Tsubakino, D., Krstic, M., Yamashita, Y.: Boundary control of a cascade of two parabolic PDEs with different diffusion coefficients. In: Proceedings of the IEEE Conference on Decision and Control, Florence, Italy, pp. 3720–3725 (2013)
36. Krstic, M.: Compensating actuator and sensor dynamics governed by diffusion PDEs. Syst. & Control. Lett. **58**, 372–377 (2009)
37. Susto, G.A., Krstic, M.: Control of PDE-ODE cascades with Neumann interconnections. J. Franklin Inst. **347**, 284–314 (2010)
38. Tang, S., Xie, C.: State and output feedback boundary control for a coupled PDE-ODE system. Syst. Control Lett. **60**, 540–545 (2011)
39. Krstic, M.: Compensating a string PDE in the actuation or sensing path of an unstable ODE. IEEE Trans. Autom. Control. **54**, 1362–1368 (2009)
40. Baudouin, L., Seuret, A., Gouaisbaut, F.: Stability analysis of a system coupled to a heat equation. Automatica **99**, 195–202 (2019)
41. Lhachemi, H., Shorten, R.: Output feedback stabilization of an ODE-reaction-diffusion PDE cascade with a long interconnection delay. Automatica **147**, 110704 (2023)
42. Kang, W., Fridman, E.: Boundary control of delayed ODE-heat cascade under actuator saturation. Automatica **83**, 252–261 (2017)
43. Kang, W., Guo, B.Z.: Stabilisation of unstable cascaded heat partial differential equation system subject to boundary disturbance. IET Control. Theory Appl. **10**, 1027–1039 (2016)
44. Wang, J.M., Liu, J.J., Ren, B.B., Chen, J.H.: Sliding mode control to stabilization of cascaded heat PDE-ODE systems subject to boundary control matched disturbance. Automatica **52**, 23–34 (2015)
45. Wang, J., Krstic, M.: Output feedback boundary control of a heat PDE sandwiched between two ODEs. IEEE Trans. Autom. Control. **64**, 4653–4660 (2019)
46. Zhang, H.W., Wang, J.M., Gu, J.J.: Exponential input-to-state stabilization of an ODE cascaded with a reaction-diffusion equation subject to disturbances. Automatica **133**, 109885 (2021)
47. Zhang, Y.L., Wang, J.M., Guo, Y.P.: Input-to-state stability of an ODE-heat cascade system with disturbances. IET Control. Theory Appl. **13**, 191–202 (2019)
48. Corbett, D.C., Fabyan, W.B., Grigoryan, B.: Thermofluidic heat exchangers for actuation of transcription in artificial tissues. Sci. Adv. **6**, 9062 (2020)
49. Da Silva, J.M.G., Tarbouriech, S.: Antiwindup design with guaranteed regions of stability: an LMI-based approach. IEEE Trans. Automat. Control **50**(1), 106–111 (2005)
50. Fridman, E., Dambrine, M.: Control under quantization, saturation and delay: a LMI approach. Automatica **45**(10), 2258–2264 (2009)
51. Hu, T., Lin, Z.: Control Systems with Actuator Saturation: Analysis and Design. Birkhäuser (2001)
52. Tarbouriech, S., Garcia, G., da Silva Jr, J.M.G., Queinnec, I.: Stability and Stabilization of Linear Systems with Saturating Actuators. Springer, New York (2011)
53. El-Farra, N.H., El-Farra, A., Christofides, P.D.: Analysis and control of parabolic PDE systems with input constraints. Automatica **39**(4), 715–725 (2003)
54. Christofides, S., Cerpa, E., Prieur, C., Andrieu, V.: Stabilization of a linear Korteweg-de Vries equation with a saturated internal control. In: Proceedings of the European Control Conference, Linz, Austria, pp. 867–872 (2015)
55. Prieur, C., Tarbouriech, S., da Silva Jr, J.M.G.: Well-posedness and stability of 1D wave equation with saturating distributed input. In: Proceedings of the IEEE Conference on Decision and Control, Los Angeles, California, USA, pp. 2846–2851 (2014)
56. Lasiecka, I., Seidman, T.I.: Strong stability of elastic control systems with dissipative saturating feedback. Syst. & Control. Lett. **48**(3), 243–252 (2003)
57. Slemrod, M.: Feedback stabilization of a linear control system in Hilbert space with an a priori bounded control. Math. Control. Signals Syst. **2**(3), 265–285 (1989)
58. Lhachemi, H., Prieur, C.: Local output feedback stabilization of a reaction-diffusion equation with saturated actuation. IEEE Trans. Autom. Control. **68**(1), 564–571 (2022)

59. Fridman, E., Bar Am, N.: Sampled-data distributed H_∞ control of transport reaction systems. SIAM J. Control. Optim. **51**, 1500–1527 (2013)
60. Fridman, E., Blighovsky, A.: Robust sampled-data control of a class of semilinear parabolic systems. Automatica **48**, 826–836 (2012)
61. Kang, W., Fridman, E.: Distributed sampled-data control of Kuramoto-Sivashinsky equation. Automatica **95**, 514–524 (2018)
62. Karafyllis, I., Krstic, M.: Sampled-data boundary feedback control of 1-D parabolic PDEs. Automatica **87**, 226–237 (2018)
63. Rathnayake, B., Diagne, M., Karafyllis, I.: Sampled-data and event-triggered boundary control of a class of reaction-diffusion PDEs with collocated sensing and actuation. Automatica **137**, 110026 (2022)
64. Ghantasala, S., El-Farra, N.H.: Active fault-tolerant control of sampled-data nonlinear distributed parameter systems. Int. J. Robust Nonlinear Control. **22**, 24–42 (2012)
65. Kang, W., Fridman, E.: Sampled-data control of 2-D Kuramoto-Sivashinsky equation. IEEE Trans. Autom. Control. **67**(3), 1314–1326 (2022)
66. Tang, J.Q., Wang, J.M., Kang, W.: Sampled-data control of an unstable cascaded heat-heat system with different reaction coefficients. Automatica **171**, 111904 (2025)
67. Selivanov, A., Fridman, E.: Delayed H_∞ control of 2-D diffusion systems under delayed pointlike measurements. Automatica **109**, 108541 (2019)
68. Bar Am, N., Fridman, E.: Network-based distributed H_∞-filtering of parabolic systems. Automatica **50**, 3139–3146 (2014)
69. Wang, P., Fridman, E.: Sampled-data finite-dimensional boundary control of 2D semilinear parabolic stochastic PDEs. IEEE Trans. Autom. Control. (2025). https://doi.org/10.1109/TAC.2024.3514520
70. Kang, W., Fridman, E.: Sampled-data control of Kuramoto-Sivashinsky equation under the point measurements. In: Proceeding of the 16th European Control Conference (ECC) (2018)
71. Jones, D., Titi, E.S.: Upper bounds on the number of determining modes, nodes, and volume elements for the Navier-Stokes equations. Indiana Univ. Math. J. **42**, 875 (1993)
72. Liberzon, D.: Switching in Systems and Control. Springer Science & Business Media (2003)
73. Deaecto, G.: Dynamic output feedback H_∞ control of continuous-time switched affine systems. Automatica **71**, 44–49 (2016)
74. Hetel, L., Fridman, E.: Robust sampled-data control of switched affine systems. IEEE Trans. Autom. Control. **58**, 2922–2928 (2013)
75. Iftime, O.V., Demetriou, M.A.: Optimal control of switched distributed parameter systems with spatially scheduled actuators. Automatica **45**, 312–323 (2009)
76. Zuazua, E.: Switching control. J. Eur. Math. Soc. **13**, 85–117 (2010)
77. Butkovskiy, A.G., Pustyl'nikov, L.M.: Mobile Control of Distributed Parameter Systems. Ellis Horwood Limited, Chichester (1987)
78. Demetriou, M.A.: Guidance of mobile actuator-plus-sensor networks for improved control and estimation of distributed parameter systems. IEEE Trans. Autom. Control. **55**, 1570–1584 (2010)
79. Demetriou, M.A.: Adaptive control of 2-D PDEs using mobile collocated actuator/sensor pairs with augmented vehicle dynamics. IEEE Trans. Autom. Control. **57**, 2977–2978 (2012)
80. Wu, H.N., Zhang, X.W.: Integrated design of switching control and mobile actuator/sensor guidance for a linear diffusion process. J. Frankl. Inst. **356**, 7246–7262 (2019)
81. Wu, H.N., Zhang, X.W.: Static output feedback stabilization for a linear parabolic PDE system with time-varying delay via mobile collocated actuator/sensor pairs. Automatica **117**, 108993 (2020)
82. Zhao, S., Lin, H., Xue, Z.: Switching control of closed quantum systems via the Lyapunov method. Automatica **48**, 1833–1838 (2012)
83. Kang, W., Fridman, E., Liu, C.X.: Stabilization by switching of parabolic PDEs with spatially scheduled actuators and sensors. Automatica **147**, 110668 (2023)
84. El-Farra, N.H., Lou, Y., Christofides, P.D.: Fault-tolerant control of fluid dynamic systems via coordinated feedback and switching. Comput. Chem. Eng. **27**, 1913–1924 (2003)

85. Espitia, N., Karafyllis, I., Krstic, M.: Event-triggered boundary control of constant-parameter reaction-diffusion PDEs: a small-gain approach. Automatica **128**, 109562 (2021)
86. Jiang, Z.P., Liu, T.: Small-gain theory for stability and control of dynamical networks: a survey. Annu. Rev. Control. **46**, 58–79 (2018)
87. Espitia, N., Tanwani, A., Tarbouriech, S.: Stabilization of boundary controlled hyperbolic PDEs via Lyapunov-based event triggered sampling and quantization. In: IEEE Conference on Decision and Control, Austria, pp. 1266–1271 (2017)
88. Selivanov, A., Fridman, E.: Distributed event-triggered control of diffusion semilinear PDEs. Automatica **68**, 344–351 (2016)
89. Seuret, A., Prieur, C., Tarbouriech, S., Zaccarian, L.: LQ-based event-triggered controller co-design for saturated linear systems. Automatica **74**, 47–54 (2016)
90. Tabuada, P.: Event-triggered real-time scheduling of stabilizing control tasks. IEEE Trans. Autom. Control. **52**, 1680–1685 (2007)
91. Tallapragada, P., Chopra, N.: Decentralized event-triggering for control of nonlinear systems. IEEE Trans. Autom. Control. **59**, 3312–3324 (2014)
92. Heemels, W.P.M.H., Donkers, M.C.F., Teel, A.R.: Periodic event-triggered control for linear systems. IEEE Trans. Autom. Control. **58**, 847–861 (2013)
93. Heemels, W.P., Johansson, K.H., Tabuada, P.: An introduction to event-triggered and self-triggered control. In: 2012 IEEE 51st IEEE Conference on Decision and Control (CDC), pp. 3270–3285. IEEE (2012)
94. Selivanov, A., Fridman, E.: Event-triggered H_∞ control: a switching approach. IEEE Trans. Autom. Control. **61**, 3221–3226 (2016)
95. Peng, C., Han, Q.L.: A novel event-triggered transmission scheme and $\mathcal{L}_2$ control co-design for sampled-data control systems. IEEE Trans. Autom. Control. **58**(10), 2620–2626 (2013)
96. Yue, D., Tian, E., Han, Q.L.: A delay system method for designing event-triggered controllers of networked control systems. In: IEEE Conference on Decision and Control and European Control Conference, vol. 58, no. 2, pp. 1668–1673 (2011)
97. Katz, R., Fridman, E., Selivanov, A.: Boundary delayed observer-controller design for reaction-diffusion systems. IEEE Trans. Autom. Control. **66**(1), 275–282 (2020)
98. Fridman, E.: Introduction to Time-Delay Systems: Analysis and Control. Birkhäuser, Basel (2014)
99. Kang, W., Baudouin, L., Fridman, E.: Event-triggered control of Korteweg-de Vries equation under averaged measurements. Automatica **123**, 109315 (2021)
100. Rosier, L.: Exact boundary controllability for the Korteweg-de Vries equation on a bounded domain. ESAIM Control Optim. Calc. Var. **2**, 33–55 (1997)
101. Rosier, L., Zhang, B.Y.: Control and stabilization of the Korteweg-de Vries equation: recent progresses. J. Syst. Sci. & Complex **22**, 647–682 (2009)
102. Wu, Z.H., Zhou, H.C., Guo, B.Z., Deng, F.: Review and new theoretical perspectives on active disturbance rejection control for uncertain finite-dimensional and infinite-dimensional systems. Nonlinear Dyn. **101**(2), 935–959 (2020)
103. Kang, W., Wang, X.N., Wu, K.N., Li, Q., Liu, Z.J.: Observer-based H_∞ control of a stochastic Korteweg-de Vries-Burgers equation. Int. J. Robust Nonlinear Control. **31**(12), 5943–5961 (2021)
104. Liu, J.J., Wang, J.M.: Stabilization of one-dimensional wave equation with nonlinear boundary condition subject to boundary control matched disturbance. IEEE Trans. Autom. Control. **64**(7), 3068–3073 (2019)
105. Han, J.: From PID to active disturbance rejection control. IEEE Trans. Ind. Electron. **56**(3), 900–906 (2009)
106. Zhou, H.C., Guo, B.Z.: Stabilization of ODE with hyperbolic equation actuator subject to boundary control matched disturbance. Int. J. Control **92**(1), 12–26 (2019)
107. Nakao, M., Ohnishi, K., Miyachi, K.: A robust decentralized joint control based on interference estimation. In: Proceedings of the IEEE International Conference on Robotics and Automation (1987)

108. Wu, Z.H., Zhou, H.C., Deng, F., Guo, B.Z.: Disturbance observer-based boundary control for an anti-stable stochastic heat equation with unknown disturbance. IEEE Trans. Autom. Control. **68**(6), 3604–3611 (2022)
109. Zhang, Y.L., Zhu, M., Li, D., Wang, J.M.: ADRC dynamic stabilization of an unstable heat equation. IEEE Trans. Autom. Control. **65**(10), 4424–4429 (2020)
110. Guo, B.Z., Jin, F.F.: The active disturbance rejection and sliding mode control approach to the stabilization of the Euler-Bernoulli beam equation with boundary input disturbance. Automatica **49**(2013), 2911–2918 (2015)
111. Guo, B.Z., Jin, F.F.: Output feedback stabilization for one-dimensional wave equation subject to boundary disturbance. IEEE Trans. Autom. Control. **60**(3), 824–830 (2015)
112. Xu, Y., Wang, C., Qiao, J., Guo, L.: Robust stabilization for a class of nonlinear positive systems with multiple disturbances. IEEE Trans. Syst. Man Cybern. Syst. **52**(7), 4611–4622 (2022)
113. Zhao, Z., Ren, Y., Mu, C., Zou, T., Hong, K.S.: Adaptive neural-network-based fault-tolerant control for a flexible string with composite disturbance observer and input constraints. IEEE Trans. Cybern. **52**(12), 12843–12853 (2021)
114. He, W., Zhang, S., Ge, S.S.: Finite dimensional disturbance observer based control for nonlinear parabolic PDE systems via output feedback. IEEE Trans. Ind. Electron. **60**(11), 5186–5194 (2013)
115. Kang, W., Zhang, J., Wang, J.M.: Disturbance rejection approaches of Korteweg-de Vries-Burgers equation under event-triggering mechanism. Automatica **169**, 111844 (2024)
116. Payne, L., Weinberger, H.: An optimal Poincaré inequality for convex domains. Arch. Rat. Mech. Anal. **5**, 286–292 (1960)
117. Wang, T.: Stability in abstract functional-differential equations. II. Applications. J. Math. Anal. Appl. **186**, 835–861 (1994)
118. Rektorys, K.: Variational Methods in Mathematics, Science and Engineering. Springer Science & Business Media (2012)
119. Brezis, H., Brézis, H.: Functional Analysis, Sobolev Spaces and Partial Differential Equations, vol. 2, no. 3, p. 5. Springer, New York (2011)
120. Robinson, J.C.: Infinite-Dimensional Dynamical Systems: An Introduction to Dissipative Parabolic PDEs and the Theory of Global Attractors. Cambridge University Press (2001)
121. Hale, J., Lunel, S.M.V.: Introduction to Functional Differential Equations. Springer (1993)
122. Kolmanovskii, V., Myshkis, A.: Applied Theory of Functional Differential Equations. Kluwer (1999)
123. Richard, J.P.: Time-delay systems: an overview of some recent advances and open problems. Automatica **39**, 1667–1694 (2003)
124. Krasovskii, N.: Stability of Motion. Stanford University Press, 1959 [Russian] (1963)
125. Razumikhin, B.: On the stability of systems with a delay. Prikl. Math. Mech. [Russian] **20**, 500–512 (1956)
126. Boyd, S., El Ghaoui, L., Feron, E., Balakrishnan, V.: Linear Matrix Inequality in Systems and Control Theory, vol. 15. Stud. Appl. Math. (1994)
127. Mikheev, Yu., Sobolev, V., Fridman, E.: Asymptotic analysis of digital control systems. Autom. Remote. Control **49**, 1175–1180 (1988)
128. Fridman, E.: Using models with aftereffect in the problem of design of optimal digital control. Autom. Remote. Control **53**(10), 1523–1528 (1992)
129. Fridman, E., Shaked, U.: Delay dependent stability and H_∞ control: constant and time-varying delays. Int. J. Control **76**(1), 48–60 (2003)
130. Fridman, E., Seuret, A., Richard, J.P.: Robust sampled-data stabilization of linear systems: an input delay approach. Automatica **40**, 1441–1446 (2004)
131. Gu, K., Kharitonov, V.L., Chen, J.: Stability of Time-Delay Systems. Birkhäuser, Boston (2003)
132. El'sgol'tz, L., Norkin, S.: Introduction to the Theory and Applications of Differential Equations with Deviating Arguments. Mathematics in Science and Engineering, vol. 105. Academic Press, New York (1973)

133. Niculescu, S.I.: Delay Effects on Stability: A Robust Control Approach. Lecture Notes in Control and Information Sciences, vol. 269. Springer, London (2001)
134. Fridman, E., Selivanov, A.: Using delay for control. Annu. Rev. Control. Robot. Auton. Syst. 8 (2025)
135. Fridman, E.: Tutorial on Lyapunov-based methods for time-delay systems. Eur. J. Control. **20**(6), 271–283 (2014)
136. Halanay, A.: Differential Equations: Stability, Oscillations. Time Lags. Academic Press, New York (1966)
137. Li, X., De Souza, C.E.: Criteria for robust stability and stabilization of uncertain linear systems with state delay. Automatica **33**(9), 1657–1662 (1997)
138. Kolmanovskii, V.B., Richard, J.P.: Stability of some linear systems with delays. IEEE Trans. Autom. Control. **44**(5), 984–989 (1999)
139. Park, P.: A delay-dependent stability criterion for systems with uncertain time-invariant delays. IEEE Trans. Autom. Control. **44**(4), 876–877 (1999)
140. Gu, K., Niculescu, S.I.: Further remarks on additional dynamics in various model transformations of linear delay systems. IEEE Trans. Autom. Control. **46**(3), 497–500 (2001)
141. Kharitonov, V.L., Melchor-Aguilar, D.: On delay-dependent stability conditions. Syst. Control. Lett. **40**(1), 71–76 (2000)
142. Fridman, E.: New Lyapunov-Krasovskii functionals for stability of linear retarded and neutral type systems. Syst. & Control. Lett. **43**, 309–319 (2001)
143. Fridman, E., Shaked, U.: Input-output approach to stability and L2-gain analysis of systems with time-varying delays. Syst. Control. Lett. **55**(12), 1041–1053 (2006)
144. Fridman, E.: Stability of linear descriptor systems with delay: a Lyapunov-based approach. J. Math. Anal. Appl. **273**(1), 24–44 (2002)
145. Suplin, V., Fridman, E., Shaked, U.: Sampled-data H^{∞} control and filtering: nonuniform uncertain sampling. Automatica **43**(6), 1072–1083 (2007)
146. Fridman, E., Orlov, Y.: Exponential stability of linear distributed parameter systems with time-varying delays. Automatica **45**(1), 194–201 (2009)
147. Park, P., Ko, J.W., Jeong, C.: Reciprocally convex approach to stability of systems with time-varying delays. Automatica **47**(1), 235–238 (2011)
148. He, Y., Wang, Q.G., Lin, C., Wu, M.: Delay-range-dependent stability for systems with time-varying delay. Automatica **43**(2), 371–376 (2007)
149. Fridman, E.: A refined input delay approach to sampled-data control. Automatica **46**, 421–427 (2010)
150. Naghshtabrizi, P., Hespanha, J.P., Teel, A.R.: Exponential stability of impulsive systems with application to uncertain sampled-data systems. Syst. Control. Lett. **57**(5), 378–385 (2008)
151. Seuret, A.: A novel stability analysis of linear systems under asynchronous samplings. Automatica **48**(1), 177–182 (2012)
152. Liu, K., Fridman, E.: Wirtinger's inequality and Lyapunov-based sampled-data stabilization. Automatica **48**(1), 102–108 (2012)
153. Liu, K., Suplin, V., Fridman, E.: Stability of linear systems with general sawtooth delay. IMA J. Math. Control. Inf. **27**(4), 419–436 (2010)
154. Liu, K., Fridman, E.: Delay-dependent methods and the first delay interval. Syst. Control Lett. **64**, 57–63 (2014)
155. Mirkin, L.: Some remarks on the use of time-varying delay to model sample-and-hold circuits. IEEE Trans. Autom. Control. **52**(6), 1109–1112 (2007)
156. Katz, R., Fridman, E.: Sampled-data finite-dimensional boundary control of 1D parabolic PDEs under point measurement via a novel ISS Halanay's inequality. Automatica **135**, 109966 (2022)
157. Wen, L.P., Yu, Y.X., Wang, W.S.: Generalized Halanay inequalities for dissipativity of Volterra functional differential equations. J. Math. Anal. Appl. **347**(1), 169–178 (2008)
158. Tarbouriech, S., Da Silva, J.M.G.: Synthesis of controllers for continuous-time delay systems with saturating controls via LMIs. IEEE Trans. Autom. Control. **45**, 105–111 (2000)

Chapter 2
Constrained Control of PDE Systems

This chapter focuses on the boundary stabilization of delayed PDE systems. This chapter explores the regional stability of the nonlinear Schrödinger equation with state delay and internal disturbances. Besides, boundary stabilization of the cascaded ODE-heat system with a time-varying state delay and actuator saturation is provided. The chapter presents control strategies based on the backstepping method and establishes stability conditions for these systems, highlighting the effectiveness of the proposed approaches through numerical examples constrained.

2.1 Boundary Constrained Control of Delayed Nonlinear Schrödinger Equation

This section studies regional boundary stabilization of nonlinear Schrödinger equation with state delay and bounded internal disturbance. The boundary constrained control law is designed by using the backstepping method. Regional input-to-state stability (ISS) of the perturbed system with time-delay is established by a Lyapunov function and a generalized Halanay's inequality. Estimates on the set of initial conditions are found starting from which the solutions are exponentially attracted to a bounded set [1].

2.1.1 Regional Stabilization: No Constraints on Control

2.1.1.1 System Description

We consider the following nonlinear Schrödinger equation with time delay and internal disturbance:

W. Kang and E. Fridman, *Delay-Robust Control of Distributed Parameter Systems*, Advances in Delays and Dynamics 13, https://doi.org/10.1007/978-981-96-6344-6_2

$$\begin{cases} \mathrm{i}z_t(x,t) = -z_{xx}(x,t) - \mu z(x,t-\tau(t))|z(x,t)|^2 + f(x,t), \\ z(0,t) \quad = 0, \\ z_x(1,t) \quad = U(t), \\ z(x,t) \quad = \psi(x,t),\ 0 \le x \le 1,\ -h \le t \le 0, \end{cases} \tag{2.1}$$

where $x \in (0,1)$, i is the imaginary unit, $\mu \in \mathbb{R}$ denotes a Landau coefficient, $z(x,t) \in \mathbb{C}$ is the state of nonlinear Schrödinger equation, $\psi(x,t)$ is the initial state, $U(t) \in \mathbb{C}$ is the control input. Here $f(x,t)$ represents an uncertain distributed disturbance source term, $\tau(t)$ is a continuously differentiable bounded delay

$$0 < h_0 \le \tau(t) \le h \tag{2.2}$$

with some constants h_0 and $h > 0$. The delay $\tau(t)$ may be unknown.

Following [2], we consider the nonlinear delay term "$\mu z(x,t-\tau(t))|z(x,t)|^2$" in system (2.1). Our results can be easily extended to another kind of nonlinear delay term "$\mu z(x,t)|z(x,t-\tau(t))|^2$" that was studied in [3, 4] (see Remark 2.2 below).

We analyze system (2.1) in the energy state space $H_L^1(0,1) = \{g \in H^1(0,1) | g(0) = 0\}$ defined in the complex number field $\mathbb{C}$. The space is naturally equipped with the H_L^1-norm given by $\|g\|^2_{H_L^1} = \int_0^1 |g'(x)|^2 dx$.

Assumption 2.1.1 The disturbances are assumed to be continuously differentiable function of class $C^1([0,\infty); H_L^1(0,1))$ subject to

$$\|f\|_{C^1([0,\infty); H_L^1(0,1))} \le \Delta_f, \tag{2.3}$$

where Δ_f is a given constant upper bound. □

Our objective is regional stabilization of (2.1). In order to find a bound on the domain of attraction, we change the state variable $v(x,t) = z_x(x,t)$ (see Remark 2.1 below for detailed explanation). Then the new state $v(x,t)$ satisfies the following PDE:

$$\begin{cases} \mathrm{i}v_t(x,t) = -v_{xx}(x,t) - \mu v(x,t-\tau(t)) \left|\int_0^x v(r,t)dr\right|^2 \\ \qquad -\mu \int_0^x v(r,t-\tau(t))dr \left[\int_0^x v(r,t)dr\overline{v(x,t)} \right. \\ \qquad \left. +v(x,t)\int_0^x \overline{v(r,t)}dr\right] + f_x(x,t), \\ v_x(0,t) \quad = 0, \\ v(1,t) \quad = U(t), \\ v(x,t) \quad = \psi_x(x,t),\ 0 \le x \le 1,\ -h \le t \le 0. \end{cases} \tag{2.4}$$

2.1.1.2 Backstepping Control

Now we consider a backstepping transformation

$$u(x,t) = v(x,t) - \int_0^x k(x,y)v(y,t)dy = (I - \mathbb{P})v, \tag{2.5}$$

that transforms the system (2.4) into the following target Schrödinger equation:

$$\begin{cases} \mathrm{i}u_t(x,t) = -u_{xx}(x,t) - \mathrm{i}cu(x,t) + (I-\mathbb{P})[F_1(v,x,t) + f_x(x,t)], \\ u_x(0,t) \ = 0, \\ u(1,t) \ \ = U(t) - \int_0^1 k(1,y)v(y,t)dy, \\ u(x,t) \ \ = \phi(x,t),\ 0 \le x \le 1,\ -h \le t \le 0, \end{cases} \tag{2.6}$$

where $c > 0$ is a constant,

$$\phi(x,t) = \psi_x(x,t) - \int_0^x k(x,y)\psi_y(y,t)dy, \tag{2.7}$$

and

$$\begin{aligned} F_1(v,x,t) = &-\mu v(x,t-\tau(t)) \left| \int_0^x v(r,t)dr \right|^2 \\ &-\mu \int_0^x v(r,t-\tau(t))dr \left[\int_0^x v(r,t)dr\overline{v(x,t)} + v(x,t) \int_0^x \overline{v(r,t)}dr \right]. \end{aligned} \tag{2.8}$$

Note that (2.5) is a modified transformation of [5] introduced for linear equation given by $v_t = -\mathrm{i}v_{xx}$. Next, we compute the kernel $k(x,y)$. Differentiation of transformation (2.5) with respect to t yields

$$\mathrm{i}u_t(x,t) = -v_{xx}(x,t) + \int_0^x k(x,y)v_{yy}(y,t)dy + (I-\mathbb{P})[F_1(v,x,t) + f_x(x,t)].$$

Substituting (2.5) into the resulting equation with some calculations that involve integration by parts, we obtain

$$\begin{aligned} \mathrm{i}u_t(x,t) = &-v_{xx}(x,t) + k(x,x)v_x(x,t) - k_y(x,x)v(x,t) + k_y(x,0)v(0,t) \\ &+ \int_0^x k_{yy}(x,y)v(y,t)dy + (I-\mathbb{P})[F_1(v,x,t) + f_x(x,t)]. \end{aligned}$$

Similarly, the first and second derivatives of $u(x,t)$ with respect to x are given by

$$u_x(x,t) = v_x(x,t) - k(x,x)v(x,t) - \int_0^x k_x(x,y)v(y,t)dy,$$

$$\begin{aligned} u_{xx}(x,t) = v_{xx}(x,t) - &\frac{d}{dx}k(x,x)v(x,t) - k(x,x)v_x(x,t) \\ &-k_x(x,x)v(x,t) - \int_0^x k_{xx}(x,y)v(y,t)dy. \end{aligned}$$

Substituting (2.5) into (2.4) and comparing with (2.6), we obtain the following set of conditions on the kernel $k(x, y)$ (cf.[5]):

$$\begin{cases} k_{xx}(x, y) = k_{yy}(x, y) - \mathrm{i}ck(x, y), \\ k_y(x, 0) = 0, \\ k(x, x) = \dfrac{\mathrm{i}c}{2}x. \end{cases} \tag{2.9}$$

The solution to (2.9) is given by

$$k(x, y) = \mathrm{i}cx\frac{I_1(\sqrt{\mathrm{i}c(y^2 - x^2)})}{\sqrt{\mathrm{i}c(y^2 - x^2)}},$$

where $I_1(\cdot)$ denotes the modified Bessel function of the first order: $I_1(x) = \sum_{n=0}^{\infty} \frac{(x/2)^{2n+1}}{n!(n+1)!}$.
The change of variable (2.5) has an inverse transformation:

$$v(x, t) = u(x, t) + \int_0^x l(x, y)u(y, t)dy = (I - \mathbb{P})^{-1}u, \tag{2.10}$$

where

$$l(x, y) = \mathrm{i}cx\frac{J_1(\sqrt{\mathrm{i}c(y^2 - x^2)})}{\sqrt{\mathrm{i}c(y^2 - x^2)}},$$

and $J_1(\cdot)$ is the Bessel function of the first order: $J_1(x) = \sum_{n=0}^{\infty} \frac{(-1)^n(x/2)^{2n+1}}{n!(n+1)!}$.

The following functionals are used to simplify the presentation (cf. (2.5), (2.10)) in the upcoming sections:

$$\mathcal{K}[v] = (I - \mathbb{P})v = v(x, t) - \int_0^x k(x, y)v(y, t)dy = u, \tag{2.11}$$

$$\mathcal{L}[u] = (I - \mathbb{P})^{-1}u = u(x, t) + \int_0^x l(x, y)u(y, t)dy = v, \tag{2.12}$$

where both $(I - \mathbb{P})$ and $(I - \mathbb{P})^{-1}$ are bounded on $L^2(0, 1)$.

We denote $|v| = |v(x, t)|$ and $|u| = |u(x, t)|$. From (2.11) and (2.12), it can be shown that

$$|\mathcal{K}[v]| \le k_1(|v| + \|v\|_{L^1}), \ |\mathcal{L}[u]| \le k_2(|u| + \|u\|_{L^1}), \tag{2.13}$$

$$\|\mathcal{L}[u]\|_{L^2} \le k_3\|u\|_{L^2}, \ \|\mathcal{K}[v]\|_{L^2} \le k_4\|v\|_{L^2}, \tag{2.14}$$

where

$$\begin{aligned} k_1 &= \max\{1, \max_{0\le y\le x\le 1} |k(x,y)|\}, & k_2 &= \max\{1, \max_{0\le y\le x\le 1} |l(x,y)|\}, \\ k_3 &= 1 + \max_{0\le y\le x\le 1} |l(x,y)|, & k_4 &= 1 + \max_{0\le y\le x\le 1} |k(x,y)|. \end{aligned} \tag{2.15}$$

By selecting the feedback controller

$$U(t) = \int_0^1 k(1,y)v(y,t)dy = k(1,1)z(1,t) - \int_0^1 k_y(1,y)z(y,t)dy, \tag{2.16}$$

from (2.6), (2.8), (2.11) and (2.12), one arrives at the target system:

$$\begin{cases} \mathrm{i}u_t(x,t) = -u_{xx}(x,t) - \mathrm{i}cu(x,t) + \mathcal{K}[F(u,x,t) + f_x(x,t)], \\ u_x(0,t) = 0, \\ u(1,t) = 0, \\ u(x,t) = \phi(x,t),\ 0 \le x \le 1,\ -h \le t \le 0, \end{cases} \tag{2.17}$$

where

$$\begin{aligned} F(u,x,t) &= F_1(v,x,t) \\ &= -\mu\mathcal{L}[u(x,t-\tau(t))]\left|\int_0^x \mathcal{L}[u(r,t)]dr\right|^2 \\ &-\mu\overline{\mathcal{L}[u(x,t)]}\int_0^x \mathcal{L}[u(r,t-\tau(t))]dr \int_0^x \mathcal{L}[u(r,t)]dr \\ &-\mu\mathcal{L}[u(x,t)]\int_0^x \mathcal{L}[u(r,t-\tau(t))]dr \int_0^x \overline{\mathcal{L}[u(r,t)]}dr. \end{aligned} \tag{2.18}$$

2.1.1.3 Well-Posedness and ISS Analysis

Consider the target system (2.17). We introduce the Hilbert space $L^2(0,1)$. For the well-posedeness, we will use the step method for solution of time-delay systems [6].

Define the system operator $\mathcal{A}_u : D(\mathcal{A}_u) \to L^2(0,1)$ for (2.17) as follows:

$$\begin{cases} \mathcal{A}_u = \mathrm{i}\dfrac{\partial^2}{\partial x^2} - c, \\ D(\mathcal{A}_u) = \left\{u \in H^2(0,1) | u'(0) = u(1) = 0\right\}. \end{cases} \tag{2.19}$$

A simple computation shows that

$$\mathrm{Re}\langle \mathcal{A}_u u, u\rangle = -c\|u\|_{L^2}^2 \le 0,\ \forall u \in D(\mathcal{A}_u).$$

Therefore, the infinitesimal operator $\mathcal{A}_u$ is dissipative. Then from the Lumer-Phillips theorem ([7, Theorem 1.4.3]), it follows that the operator $\mathcal{A}_u$ generates an exponentially stable semigroup $e^{\mathcal{A}_u t}$ on $L^2(0,1)$ (see also [5] that claimed this fact).

While being viewed over the time segment $[0, h_0]$, with the operator $\mathcal{A}_u$ at hand, the system (2.17) can be rewritten as the differential equation in $L^2(0,1)$:

$$\frac{d}{dt}\zeta(t) = \mathcal{A}_u\zeta(t) + \hat{F}(\zeta, t), \tag{2.20}$$

$$\zeta(0) = \phi(\cdot, 0), \tag{2.21}$$

where

$$\zeta(t) = u(\cdot, t), \ \ \hat{F}(\zeta, t) = -\mathrm{i}(I - \mathbb{P})\bar{F}(\zeta, t) \tag{2.22}$$

subject to

$$\begin{aligned}\bar{F}(\zeta, t) = &-\mu\mathcal{L}[\phi(t - \tau(t))]\left|\int_0^x \mathcal{L}[\zeta]dr\right|^2 \\ &-\mu\overline{\mathcal{L}[\zeta]}\int_0^x \mathcal{L}[\phi(t - \tau(t))]dr \int_0^x \mathcal{L}[\zeta]dr \\ &-\mu\mathcal{L}[\zeta]\int_0^x \mathcal{L}[\phi(t - \tau(t))]dr \int_0^x \overline{\mathcal{L}[\zeta]}dr + f_x(x, t).\end{aligned}$$

Definition 2.1 A function $\zeta \in L^1([0, T]; D(\mathcal{A}_u)) \cap C([0, T]; L^2(0, 1))$ such that $\dot{\zeta} \in L^1([0, T]; L^2(0, 1))$ is called a strong solution of initial value problem (2.20), (2.21) if (2.21) is satisfied and (2.20) holds a.e. on $[0, T]$.

Our definition of strong solution follows [7] (see page 109).

Consider the following initial space for the target system:

$$Z \triangleq C\big([-h, 0]; D(\mathcal{A}_u)\big) \cap C^1\big([-h, 0]; L^2(0, 1)\big), \tag{2.23}$$

which is endowed with the norm

$$\|\phi\|_Z = \|\phi\|_{C([-h,0];D(\mathcal{A}_u))} + \|\phi\|_{C^1([-h,0];L^2(0,1))}.$$

Lemma 2.1 *Given any $h_0 > 0$ and $\phi \in Z$, let $\hat{F}(\zeta, t)$ be defined by (2.22), where $t \in [0, h_0]$ and $f \in C^1([0, h_0]; H_L^1(0, 1))$. Then*

(i) $\hat{F}$ is Lipschitz continuous in both variables locally in ζ, i.e. for any $t_1, t_2 \in [0, h_0]$ and any $R > 0$, there exists a positive constant $K(R)$ such that the inequality

$$\|\hat{F}(\zeta_1, t_1) - \hat{F}(\zeta_2, t_2)\|_{L^2} \leq K(R)\left(|t_1 - t_2| + \|\zeta_1 - \zeta_2\|_{L^2}\right) \tag{2.24}$$

holds for all $\zeta_1, \zeta_2 \in L^2(0, 1)$ with $\|\zeta_1\|_{L^2} \leq R$, $\|\zeta_2\|_{L^2} \leq R$.

(ii) There exists a unique strong solution of (2.20) on some interval $[0, t'] \subset [0, h_0]$ (here $t' > 0$ depends on ϕ). Moreover, if this solution admits a priori estimate

$$\|\zeta\|_{L^2} \leq C(\phi),$$

where $C(\phi)$ is constant depending on initial state ϕ, then the solution exists on the entire interval $[0, h_0]$. This solution is Lipschitz continuous on $[0, h_0]$ meaning that $\|\zeta(t_1) - \zeta(t_2)\|_{L^2} \leq L(\phi)|t_1 - t_2|$ (here $L(\phi)$ depends on ϕ).

Proof (i) Since $\phi \in Z$, τ is continuously differentiable, and $\mathcal{L}$ is bounded (cf. (2.14)), we have

$$
\begin{aligned}
&\|\mathcal{L}[\phi(t_1-\tau(t_1))-\phi(t_2-\tau(t_2))]\|_{L^2} \\
&\leq \|\mathcal{L}[\phi(t_1-\tau(t_1))-\phi(t_1-\tau(t_2))]\|_{L^2}+\|\mathcal{L}[\phi(t_1-\tau(t_2))-\phi(t_2-\tau(t_2))]\|_{L^2} \\
&\leq k_3\|\phi\|_Z\left(1+\max_{0\leq s\leq h_0}|\dot{\tau}(s)|\right)|t_1-t_2|,\ \forall t_1,t_2\in[0,h_0].
\end{aligned}
\tag{2.25}
$$

Here k_3 and k_4 are given by (2.15).

Since $I-\mathbb{P}$ is bounded (cf. (2.14)), application of the Minkowski's and Cauchy-Schwarz's inequalities leads to

$$
\begin{aligned}
&\|\hat{F}(\zeta_1,t_1)-\hat{F}(\zeta_2,t_2)\|_{L^2} \\
&\leq \|(I-\mathbb{P})\|\cdot\|\bar{F}(\zeta_1,t_1)-\bar{F}(\zeta_2,t_2)\|_{L^2}\leq k_4\|\bar{F}(\zeta_1,t_1)-\bar{F}(\zeta_2,t_2)\|_{L^2} \\
&\leq k_4\|\bar{F}(\zeta_1,t_1)-\bar{F}(\zeta_1,t_2)\|_{L^2}+k_4\|\bar{F}(\zeta_1,t_2)-\bar{F}(\zeta_2,t_2)\|_{L^2} \\
&\leq 3k_4\mu\|\mathcal{L}[\phi(t_1-\tau(t_1))-\phi(t_2-\tau(t_2))]\|_{L^2}\cdot\|\mathcal{L}[\zeta_1]\|_{L^2}^2 < \\
&+3k_4\mu\|>L[\phi(t_2-\tau(t_2))]\|_{L^2}\cdot\|\mathcal{L}[\zeta_1-\zeta_2]\| \\
&\times\left[\|\mathcal{L}[\zeta_1]\|_{L_2}+\mathcal{L}[\zeta_2]\|_{L_2}\right]+k_4\|f(\cdot,t_1)-f(\cdot,t_2)\|_{H_L^1}, \\
&\qquad\qquad \forall t_1,t_2\in[0,h_0],\ \forall\zeta_1,\zeta_2\in L^2(0,1).
\end{aligned}
\tag{2.26}
$$

The fact that $f\in C^1([0,h_0];H_L^1(0,1))$ implies the local Lipschitz condition holds, i.e.$\|f(\cdot,t_1)-f(\cdot,t_2)\|_{H_L^1}\leq K_1|t_1-t_2|$, $\forall t_1,t_2\in[0,h_0]$ for some $K_1>0$.

Taking into account (2.25) and the latter relation, we conclude that there exists a positive constant $K(R)$ such that (2.24) holds for all $\zeta_1,\zeta_2\in L^2(0,1)$ with $\|\zeta_1\|_{L^2}\leq R$, $\|\zeta_2\|_{L^2}\leq R$.

(ii) The existence and uniqueness of local strong solution of (2.20) on some interval $[0,t']\subset[0,h_0]$ follow now from Theorems 6.1.4 and 6.1.6 of [7].

Continuation of this solution under a priori bound to entire interval $[0,h_0]$ follows from Theorem 6.23.5 of [8]. The fact that solution is Lipschitz follows from the proof of Theorem 6.1.6 of [7] (cf. (6.1.20)).

Definition 2.2 By strong solution of the initial value problem for the target system (2.17) with the initial function $\phi\in Z$, we understand a function $u\in L^1([0,\infty);D(\mathcal{A}_u))\cap C([0,\infty);L^2(0,1))$ such that $\dot{u}\in L^1([0,\infty);L^2(0,1))$ and the corresponding evolution equation holds a.e. on $[0,\infty)$.

Now we are in a position to formulate the conditions that guarantee the existence of strong solution of (2.17) for all $t\geq 0$ with small enough initial data $\max_{-h\leq\theta\leq 0}\|\phi(\cdot,\theta)\|_{L^2}$ as well as for regional ISS of (2.17):

Proposition 2.1 *Consider the target system (2.17) under Assumption 1. Given* $h_0 > 0$, $c > 0$, $\Delta_f > 0$, $C_1 > 0$ *and* $\mu \in \mathbb{R}$, *assume that there exist* $0 < \delta_1 < \delta_0$ *and* $\gamma > 0$ *such that the following LMIs hold:*

$$(\delta_0 - \delta_1)C_1 > \gamma \Delta_f^2, \tag{2.27}$$

$$\Theta_0 = \begin{bmatrix} -c + \delta_0 & 6k_1k_2k_3^2|\mu|C_1 & k_1 \\ * & -\delta_1 & 0 \\ * & * & -\gamma \end{bmatrix} < 0, \tag{2.28}$$

where $k_i > 0$ $(i = 1, 2, 3)$ *are defined by (2.15). Then for any initial condition* ϕ *from the set*

$$\mathcal{X}_\phi = \{\phi \in Z : \max_{-h \le \theta \le 0} \|\phi(\cdot, \theta)\|_{L^2}^2 \le C_1 - \frac{\gamma}{\sigma}\Delta_f^2\}, \tag{2.29}$$

where $\sigma = \delta_0 - \delta_1$, *the following holds:*

(i) [Well-posedness] There exists a unique strong solution of the initial value problem for the target system (2.17) for all $t \ge 0$.

(ii) [Regional ISS] The strong solution of (2.17) satisfies

$$\int_0^1 |u(x,t)|^2 dx \le e^{-2\delta t} \max_{\theta \in [-h,0]} \int_0^1 |u(x,\theta)|^2 dx + \frac{\gamma}{\sigma}\Delta_f^2 \tag{2.30}$$

for all $t \ge 0$, *where* δ *is a unique solution of (1.45).*

(iii) Given $C_1 > 0$ *and any* $\delta > 0$, *the decay rate* δ *in the inequality (2.30) can be achieved by appropriate choice of the design parameter* c.

Proof (i)–(ii): Step 1: From (ii) of Lemma 2.1, it follows that there exists a strong solution of (2.20) initialized with $\phi \in \mathcal{X}_\phi (\subset Z)$ on some interval $[0, t'] \subset [0, h_0]$. Therefore, we first employ the Lyapunov functional for $[0, t']$,

$$V(t) = \frac{1}{2}\int_0^1 |u(x,t)|^2 dx. \tag{2.31}$$

Taking the time derivative of the Lyapunov function along (2.17), we obtain

$$\begin{aligned} &\dot{V}(t) + 2\delta_0 V(t) - 2\delta_1 \sup_{-h \le \theta \le 0} V(t+\theta) - \gamma \|f(\cdot,t)\|_{H_L^1}^2 \\ &\le -c\int_0^1 |u(x,t)|^2 dx + \delta_0 \int_0^1 |u(x,t)|^2 dx - \delta_1 \int_0^1 |u(x,t-\tau(t))|^2 dx \\ &-\gamma \|f(x,t)\|_{H_L^1}^2 + \int_0^1 |u(x,t)| \cdot |\mathcal{K}[F(u,x,t) + f_x(x,t)]| dx. \end{aligned} \tag{2.32}$$

From (2.13) and (2.14), we have

$$\int_0^1 |u(x,t)| \cdot |\mathcal{K}[F(u,x,t)]|dx \le k_1 \int_0^1 |u(x,t)|[|F(u,x,t)| + \|F(u,x,t)\|_{L^1}]dx,$$

$$\int_0^1 |\mathcal{L}[u(x,t)]|^2 dx \le k_3^2 \int_0^1 |u(x,t)|^2 dx. \tag{2.33}$$

By the Cauchy-Schwartz inequality and (2.33), we obtain

$$\begin{aligned}
&\int_0^1 |u(x,t)| \cdot |F(u,x,t)|dx \\
&\le |\mu| \int_0^1 |u(x,t)||\mathcal{L}[u(x,t-\tau(t))]| \left|\int_0^x \mathcal{L}[u(r,t)]dr\right|^2 dx \\
&+2|\mu| \int_0^1 |u(x,t)||\mathcal{L}[u(x,t)]| \int_0^x |\mathcal{L}[u(r,t-\tau(t))]|\, dr \times \int_0^x |\mathcal{L}[u(r,t)]|drdx \\
&\le 6|\mu|k_2k_3^2\|u(\cdot,t)\|_{L^2}^2[\|u(\cdot,t-\tau(t))\|_{L^2}\|u(\cdot,t)\|_{L^2}].
\end{aligned} \tag{2.34}$$

Similarly, by the Minkowski inequality, we find

$$\int_0^1 |u(x,t)|\|F(u,x,t)\|_{L^1}dx \le 6|\mu|k_2k_3^2\|u(\cdot,t)\|_{L^2}^2[\|u(\cdot,t-\tau(t))\|_{L^2}\|u(\cdot,t)\|_{L^2}].$$

Set $\eta(t) = \mathrm{col}\left\{\|u(\cdot,t)\|_{L^2}, \|u(\cdot,t-\tau(t))\|_{L^2}, \|f(\cdot,t)\|_{H_L^1}\right\}$. If (2.27), (2.28) are feasible, then for any initial state $\phi \in \mathcal{X}_\phi$ we obtain

$$\dot{V}(t) + 2\delta_0 V(t) - 2\delta_1 \sup_{-h\le\theta\le 0} V(t+\theta) - \gamma\|f(\cdot,t)\|_{H_L^1}^2 \le \eta(t)^\top \Theta_0 \eta(t) \le 0.$$

By the generalized Halanay inequality, (2.30) holds on $[0, t']$.

Step 2: Given $\phi \in \mathcal{X}_\phi$. The solution of (2.17) initialized with $\phi \in \mathcal{X}_\phi$ is uniformly bounded whenever it exists meaning that

$$\|u(\cdot,t)\|_{L^2} \le e^{-2\delta t} \max_{\theta\in[-h,0]} \|\phi(\cdot,\theta)\|_{L^2}^2 + \frac{\gamma}{\sigma}\Delta_f^2 \le C_1,$$

where δ is the solution of (1.45).

Thus, from (ii) of Lemma 2.1, we obtain that the strong solution of (2.17) initialized with $\phi \in \mathcal{X}_\phi$ exists for all $t \in [0, h_0]$.

Step 3: We apply the same arguments step-by-step for $[h_0, 2h_0]$, $[2h_0, 3h_0]$, $\cdots$. For $t \in [h_0, 2h_0]$, the system (2.17) can be also rewritten as an evolution equation (2.20) with $\zeta(h_0) \in D(\mathcal{A}_u)$, where $\hat{F}(\zeta, t)$ is given by (2.22) with a modified ϕ:

$$\phi(\theta) = \begin{cases} \phi(\theta), & \theta \le 0, \\ \zeta(\theta), & \theta \in [0, h_0]. \end{cases}$$

Since the modified ϕ is Lipschitz, inequalities (2.25), (2.26), and thus, (2.24) hold for any $t_1, t_2 \in [h_0, 2h_0]$, any $R > 0$ and for all $\zeta_1, \zeta_2 \in L^2(0,1)$ such that $\|\zeta_1\|_{L^2} \le R$, $\|\zeta_2\|_{L^2} \le R$. Therefore, by applying arguments of Steps 1 and 2 we conclude that there exists a strong solution for $t \in [h_0, 2h_0]$. Then by step method, the strong solution exists for all $t \ge 0$. Furthermore, the generalized Halanay's inequality implies (2.30) for all $t \ge 0$. The proof of (i) and (ii) is completed.

(iii) The LMIs (2.27) and (2.28) are always feasible for appropriate choice of c. Indeed, applying Schur complement theorem, we have

$$\Theta_0 < 0 \Longleftrightarrow -c + \delta_0 + k_1^2\gamma^{-1} - (6k_1k_2k_3^2|\mu|C_1)^2\delta_1^{-1} < 0. \tag{2.35}$$

For given $C_1 > 0$, $\delta_1 > 0$ and large enough $\gamma > 0$, we can always choose $\delta_0 > \delta_1$ as large as we want that satisfy (2.27), (2.35) by appropriate choice of c. The latter means that the decay rate bound can be improved by appropriate choice of design parameter c.

Returning to the original system (2.1) with the control law (2.16), from the definition of $D(\mathcal{A}_u)$ (cf. (2.19)), $v = z_x$ and the backstepping transformation (2.11), we define

$$D(\mathcal{A}_z) = \left\{ z \in H^3(0,1) | z(0) = 0, z'(1) = k(1,1)z(1) - \int_0^1 k_y(1,y)z(y)dy \right\},$$

and initial space for the original system:

$$\Omega = C([-h,0]; D(\mathcal{A}_z)) \cap C^1([-h,0]; H_L^1(0,1)). \tag{2.36}$$

Definition 2.3 By strong solution of the initial value problem for the closed-loop system (2.1), (2.16) with the initial function $\psi \in \Omega$, we understand a function

$$z \in L^1([0,\infty); D(\mathcal{A}_z)) \cap C([0,\infty); H_L^1(0,1)),$$

such that $\dot{z} \in L^1([0,\infty); H_L^1(0,1))$ and the corresponding evolution equation holds a.e. on $[0,\infty)$.

The following theorem states the well-posedness as well as regional ISS for the closed-loop system (2.1), (2.16):

Theorem 2.1 *Consider the original system (2.1) with the control law (2.16) under Assumption 2.1.1. Given $h_0 > 0$, $c > 0$, $\Delta_f > 0$, $C_1 > 0$ and $\mu \in \mathbb{R}$, assume that there exist $0 < \delta_1 < \delta_0$ and $\gamma > 0$ such that LMIs (2.27) and (2.28) hold, where $k_i > 0$ $(i = 1,2,3,4)$ are defined by (2.15). The initial state ψ is assumed to be in the set*

$$\mathcal{X}_\psi = \left\{ \psi \in \Omega : k_4 \max_{-h\leq\theta\leq 0} \|\psi(\cdot,\theta)\|_{H_L^1} \leq \sqrt{C_1 - \frac{\gamma}{\sigma}\Delta_f^2} \right\}, \tag{2.37}$$

where $\sigma = \delta_0 - \delta_1$ and Ω is defined by (2.36). Then the following holds:

(i) [Well-posedness] A strong solution of the closed-loop system exists for all $t \geq 0$.

(ii) [Regional ISS] The closed-loop system is exponentially ISS:

$$\|z(\cdot,t)\|_{H_L^1}^2 \leq k_3^2k_4^2e^{-2\delta t} \max_{\theta\in[-h,0]} \|z(\cdot,\theta)\|_{H_L^1}^2 + \frac{\gamma k_3^2}{\sigma}\Delta_f^2, \forall t \geq 0. \tag{2.38}$$

Proof By employing (2.11) and (2.14) with $v = z_x$, we have

$$\|u(\cdot,t)\|_{L^2} = \|\mathcal{K}[z_x(\cdot,t)]\|_{L^2} \le k_4 \|z_x(\cdot,t)\|_{L^2},$$

Hence, if the initial state ψ of (2.1), (2.16) starts from the set $\mathcal{X}_\psi$, then the initial state ϕ of the target system (2.17) belongs to the set $\mathcal{X}_\phi$, where $\mathcal{X}_\phi$ is given by (2.29), that is to say,

$$\max_{-h\le\theta\le 0} \|\phi(\cdot,\theta)\|_{L^2}^2 \le k_4^2 \max_{-h\le\theta\le 0} \|\psi(\cdot,\theta)\|_{H_L^1}^2 \le C_1 - \frac{\gamma}{\sigma}\Delta_f^2. \tag{2.39}$$

Therefore, from Proposition 2.1 and the model transformation formulas (3.257) and (2.12) with bounded operators $(I - \mathbb{P})$ and its inverse, we obtain that if the LMIs (2.27) and (2.28) are feasible, then the strong solution of (2.1), (2.16) with initial condition $\psi \in \mathcal{X}_\psi$ exists. Moreover, the strong solution of target system (2.17) initialized with $\phi \in \mathcal{X}_\phi$ satisfies (2.30).

By using (2.14), (2.30), together with (2.39), we arrive at (2.38):

$$\begin{aligned}
&\|z(\cdot,t)\|_{H_L^1}^2 = \|z_x(\cdot,t)\|_{L^2}^2 = \|\mathcal{L}[u(\cdot,t)]\|_{L^2}^2 \\
&\le k_3^2 \|u(\cdot,t)\|_{L^2}^2 \le k_3^2 \left[e^{-2\delta t} \max_{\theta\in[-h,0]} \|\phi(\cdot,\theta)\|_{L^2}^2 + \frac{\gamma}{\sigma}\Delta_f^2 \right] \\
&\le k_3^2 \left[k_4^2 e^{-2\delta t} \max_{\theta\in[-h,0]} \|\psi(\cdot,\theta)\|_{H_L^1}^2 + \frac{\gamma}{\sigma}\Delta_f^2 \right], \quad \forall t \ge 0.
\end{aligned}$$

Remark 2.1 Here we explain why we introduce the change of variable: $v(x,t) = z_x(x,t)$. If we use the backstepping transformation directly, without this change of variable, then by applying arguments of Proposition 2.1, instead of (2.34) we arrive at the following inequality:

$$\int_0^1 |u(x,t)| \cdot |F(u,x,t)| dx \le 3|\mu| \int_0^1 |u(x,t)||\mathcal{L}[u(x,t-\tau(t))]||\mathcal{L}[u(x,t)]|^2 dx,$$

where $\mathcal{L}$ is a bounded linear operator which is related to the backstepping transformation. It is seen that the right-hand side of the latter inequality cannot be bounded in terms of $\|u\|_{L^2}$.

Consider the system (2.17) without the disturbance. Then for the unperturbed system we obtain the following result:

Corollary 2.1 *Consider the system (2.17) with $f(x,t) \equiv 0$. Given scalars $h_0 > 0$, $C_1 > 0$ and $\mu \in \mathbb{R}$, let $c > 0$ be a scalar such that the following inequality*

$$c > 12k_1k_2k_3^2|\mu|C_1 \tag{2.40}$$

holds, where $k_i > 0$ $(i = 1, 2, 3)$ are defined by (2.15). Then the solutions of unperturbed system starting from the initial set $\mathcal{X}_\phi$ given by (2.29) with $\Delta_f = 0$ converge to zero with a small enough decay rate.

Proof From (2.28) we conclude that the unperturbed system is exponentially stable if the LMI

$$\begin{bmatrix} -c + \delta_0 & 6k_1k_2k_3^2|\mu|C_1 \\ * & -\delta_1 \end{bmatrix} < 0 \tag{2.41}$$

is satisfied. The feasibility of strict LMI (2.41) with $\delta_0 = \delta_1 > 0$ implies its feasibility for $\bar{\delta}_0$ and $\bar{\delta}_1$ given by $\bar{\delta}_0 = \delta_0 + \varepsilon > \delta_0 = \bar{\delta}_1$ for small enough $\varepsilon > 0$. Since Halanay's inequality holds with $\bar{\delta}_0$ and $\bar{\delta}_1$, the system is exponentially stable with a small enough decay rate.

By Schur complements, we show next that (2.40) guarantees the feasibility of (2.41) with $\delta_0 = \delta_1 > 0$. Indeed, the LMI (2.41) is feasible with $\delta_0 = \delta_1 = \bar{\delta}$ iff

$$-(c - \bar{\delta})\bar{\delta} + [6k_1k_2k_3^2|\mu|C_1]^2 < 0. \tag{2.42}$$

It is easily seen that $\bar{\delta} = \frac{c}{2}$ minimizes the left-hand side of (2.42) in $\bar{\delta}$. Substituting $\bar{\delta} = \frac{c}{2}$ into (2.42), we obtain

$$-\frac{c^2}{4} + [6k_1k_2k_3^2|\mu|C_1]^2 < 0,$$

i.e. (2.40) guarantees exponential stability.

Remark 2.2 As in [3, 4], we can consider the nonlinear delay term "$\mu z(x,t)|z(x,t-\tau(t))|^2$". In this case, we can apply the same transformation $v = z_x$ and further the same backstepping transformation that leads to the target system (2.17) with a correspondingly modified nonlinearity F:

$$\begin{aligned} F(u,x,t) = & -\mu\mathcal{L}[u(x,t)]\left|\int_0^x \mathcal{L}[u(r,t-\tau(t))]dr\right|^2 \\ & -\mu\overline{\mathcal{L}[u(x,t-\tau(t))]}\int_0^x \mathcal{L}[u(r,t)]dr \int_0^x \mathcal{L}[u(r,t-\tau(t))]dr \\ & -\mu\mathcal{L}[u(x,t-\tau(t))]\int_0^x \mathcal{L}[u(r,t)]dr \int_0^x \overline{\mathcal{L}[u(r,t-\tau(t))]}dr. \end{aligned}$$

Lemma 2.1 for the corresponding target system holds. We can apply the same V to the target system. Here we do not need Halanay's inequality since the delayed state is treated as bounded in L^2-norm. The LMI of Theorem 2.1 here has a form

$$\Theta_0 = \begin{bmatrix} -c + \delta + 12k_1k_2k_3^2|\mu|C_1 & k_1 \\ * & -\gamma \end{bmatrix} < 0. \tag{2.43}$$

Let $\mathcal{X}_\psi$ be defined by (2.37), where σ is replaced by δ. Then solution of the closed-loop system starting from any initial state $\psi \in \mathcal{X}_\psi$ exponentially converges in the sense that (2.38) holds, which implies that the closed-loop system is ISS.

Remark 2.3 In Corollary 2.1 we have presented an explicit condition (2.40) in terms of system parameters that guarantees the feasibility of (2.28), (2.43) with a small enough δ and large enough γ. The delay-independent stability condition (2.40) (and thus (2.28) and (2.43)) are not conservative because the same condition (2.40) follows from the direct Lyapunov analysis of the non-delayed target system (cf. (2.43)).

2.1.2 Constrained Control Law

In this section, we consider the system (2.1) with the control law which is subject to the following amplitude constraint:

$$|U(t)| \le \bar{u}.$$

We design the state feedback controller in the following form:

$$U_{sat}(t) = \text{sat}_{\mathbb{C}}(U(t), \bar{u}), \tag{2.44}$$

where $U(t)$ is given by (2.16), and the saturation function is defined by

$$\text{sat}_{\mathbb{C}}(s, \bar{u}) := \begin{cases} s, & |s| \le \bar{u}, \\ \dfrac{s}{|s|}\bar{u}, & |s| > \bar{u}. \end{cases}$$

Applying the latter control law (2.44), we represent the saturated closed-loop system as (2.6) with the following boundary condition:

$$u(1, t) = \text{sat}_{\mathbb{C}}(U(t), \bar{u}) - U(t). \tag{2.45}$$

We now provide sufficient conditions for saturation avoidance:

Proposition 2.2 *Consider the target system (2.6) with the constrained control law (2.44) under Assumption 1. Given scalars $h_0 > 0$, $\Delta_f > 0$, $c > 0$, $C_1 > 0$ and $\mu \in \mathbb{R}$,* *assume that there exist* $0 < \delta_1 < \delta_0$ *and* $\gamma > 0$ *such that LMIs (2.27) and (2.28) hold, where* $k_i > 0$ $(i = 1, 2, 3)$ *are given by (2.15). Let* $\beta > 0$ *satisfy the following inequalities*

$$\frac{\gamma}{\sigma}\Delta_f^2 < \beta \leq C_1, \ \sigma = \delta_0 - \delta_1, \tag{2.46}$$

$$\bar{u}^2 \geq \beta \left[\max_{0 \leq y \leq 1} |l(1, y)| \right]^2. \tag{2.47}$$

Consider the set of initial states ϕ*:*

$$\mathcal{X}_\phi^\beta = \left\{ \phi \in Z : \max_{-h \leq \theta \leq 0} \|\phi(\cdot, \theta)\|_{L^2}^2 \leq \beta - \frac{\gamma}{\sigma}\Delta_f^2 \right\},$$

which is a subset of $\mathcal{X}_\phi$ *given by (2.29). Then all strong solutions of (2.6), (2.44) starting from* $\mathcal{X}_\phi^\beta$ *satisfy (2.30).*

Proof From the backstepping transformations (2.5) and (2.10), $U(t)$ admits the following representation:

$$U(t) = \int_0^1 k(1, y)v(y, t)dy = \int_0^1 l(1, y)u(y, t)dy.$$

Due to (2.10), $l(x, y)$ is a continuous function bounded on any compact. Then Jensen's inequality implies

$$|U(t)| \leq \max_{0 \leq y \leq 1} |l(1, y)| \cdot \|u(\cdot, t)\|_{L^2}. \tag{2.48}$$

Denote $K \triangleq \max_{0 \leq y \leq 1} |l(1, y)|$. Given $\bar{u} > 0$, we define the following set:

$$\mathcal{J}(K, \bar{u}) = \left\{ u(\cdot, t) \in L^2(0, 1) : K\|u(\cdot, t)\|_{L^2} \leq \bar{u} \right\}. \tag{2.49}$$

Thus, we can obtain that if $u \in \mathcal{J}(K, \bar{u})$, then $|U(t)| \leq \bar{u}$, and the saturation is avoided. Thus, the system (2.6) subject to (2.45) admits the linear representation (2.17).

By Proposition 2.1, we find that if the LMIs (2.27), (2.28) and the inequality (2.46) are satisfied, then we have

$$\|u(\cdot, t)\|_{L^2} \leq e^{-2\delta t} \max_{\theta \in [-h, 0]} \|\phi(\cdot, \theta)\|_{L^2}^2 + \frac{\gamma}{\sigma}\Delta_f^2, \quad \forall t \geq 0. \tag{2.50}$$

Moreover, the inequality (2.50) guarantees that the trajectories $u(x, t)$ starting from $\phi \in \mathcal{X}_\phi^\beta$ remain within the reachable set $\mathcal{X}_\beta$, where

$$\mathcal{X}_\beta = \left\{ u(\cdot, t) \in L^2(0, 1) : \int_0^1 |u(x, t)|^2 dx \leq \beta \right\}.$$

The "ball" $\mathcal{X}_\beta$ is contained in $\mathcal{J}(K, \bar{u})$, if the following implication holds

$$\|u(\cdot, t)\|_{L^2}^2 \leq \beta \Rightarrow K \|u(\cdot, t)\|_{L^2} \leq \bar{u}$$

for all $u(x, t)$, i.e. if

$$K^2 \|u(\cdot, t)\|_{L^2}^2 \leq \beta^{-1} \bar{u}^2 \|u(\cdot, t)\|_{L^2}^2.$$

The latter inequality is guaranteed if the inequality (2.47) holds. Thus, the inequality (2.47) guarantees the saturation avoidance, and together with Proposition 2.1 imply (2.30). □

From Proposition 2.2, by arguments of Theorem 2.1 we arrive at:

Theorem 2.2 *Consider the original system (2.1) with the constrained control law (2.44) under Assumption 1. Given scalars* $h_0 > 0$, $\Delta_f > 0$, $c > 0$, $C_1 > 0$ *and* $\mu \in \mathbb{R}$, *assume that there exist* $0 < \delta_1 < \delta_0$ *and* $\gamma > 0$ *such that LMIs (2.27) and (2.28) hold, where* $k_i > 0$ $(i = 1, 2, 3, 4)$ *are given by (2.15). Let* $\beta > 0$ *satisfy the inequalities (2.46), (2.47). Then all the strong solutions of the closed-loop system (2.1), (2.44) starting from the initial set*

$$\mathcal{X}_\psi^\beta = \left\{ \psi \in \Omega : k_4 \max_{-h \leq \theta \leq 0} \|\psi(\cdot, \theta)\|_{H_L^1} \leq \sqrt{\beta - \frac{\gamma}{\sigma} \Delta_f^2} \right\}$$

satisfy (2.38).

2.1.3 Example

Consider the system (2.1) and (2.6) with the parameters $c = 8$, $\mu = \pm 0.01$, $\bar{u} = 10$ and $\Delta_f = 0.01$. Here $k_1 = k_2 = 4.6259$, $k_3 = k_4 = 5.6259$.

In order to increase an initial ball $\mathcal{X}_\psi^\beta$ inside of domain of initial set, we minimize γ (i.e. maximize $M_0 \triangleq \beta - \gamma \Delta_f^2 / \sigma$) subject to the LMIs of Theorem 2.2. Fixing $C_1 = 0.064$, we find that the LMIs of Theorem 2.2 are feasible with $\delta_0 = 2$, $\delta_1 = 1.5$ and $\gamma = 14.3570$ leading to max $M_0 = 0.0612$. Hence, we obtain the following ball inside of initial set for original system:

$$\mathcal{X}_\psi^\beta = \left\{ \psi \in \Omega : 5.6259 \max_{\theta \in [-h, 0]} \|\psi(\cdot, \theta)\|_{H_L^1} \leq 0.247 \right\}.$$

Next, a finite difference method is applied to compute the state of the closed-loop perturbed system (2.1) with the constrained control law (2.44). The steps of

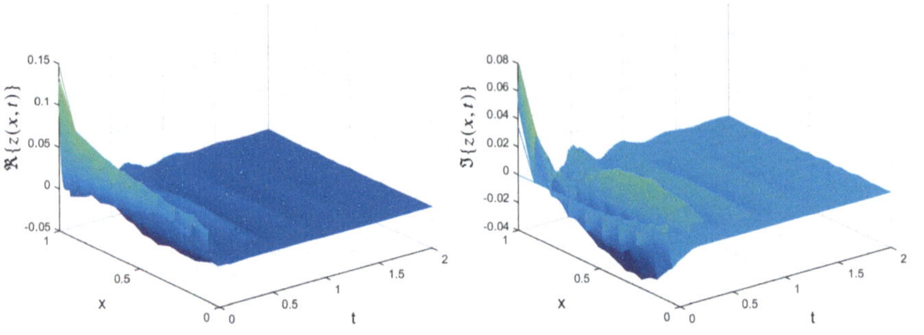

(a) State $\Re\{z(x,t)\}$ (b) State $\Im\{z(x,t)\}$

Fig. 2.1 Input-to-state stability of the perturbed system with time delay

space and time are taken as 0.1 and 0.00001, respectively. Here we choose the time-varying delay $\tau(t) = 0.2(\sin t + 1)$, the space-and time-varying perturbation $f(x,t) = 0.01\sin(x)\sin(t)$, and the following initial conditions:

$$z(x,\theta) = \psi(x,\theta) = 0.88\sin(\frac{x}{10}),\ \theta \in [-0.4, 0]. \tag{2.51}$$

Hence, $5.6259 \max_{\theta\in[-0.4,0]} \|\psi(\cdot,\theta)\|_{H_L^1} = 0.245 < 0.247$. Figure 2.1 demonstrates the state of the closed-loop system of (2.1) with constrained control (2.44). It is seen that the solution of the closed-loop system converges to a neighborhood of the origin as the time increases.

In Fig. 2.2, we choose the constant delay $\tau(t) \equiv 0.2$ and the initial state outside $\mathcal{X}_\psi^\beta$: $z(x,\theta) = \psi(x,\theta) = 15\sinh x + 5ix,\ \theta \in [-0.2, 0]$. Here $\max_{\theta\in[-0.2,0]} \|\psi(\cdot,\theta)\|_{H_L^1} = 18.47$. It is seen that the state of nonlinear Schrödinger

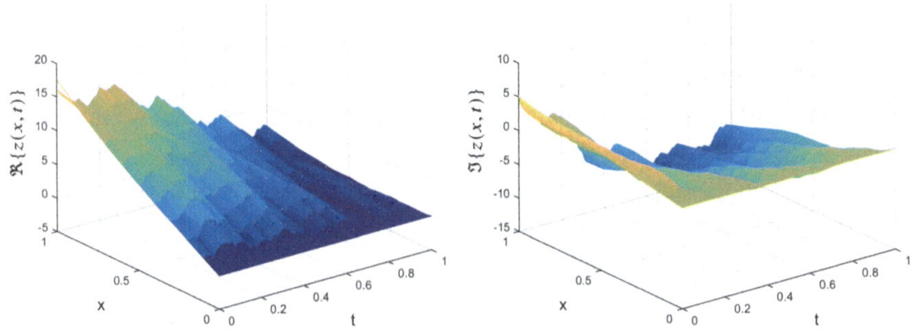

(a) State $\Re\{z(x,t)\}$ (b) State $\Im\{z(x,t)\}$

Fig. 2.2 Instability when the initial state is chosen outside the ball $\mathcal{X}_\psi^\beta$

equation is vibrating. The simulations of the solutions confirm the theoretical results.

2.2 Boundary Control of Delayed ODE-Heat Cascade Under Actuator Saturation

This section focuses on the study of boundary stabilization for a cascade of ODE-heat system with a time-varying state delay under actuator saturation. To stabilize the system, we design a state feedback controller via the backstepping method and find a bound on the domain of attraction. The latter bound is based on Lyapunov method, whereas the exponential stability conditions for the delayed cascaded system are derived by using Halanay's inequality. Numerical examples illustrate the efficiency of the method [18].

2.2.1 Backstepping Control for Cascaded ODE-Heat Equations with Delay

We consider the following coupled ODE-reaction diffusion system:

$$\begin{cases} \dot{X}(t) = AX(t) + A_1X(t - \tau(t)) + Bz(0, t), \\ z_t(x, t) = z_{xx}(x, t) + a_2z(x, t - \tau(t)) + az(x, t), \\ z_x(0, t) = 0, \\ (X(t), z(x, t)) = (f(t), \psi(x, t)), \ -h \le t \le 0, \end{cases} \tag{2.52}$$

with Dirichlet boundary actuator:

$$z(1, t) = U(t), \ t > 0, \tag{2.53}$$

or Neumann boundary actuator:

$$z_x(1, t) = U(t), \ t > 0. \tag{2.54}$$

Here $x \in (0, 1)$, $A, A_1 \in \mathbb{R}^{n\times n}$, $B \in \mathbb{R}^{n\times 1}$, $a, a_2 \in \mathbb{R}$ denotes a constant coefficient, $\tau(t)$ corresponds to a time varying delay, and $(f(t), \psi(x, t))$ is the initial state. $X(t) \in \mathbb{R}^n$ is the state of ordinary differential equation, $z(x, t) \in \mathbb{R}$ is the displacement of heat equation, and $U(t) \in \mathbb{R}$ is the control actuation.

We assume that (A, B) is controllable. Assume that the time-varying delay $\tau(t)$ is a continuously differentiable function of t that satisfies

$$0 < h_0 \le \tau(t) \le h \tag{2.55}$$

with some constants h_0 and $h > 0$. Note that the assumption $h_0 > 0$ is used for simplification of the proof of well-posedness. The delay and its bounds may be unknown for the exponential stability conditions (without finding a decay rate) and for the domain of attraction in the presence of actuator saturation. However, the upper bound h on the delay should be known for finding a bound on the decay rate of the exponential stability.

The first equation of (2.52) is ODE with delay or a difference-differential equation. So, we call it ODE in order to distinguish it from PDE.

First, we look for a coordinate transformation

$$\begin{cases} X(t) = X(t), \\ w(x,t) = z(x,t) - \int_0^x k(x,y)z(y,t)dy - \gamma(x)X(t), \end{cases} \tag{2.56}$$

that transforms the system (2.52) into the following intermediate ODE-heat cascade:

$$\begin{cases} \dot{X}(t) = (A+BK)X(t) + A_1X(t-\tau(t)) + Bw(0,t), \\ w_t(x,t) = w_{xx}(x,t) + a_2w(x,t-\tau(t)) + aw(x,t) \\ \qquad -\gamma(x)[A_1 - a_2I]X(t-\tau(t)), \\ w_x(0,t) = 0, \\ (X(t), w(x,t)) = (f(t), \phi(x,t)), \ -h \le t \le 0, \end{cases} \tag{2.57}$$

where K is chosen such that

$$\dot{X}(t) = (A+BK)X(t) + A_1X(t-\tau(t))$$

is asymptotically stable, and

$$\phi(x,t) = \psi(x,t) - \int_0^x k(x,y)\psi(y,t)dy - \gamma(x)f(t). \tag{2.58}$$

Boundary actuation (2.53) is transformed into

$$w(1,t) = U(t) - \int_0^1 k(1,y)z(y,t)dy - \gamma(1)X(t), \tag{2.59}$$

and (2.54) is transformed into

$$w_x(1,t) = U(t) - k(1,1)z(1,t) - \int_0^1 k_x(1,y)z(y,t)dy - \gamma'(1)X(t). \tag{2.60}$$

Second, a further transformation, where $(X, w) \mapsto (X, u)$, can be given by

$$\begin{cases} X(t) = X(t), \\ u(x,t) = w(x,t) - \int_0^x q(x,y)w(y,t)dy. \end{cases} \tag{2.61}$$

Here the kernel $q(x, y)$ should be chosen to transform the system (2.57) into the target ODE-heat cascade:

$$\begin{cases} \dot{X}(t) = (A + BK)X(t) + A_1 X(t - \tau(t)) + Bu(0, t), \\ u_t(x, t) = u_{xx}(x, t) - cu(x, t) + a_2 u(x, t - \tau(t)) \\ -[\gamma(x) - \int_0^x q(x, y)\gamma(y)dy](A_1 - a_2 I)X(t - \tau(t)), \\ u_x(0, t) = 0, \\ (X(t), u(x, t)) = (f(t), \varphi(x, t)), \ -h \le t \le 0, \end{cases} \tag{2.62}$$

where $c > 0$ is a constant, and

$$\varphi(x, t) = \phi(x, t) - \int_0^x q(x, y)\phi(y, t)dy. \tag{2.63}$$

Boundary actuation (2.59) is transformed into

$$u(1, t) = U(t) - \int_0^1 k(1, y)z(y, t)dy - \gamma(1)X(t) - \int_0^1 q(1, y)w(y, t)dy, \tag{2.64}$$

and (2.60) is transformed into

$$\begin{aligned} u_x(1, t) = U(t) - k(1, 1)z(1, t) - \int_0^1 k_x(1, y)z(y, t)dy - \gamma'(1)X(t) \\ -q(1, 1)w(1, t) - \int_0^1 q_x(1, y)w(y, t)dy. \end{aligned} \tag{2.65}$$

Next, we compute the kernels of $k(x, y)$, $\gamma(x)$ and $q(x, y)$. Motivated by [9], we will show that the transformation for undelayed equations (see [10]) still works for the above class of delayed equations.

Differentiation of (2.56) with respect to t yields

$$\begin{aligned} w_t(x, t) = z_{xx}(x, t) + a_2 z(x, t - \tau(t)) + a[z(x, t) - \int_0^x k(x, y)z(y, t)dy] \\ - \int_0^x k(x, y)[z_{yy}(y, t) + a_2 z(y, t - \tau(t))]dy \\ -\gamma(x)[AX(t) + A_1 X(t - \tau(t)) + Bz(0, t)]. \end{aligned}$$

Substitution of (2.56) into the resulting equation implies

$$\begin{aligned} w_t(x, t) = z_{xx}(x, t) + a_2 w(x, t - \tau(t)) + aw(x, t) - k(x, x)z_x(x, t) + k_y(x, x)z(x, t) \\ -k_y(x, 0)z(0, t) - \int_0^x k_{yy}(x, y)z(y, t)dy \\ -\gamma(x)[(A - aI)X(t) + (A_1 - a_2 I)X(t - \tau(t))] - \gamma(x)Bz(0, t). \end{aligned}$$

Similarly, the first and second derivatives of $w(x, t)$ with respect to x are given by

$$\begin{aligned} w_x(x, t) &= z_x(x, t) - k(x, x)z(x, t) \\ &\quad - \int_0^x k_x(x, y)z(y, t)dy - \gamma'(x)X(t), \\ w_{xx}(x, t) &= z_{xx}(x, t) - \frac{d}{dx}k(x, x)z(x, t) \\ &\quad -k(x, x)z_x(x, t) - k_x(x, x)z(x, t) \\ &\quad - \int_0^x k_{xx}(x, y)z(y, t)dy - \gamma''(x)X(t). \end{aligned}$$

Substituting (2.56) into (2.52) and comparing with (2.57), we obtain the following set of conditions on the kernels $k(x, y)$ and $\gamma(x)$ (see e.g.[11]):

$$\begin{cases} k_{xx}(x, y) = k_{yy}(x, y), \\ k_y(x, 0) = -\gamma(x)B, \\ k(x, x) = 0, \end{cases} \tag{2.66}$$

$$\begin{cases} \gamma''(x) = \gamma(x)(A - aI), \\ \gamma(0) = K, \\ \gamma'(0) = 0. \end{cases} \tag{2.67}$$

The solution to (2.66) and (2.67) is given by

$$\begin{aligned} k(x, y) &= \int_0^{x-y} \gamma(\sigma)Bd\sigma, \\ \gamma(x) &= \begin{bmatrix} K & 0 \end{bmatrix} e^{\begin{bmatrix} 0 & A-aI \\ I & 0 \end{bmatrix}x} \begin{bmatrix} I \\ 0 \end{bmatrix}. \end{aligned} \tag{2.68}$$

In the similar manner, the change of variable (2.56) has an inverse transformation:

$$z(x, t) = w(x, t) + \int_0^x n(x, y)w(y, t)dy + \psi(x)X(t), \tag{2.69}$$

where

$$\begin{aligned} n(x, y) &= \int_0^{x-y} \psi(\sigma)Bd\sigma, \\ \psi(x) &= \begin{bmatrix} K & 0 \end{bmatrix} e^{\begin{bmatrix} 0 & A+BK-aI \\ I & 0 \end{bmatrix}x} \begin{bmatrix} I \\ 0 \end{bmatrix}. \end{aligned} \tag{2.70}$$

By the standard procedures (see [12]), we differentiate (2.61) with respect to t and x respectively to obtain

$$\begin{aligned} u_t(x, t) &= w_{xx}(x, t) + a_2 u(x, t - \tau(t)) + aw(x, t) \\ &\quad -q(x, x)w_x(x, t) + q_y(x, x)w(x, t) \\ &\quad -q_y(x, 0)w(0, t) - \int_0^x q_{yy}(x, y)w(y, t)dy \\ &\quad -a \int_0^x q(x, y)w(y, t)dy \\ &\quad -[\gamma(x) - \int_0^x q(x, y)\gamma(y)dy](A_1 - a_2 I) \\ &\quad \times X(t - \tau(t)), \\ u_x(x, t) &= w_x(x, t) - q(x, x)w(x, t) \\ &\quad - \int_0^x q_x(x, y)w(y, t)dy, \\ u_{xx}(x, t) &= w_{xx}(x, t) - \frac{d}{dx}q(x, x)w(x, t) \\ &\quad -q(x, x)w_x(x, t) - q_x(x, x)w(x, t) \\ &\quad - \int_0^x q_{xx}(x, y)w(y, t)dy. \end{aligned} \tag{2.71}$$

Comparing (2.71) with the second equation of (2.62), we obtain that $q(x, y)$ satisfies

$$\begin{cases} q_{xx}(x, y) = q_{yy}(x, y) + (a + c)q(x, y), \\ q_y(x, 0) = 0, \\ q(x, x) = -\dfrac{a + c}{2}x. \end{cases} \tag{2.72}$$

The solution to (2.72) is given by

$$q(x, y) = -(a + c)x\frac{I_1(\sqrt{(a + c)(x^2 - y^2)})}{\sqrt{(a + c)(x^2 - y^2)}},$$

where $I_1(\cdot)$ denotes the modified Bessel function of the first order:

$$I_1(x) = \sum_{n=0}^{\infty}\frac{(x/2)^{2n+1}}{n!(n + 1)!}.$$

In the similar manner, the change of variable (2.61) has an inverse transformation:

$$w(x, t) = u(x, t) + \int_0^x l(x, y)u(y, t)dy, \tag{2.73}$$

where

$$l(x, y) = -(a + c)x\frac{J_1(\sqrt{(a + c)(x^2 - y^2)})}{\sqrt{(a + c)(x^2 - y^2)}}, \tag{2.74}$$

and $J_1(\cdot)$ is Bessel function of the first order:

$$J_1(x) = \sum_{n=0}^{\infty}\frac{(-1)^n(x/2)^{2n+1}}{n!(n + 1)!}.$$

2.2.1.1 Dirichlet Actuation

We design the state feedback controller for the target system (2.62). By selecting the following feedback controller:

$$\begin{aligned} &U(t) = \textstyle\int_0^1 k(1, y)z(y, t)dy + \gamma(1)X(t) \\ &+ \textstyle\int_0^1 q(1, y)\left[z(y, t) - \int_0^y k(y, s)z(s, t)ds - \gamma(y)X(t)\right]dy, \end{aligned} \tag{2.75}$$

one arrives to the closed-loop system of (2.62) with boundary actuation (2.64) as follows:

$$\begin{cases} \dot{X}(t) = (A + BK)X(t) + A_1 X(t - \tau(t)) + Bu(0, t), \\ u_t(x, t) = u_{xx}(x, t) - cu(x, t) + a_2 u(x, t - \tau(t)) \\ -[\gamma(x) - \int_0^x q(x, y)\gamma(y)dy](A_1 - a_2 I)X(t - \tau(t)), \\ u_x(0, t) = 0, \\ (X(t), u(x, t)) = (f(t), \varphi(x, t)), \ -h \le t \le 0, \end{cases} \tag{2.76}$$

subject to

$$u(1, t) = 0. \tag{2.77}$$

Remark 2.4 Differently from the non-delayed case [11], the resulting target system (2.76), (2.77) is coupled. However, each differential equation (for X and for z) contains the design parameter (either K or c). This allows to stabilize the target system by appropriate choice of K and c (see (ii) of Propositions 2.3, 2.4 and Remark 2.6 below).

2.2.1.2 Neumann Actuation

The Neumann controller is obtained using the same exact transformation as in the case of the Dirichlet actuation, but with the appropriate change in the boundary condition of the target system. In this case, the backstepping approach yields the following controller for the target system (2.62):

$$\begin{aligned} U(t) = &\int_0^1 k_x(1, y)z(y, t)dy + \gamma'(1)X(t) \\ &+q(1, 1)w(1, t) + \int_0^1 q_x(1, y)w(y, t)dy. \end{aligned} \tag{2.78}$$

Here we use the fact that $k(1, 1) = 0$.

Under (2.78), the closed-loop system of (2.62) with boundary actuation (2.65) becomes (2.76) subject to

$$u_x(1, t) = 0. \tag{2.79}$$

2.2.2 *Well-Posedness of the Closed-Loop Systems*

We start with the Dirichlet actuation. Consider the closed-loop target system (2.76) and (2.77). We introduce the Hilbert space $H_R^1(0, 1) = \{f \in H^1(0, 1) | f(1) = 0\}$. Let $\mathcal{H} = \mathbb{R}^n \times L^2(0, 1)$ be the Hilbert space with the norm: $\|(f, g)\|_{\mathcal{H}}^2 = |f|_{\mathbb{R}^n}^2 + \|g\|_{L^2(0,1)}^2$.

While being viewed over the time segment $[0, h_0]$, the system can be rewritten as the differential equation:

$$\begin{cases} \frac{d}{dt}Y(\cdot,t) = \mathcal{A}_u Y(\cdot,t) + \mathcal{A}_1 Y(\cdot, t-\tau(t)), \\ Y(\cdot,\theta) = (f(\theta), \varphi(\cdot,\theta)), \theta \in [-h, 0] \end{cases} \tag{2.80}$$

in $\mathcal{H}$, where the system operator $\mathcal{A}_u : D(\mathcal{A}_u) \subset \mathcal{H} \to \mathcal{H}$ is defined by

$$\begin{cases} \mathcal{A}_u(X, u) = [(A + BK)X + Bu(0), u_{xx} - cu], \\ D(\mathcal{A}_u) = \quad \{(X, u) \in \mathbb{R}^n \times (H^2(0,1) \cap H_R^1(0,1))| \\ \qquad\qquad\quad u_x(0) = 0\}, \end{cases} \tag{2.81}$$

and the bounded operator $\mathcal{A}_1 : \mathcal{H} \to \mathcal{H}$ is defined by

$$\mathcal{A}_1(X, u) = [A_1 X, a_2 u - g(\cdot)(A_1 - a_2 I)X],$$

where $g(x) = \gamma(x) - \int_0^x q(x, y)\gamma(y)dy$.

A straightforward computation gives

$$\begin{cases} \mathcal{A}_u^*(X^*, u^*) = [(A + BK)^\top X, u_{xx}^* - cu], \\ D(\mathcal{A}_u^*) = \quad \{(X^*, u^*) \in \mathbb{R}^n \times (H^2(0,1) \cap H_R^1(0,1))| \\ \qquad\qquad\quad u_x^*(0) = -B^\top X^*\}, \end{cases} \tag{2.82}$$

where $\mathcal{A}_u^*$ is the adjoint operator of $\mathcal{A}_u$.

By the arguments of [13], it can be shown that there is a sequence of eigenfunctions of $\mathcal{A}_u^*$ which forms a Riesz basis for $\mathcal{H}$ and hence $\mathcal{A}_u^*$ generates an exponentially stable semigroup. Then by Proposition 2.8.1 and Proposition 2.8.5 of [14], we obtain that $\mathcal{A}_u$ generates a C_0-semigroup.

Define the initial conditions in the space

$$W \triangleq C([-h, 0], D(\mathcal{A}_u)) \cap C^1([-h, 0], \mathcal{H}).$$

The inhomogeneous term $\mathcal{A}_1 Y(\cdot, t - \tau(t))$ is of class C^1 on $[0, h_0]$. By Theorem 3.1.3 of [15], for any initial value $(X(\theta), u(\cdot,\theta)) \in W$, the closed-loop target system admits a unique classical solution $(X(t), u(\cdot, t))$ for all $t \in [0, h_0]$.

The same line of reasoning is step-by-step applied to the time segments $[h_0, 2h_0]$, $[2h_0, 3h_0]$, $[3h_0, 4h_0]$, ... Following this procedure, we obtain that there exists a unique classical solution $(X(t), u(\cdot, t))$ for all $t \geq 0$ with the initial condition $(X(\theta), u(\cdot, \theta)) \in W$ (see e.g. [16]).

Consider next the closed-loop target system (2.76), (2.79) under the Neumann actuation. Let $\mathcal{H}_1 = \mathbb{R}^n \times H^1(0, 1)$ be the Hilbert space with the norm:

$$\|(f, g)\|_{\mathcal{H}_1}^2 = |f|_{\mathbb{R}^n}^2 + \|g\|_{L^2(0,1)}^2 + \|g'\|_{L^2(0,1)}^2.$$

The existence and uniqueness of the solution of the system (2.76) subject to (2.79) can be easily obtained by applying the same procedure. But the expression of the domain $D(\mathcal{A}_u)$ and initial space W should be changed into

$$D(\mathcal{A}_u) = \{(X, u) \in \mathbb{R}^n \times H^2(0, 1)|u'(0) = u'(1) = 0\},$$

and

$$W = C([-h, 0], D(\mathcal{A}_u)) \cap C^1([-h, 0], \mathcal{H}_1).$$

Remark 2.5 By using the transformation (2.56) and (2.61), we establish the well-posedness of the closed-loop original system (2.52) under the Dirichlet or Neumann actuation.

For the case of Dirichlet actuation, we define

$$\begin{aligned}
&D(\mathcal{A}_z) = \Big\{(X, z) \in \mathbb{R}^n \times H^2(0, 1)|z'(0) = 0,\\
&z(1) = \textstyle\int_0^1 k(1, y)z(y)dy + \gamma(1)X\\
&+ \textstyle\int_0^1 q(1, y)[z(y) - \int_0^y k(y, s)z(s)ds - \gamma(y)X]dy\Big\},\\
&W_1 \triangleq C([-h, 0], D(\mathcal{A}_z)) \cap C^1([-h, 0], \mathcal{H}).
\end{aligned}$$

Thus, for any initial value $(X(\theta), z(\cdot, \theta)) \in W_1$, the closed-loop original system admits a unique classical solution $(X(t), z(\cdot, t))$ for all $t \geq 0$.

For the case of Neumann actuation, we define

$$\begin{aligned}
&D(\mathcal{A}_z) = \Big\{(X, z) \in \mathbb{R}^n \times H^2(0, 1)|z'(0) = 0,\\
&z'(1) = \textstyle\int_0^1 k_x(1, y)z(y)dy + \gamma'(1)X\\
&+q(1, 1)[z(1) - \textstyle\int_0^1 k(1, y)z(y)dy - \gamma(1)X]\\
&+ \textstyle\int_0^1 q_x(1, y)[z(y) - \int_0^y k(y, s)z(s)ds - \gamma(y)X]dy\Big\},\\
&W_1 = C([-h, 0], D(\mathcal{A}_z))) \cap C^1([-h, 0], \mathcal{H}_1).
\end{aligned}$$

Thus, well-posedness of the closed-loop original system can be obtained.

2.2.3 *Stability Analysis*

In Theorem 2 of [9], a delay-independent condition for the exponential stability of target system, which is described by reaction diffusion equation with state delay, has been shown by applying Lyapunov-Razumikhin theory. In this section, we will derive an exponential bound on the solution of the target coupled system via Halanay's inequality. This solution bound will allow to find a domain of attraction in the case of actuator saturation.

2.2.3.1 Stability of System (2.76) Subject to (2.77)

For the case of Dirichlet actuation, we choose the Lyapunov functions of the form

$$V(t) = X^\top PX + p_1 \int_0^1 u^2(x,t)dx, \tag{2.83}$$

Proposition 2.3 (i) *Given gains K and c, and tuning parameters $r > 0, 0 < \delta_1 < \delta_0$, let there exist scalars $p_1 > 0$, $0 < \lambda \le 2p_1$ and an $n \times n$ matrix $P > 0$ that satisfy the following linear matrix inequalities:*

$$\Theta_1 \triangleq \Xi + p_1 r^{-1} R < 0, \tag{2.84}$$

$$\Theta_2 \triangleq \begin{bmatrix} (-2c + 2\delta_0 + r - \frac{\pi^2}{2})p_1 + \frac{\pi^2}{4}\lambda & a_2 p_1 \\ * & -2\delta_1 p_1 \end{bmatrix} < 0, \tag{2.85}$$

where

$$\Xi = \begin{bmatrix} \theta_{11} & PA_1 & PB \\ * & -2\delta_1 P & 0 \\ * & * & -\lambda \end{bmatrix}, \tag{2.86}$$

$$R = \mathrm{diag}\{0, \zeta(A_1 - a_2 I)^\top (A_1 - a_2 I), 0\}, \tag{2.87}$$

$$\theta_{11} = P(A + BK) + (A + BK)^\top P + 2\delta_0 P, \tag{2.88}$$

$$\zeta \triangleq (1 + \max_{0 \le y \le x \le 1} |q(x,y)|)^2 (\max_{0 \le x \le 1} |\gamma(x)|)^2. \tag{2.89}$$

Then, for all $h_0 > 0$, $h > 0$ and $\tau(t) \in [h_0, h]$, the system (2.76) subject to (2.77) with initial conditions $(f, \varphi) \in W$ is exponentially stable with a decay rate δ in the sense that (1.43) holds, where δ is a unique solution of $\delta = \delta_0 - \delta_1 e^{2\delta h}$. Moreover, if the strict LMIs (2.84) and (2.85) with $\delta_0 = \delta_1 > 0$ hold, then for all $h_0 > 0$, $h > 0$ and $\tau(t) \in [h_0, h]$, the system (2.76) subject to (2.77) is exponentially stable with a small enough decay rate.

(ii)*Assume now that A_1 is a scalar matrix, i.e. $A_1 = a_1 I$, where a_1 is some constant. Then given any $\delta > 0$, the exponential stability of the system (2.76) subject to (2.77) with the decay rate $\delta > 0$ can be achieved by appropriate choice of design parameters c and K.*

Proof (i) Differentiating V along (2.76) and (2.77) we find

$$\dot{V}(t) = 2p_1 \int_0^1 u(x,t)u_t(x,t)dx + X^\top(t)P\dot{X}(t) + \dot{X}^\top(t)PX(t).$$

Integration by parts and substitution of the boundary conditions $u_x(0,t) = u(1,t) = 0$ lead to

$$\begin{aligned}
&\dot{V}(t) + 2\delta_0 V(t) - 2\delta_1 \sup_{-h\le\theta\le 0} V(t+\theta) \\
&\le -2p_1 \textstyle\int_0^1 u_x^2(x,t)dx + 2a_2 p_1 \int_0^1 u(x,t)u(x,t-\tau(t))dx \\
&-2p_1 c \textstyle\int_0^1 u^2(x,t)dx - 2\delta_1 p_1 \int_0^1 u^2(x,t-\tau(t))dx \\
&-2p_1 \textstyle\int_0^1 u(x,t)[\gamma(x) - \int_0^x q(x,y)\gamma(y)dy]dx(A_1 - a_2 I) \\
&\times X(t-\tau(t)) + X^\top(t)[P(A+BK) + (A+BK)^\top P]X(t) \\
&+2X^\top(t)PBu(0,t) + 2X^\top(t)PA_1X(t-\tau(t)) \\
&+2\delta_0 p_1 \textstyle\int_0^1 u^2(x,t)dx + 2\delta_0 X^\top(t)PX(t) \\
&-2\delta_1 X^\top(t-\tau(t))PX(t-\tau(t)).
\end{aligned} \tag{2.90}$$

Sobolev's inequality and Wirtinger's inequality imply

$$-\int_0^1 u_x^2(x,t)dx \le -u^2(0,t), \tag{2.91}$$

$$-\int_0^1 u_x^2(x,t)dx \le -\frac{\pi^2}{4}\int_0^1 u^2(x,t)dx. \tag{2.92}$$

Multiplying the inequality (2.91) by a constant $\lambda \in (0, 2p_1]$ and multiplying the inequality (2.92) by $2p_1 - \lambda$ on both sides and summing, we obtain that

$$-2p_1 \int_0^1 u_x^2(x,t)dx \le -\frac{\pi^2}{4}(2p_1 - \lambda)\int_0^1 u^2(x,t)dx - \lambda u^2(0,t). \tag{2.93}$$

As $\gamma(x)$, $q(x,y)$ are continuous functions bounded on any compact, the following inequality can be obtained:

$$\int_0^1 \left|\gamma(x) - \int_0^x q(x,y)\gamma(y)dy\right|^2 dx \le (1 + \max_{0\le y\le x\le 1} |q(x,y)|)^2 (\max_{0\le x\le 1} |\gamma(x)|)^2 = \zeta, \tag{2.94}$$

which together with Young's inequality implies

$$\begin{aligned}
&-2p_1 \textstyle\int_0^1 u(x,t)[\gamma(x) - \int_0^x q(x,y)\gamma(y)dy]dx(A_1 - a_2 I)X(t-\tau(t)) \\
&\quad \le p_1 \left[r \textstyle\int_0^1 u^2(x,t)dx + r^{-1}X^\top(t-\tau(t))SX(t-\tau(t))\right],
\end{aligned} \tag{2.95}$$

where

$$S = \zeta(A_1 - a_2 I)^\top (A_1 - a_2 I). \tag{2.96}$$

Set $\eta_1(t) = \mathrm{col}\{X(t), X(t-\tau(t)), u(0,t)\}$, $\eta_2(t) = \mathrm{col}\{u(x,t), u(x,t-\tau(t))\}$. Then substituting (2.93), (2.95) into (2.90) yields

$$\dot{V}(t) + 2\delta_0 V(t) - 2\delta_1 \sup_{-h\le\theta\le 0} V(t+\theta) \le \sum_{i=1}^{2}\int_0^1 \eta_i^\top(t)\Theta_i\eta_i(t)dx \le 0$$

if the LMIs $\Theta_1 < 0$ and $\Theta_2 < 0$ hold. Therefore, the feasibility of $\Theta_1 < 0$ and $\Theta_2 < 0$ guarantees that the Halanay inequality (1.43) holds meaning that the system (2.76) subject to (2.77) is exponentially stable.

The feasibility of strict inequalities (2.84) and (2.85) with $\delta_1 = \delta_0 > 0$ implies feasibility of these inequalities with $\bar{\delta}_0$ and $\bar{\delta}_1$ given by $\bar{\delta}_0 = \delta_0 + \epsilon > \delta_0 = \bar{\delta}_1$ for small enough $\epsilon > 0$. Since Halanay's inequality holds with $\bar{\delta}_0$ and $\bar{\delta}_1$, the system is exponentially stable with a small enough decay rate.

(ii) The decay rate bound can be enlarged if for given $\delta_1 > 0$ we can enlarge $\delta_0 > \delta_1$ subject to $\Theta_1 < 0,\ \Theta_2 < 0$. Applying Schur complement theorem, we obtain

$$\Xi < 0 \Longleftrightarrow P(A+BK) + (A+BK)^\top P + \lambda^{-1} PBB^\top P + [2\delta_0 + (2\delta_1)^{-1} a_1^2] P < 0. \tag{2.97}$$

Multiplying the last inequality by $Q = P^{-1}$ from left and right we arrive at

$$\Xi < 0 \Longleftrightarrow (A+BK)Q + Q(A+BK)^\top + \lambda^{-1} BB^\top + [2\delta_0 + (2\delta_1)^{-1} a_1^2] Q < 0. \tag{2.98}$$

Since (A, B) is controllable, for any $0 < \delta_1 < \delta_0$ and $0 < \lambda \leq 2$, we can choose K such that Lyapunov inequality (2.98) has a solution $Q > 0$. Then there exist large enough $r > 0$ and $p_1 = 1$ such that (2.84) holds.

By Schur complement theorem,

$$\Theta_2 < 0 \Longleftrightarrow -2c + 2\delta_0 - \frac{\pi^2}{4}(2 - \lambda p_1^{-1}) + r + (2\delta_1)^{-1} a_2^2 < 0. \tag{2.99}$$

With the chosen above parameters δ_0, δ_1, p_1, λ and r, (2.99) always holds for large enough c. Thus, given h, any decay rate bound may be achieved by appropriate choice of design parameters c and K. □

Remark 2.6 Less conservative delay-dependent stability conditions for system (2.76) subject to (2.77) with fast varying delays can be derived by using Lypunov-Krasovskii approach similar to [6, 17]. In fact, one can consider the following Lypunov-Krasovskii functional

$$\begin{aligned} V(t) = {} & X^\top(t)PX(t) + \int_{t-h}^{t} e^{-2\delta_0(t-s)} X^\top(s)SX(s)ds \\ & + h\int_{-h}^{0}\int_{t+\theta}^{t} e^{-2\delta_0(t-s)} \dot{X}^\top(s)R\dot{X}(s)ds d\theta + p_1\int_0^1 u^2(x,t)dx \end{aligned}$$

combined with the Halanay inequality, where $P, S, R > 0$ are some matrices, and $p_1 > 0$ is a constant. The resulting conditions will be always feasible for small enough h provided $(A + A_1, B)$ is controllable.

Remark 2.7 The original system (2.52) is equivalent to system (2.62) under the invertible transformation (2.56), and (2.47). Therefore, under the conditions of Proposition 2.3, for the original system (2.52), the same decay rate can be guaranteed by the controller $U(t)$ given by (2.75) .

2.2.3.2 Stability of System (2.76) Subject to (2.79)

For the case of Neumann actuation, we choose the Lyapunov function

$$V_1(t) = V(t) + p_2\|u_x\|^2 = X^\top PX + p_1\|u\|^2 + p_2\|u_x\|^2.$$

where the $n \times n$ matrix $P = P^\top > 0$, the parameters $p_1 > 0$ and $p_2 > 0$ will be chosen later, and $V(t)$ is defined by (2.83).

Remark 2.8 Similar to the case of Dirichlet actuation, for the proof of the stability of system (2.76) subject to (2.79), we can choose $p_2 = 0$. For finding a domain of attraction under Neumann actuation in the presence of actuator saturation, we need $p_2 > 0$ (see Sect. 2.2.4).

Proposition 2.4 (i) *Given gains K and c, and tuning parameters $r > 0$, $0 < r_1 < 2$, $0 < \delta_1 < \delta_0$, let there exist an $n \times n$ matrix $P > 0$, and scalars $p_1 > 0$, $p_2 > 0$, $\lambda > 0$ and $\lambda_1 \geq 0$ that satisfy the LMIs*

$$\bar{\Theta}_1 \triangleq \Theta_1 + p_2 r_1^{-1} R = \Xi + (p_1 r^{-1} + p_2 r_1^{-1})R < 0, \tag{2.100}$$

$$\bar{\Theta}_2 \triangleq \begin{bmatrix} (-2c + 2\delta_0 + r)p_1 + 2\lambda & a_2 p_1 & 0 \\ * & -2\delta_1 p_1 & -a_2 p_2 \\ * & * & \theta_{33} \end{bmatrix} < 0, \tag{2.101}$$

and the inequality

$$-2p_1 - 2p_2 c + \lambda + 2\delta_0 p_2 - \frac{\pi^2}{4}\lambda_1 \leq 0, \tag{2.102}$$

where Ξ, R are defined by (2.86) and (2.87) respectively,

$$\theta_{33} = -(2 - r_1)p_2 + \lambda_1.$$

Then, for all $h_0 > 0$, $h > 0$ and $\tau(t) \in [h_0, h]$, the system (2.76) subject to (2.79) with initial condition $(f, \varphi) \in W$ is exponentially stable with a decay rate δ, where δ is a unique solution of $\delta = \delta_0 - \delta_1 e^{2\delta h}$. Moreover, if (2.100), (2.101) and (2.102) hold with $\delta_0 = \delta_1 > 0$, then for all $h_0 > 0$ and $h > 0$, the system (2.76) subject to (2.79) is exponentially stable with a small enough decay rate for all $\tau(t) \in [h_0, h]$.

(ii) *Assume now that A_1 is a scalar matrix, i.e. $A_1 = a_1 I$, where a_1 is some constant. Then given any $\delta > 0$, the exponential stability of the system (2.76) subject to (2.79) with the decay rate $\delta > 0$ can be achieved.*

Proof (i) Taking the time derivative of the Lyapunov function along the solution of (2.76) subject to (2.79), and from (2.90) we get

$$
\begin{array}{l}
\dot{V}_1(t) + 2\delta_0 V_1(t) - 2\delta_1 \sup\limits_{-h\le\theta\le 0} V_1(t+\theta) \\
\le -2p_1 \int_0^1 u_x^2(x,t)dx + 2a_2 p_1 \int_0^1 u(x,t)u(x,t-\tau(t))dx \\
-2p_1 c \int_0^1 u^2(x,t)dx - 2p_2 c \int_0^1 u_x^2(x,t)dx \\
-2p_2 \int_0^1 u_{xx}^2(x,t)dx - 2a_2 p_2 \int_0^1 u_{xx}(x,t)u(x,t-\tau(t))dx \\
-2\int_0^1 [p_1 u(x,t) - p_2 u_{xx}(x,t)][\gamma(x) - \int_0^x q(x,y)\gamma(y)dy]dx(A_1 - a_2 I)X(t-\tau(t)) \\
+X^\top(t)[P(A+BK) + (A+BK)^\top P]X(t) \\
+2X^\top(t)PBu(0,t) + 2X^\top(t)PA_1 X(t-\tau(t)) \\
+2\delta_0 p_1 \int_0^1 u^2(x,t)dx + 2\delta_0 p_2 \int_0^1 u_x^2(x,t)dx \\
+2\delta_0 X^\top(t)PX(t) - 2\delta_1 X^\top(t-\tau(t))PX(t-\tau(t)) \\
-2\delta_1 p_2 \int_0^1 u_x^2(x,t-\tau(t))dx - 2\delta_1 p_1 \int_0^1 u^2(x,t-\tau(t))dx.
\end{array}
\tag{2.103}
$$

From Young's inequality, we have (2.95) and

$$
\begin{array}{l}
2\int_0^1 p_2 u_{xx}(x,t)[\gamma(x) - \int_0^x q(x,y)\gamma(y)dy]dx(A_1 - a_2 I)X(t-\tau(t)) \\
\le p_2[r_1 \int_0^1 u_{xx}^2(x,t)dx + r_1^{-1} X^\top(t-\tau(t))SX(t-\tau(t))],
\end{array}
\tag{2.104}
$$

where $r_1 > 0$ and S is defined by (2.96).

By using Agmon's and Wirtinger's inequalities, we have

$$
|u(0,t)|^2 \le 2\|u\|^2 + \|u_x\|^2, \quad \|u_x\|^2 \le \frac{4}{\pi^2}\|u_{xx}\|^2.
$$

Hence,

$$
0 \le \lambda[\|u_x\|^2 + 2\|u\|^2 - |u(0,t)|^2], \tag{2.105}
$$

$$
0 \le \lambda_1[\|u_{xx}\|^2 - \frac{\pi^2}{4}\|u_x\|^2], \tag{2.106}
$$

where $\lambda, \lambda_1 > 0$ are some constants.

Set $\eta_1(t) = \mathrm{col}\{X(t), X(t-\tau(t)), u(0,t)\}$, $\eta_2(t) = \mathrm{col}\{u(x,t), u(x,t-\tau(t)), u_{xx}(x,t)\}$. Let Θ_1 be defined by (2.84) and R by (2.87). We add (2.105) and (2.106) to (2.103). Then we obtain

$$
\begin{array}{l}
\dot{V}_1(t) + 2\delta_0 V_1(t) - 2\delta_1 \sup\limits_{-h\le\theta\le 0} V_1(t+\theta) \\
\le \sum\limits_{i=1}^{2} \int_0^1 \eta_i^\top(t)\bar{\Theta}_i \eta_i(t)dx \\
-(2p_1 + 2p_2 c - \lambda - 2\delta_0 p_2 + \frac{\pi^2}{4}\lambda_1) \int_0^1 u_x^2(x,t)dx \le 0
\end{array}
\tag{2.107}
$$

if the LMIs $\bar{\Theta}_1 < 0, \bar{\Theta}_2 < 0$ are feasible and the inequality (2.102) holds. Application of Halanay's inequality, completes the proof of (i).

(ii) By (ii) of Proposition 2.4, $\Theta_1 < 0$ is feasible for given $0 < \delta_1 < \delta_0$ and appropriate K. Then for $r_1 = 1$ and small enough $p_2 > 0$, $\bar{\Theta}_1 < 0$ is feasible.

Now given $0 < \delta_1 < \delta_0$, $\lambda > 0$, $p_1 = 1$, $p_2 > 0$, and $\lambda_1 \ge 0$ such that $\theta_{33} < 0$, we show that (2.101) and (2.102) are feasible for appropriate choice of large enough

$c > 0$. For (2.102), this is evident. For (2.101), this is true by Schur complements theorem. □

Remark 2.9 For simplicity only, in the cascade model we consider a constant coefficient a of the undelayed term $az(x,t)$. For the variable $a(s)$, one have to modify kernels of the transformations similarly to [9]. Halanay's inequality is applicable for the resulting target system.

2.2.4 Control Under Saturation: Regional Stabilization

In this section, we consider (2.52) with the control law which is subject to the following amplitude constraint:

$$|U(t)| \leq \bar{u}. \tag{2.108}$$

Denoting the state trajectory of (2.52) subject to Dirichlet or Neumann boundary actuation with the initial condition $(X_0, z_0) \triangleq (f(\theta), \psi(\cdot,\theta)) \in W_1$ by $(X(t; X_0), z(x,t; z_0))$.

For the case of Dirichlet actuation, the domain of attraction of the closed-loop original system is then the set

$$\tilde{S} = \left\{(X_0, z_0) \in W_1 : \lim_{t\to\infty} \|(X(t; X_0), z(x,t; z_0))\|_{\mathcal{H}} = 0\right\}. \tag{2.109}$$

For the case of Neumann actuation, the domain of attraction of the closed-loop original system is given by (2.109), where $\mathcal{H}$ is replaced by $\mathcal{H}_1$.

2.2.4.1 Dirichlet Control Under Saturation

We first find domain of attraction for the closed-loop target system. Denoting the state trajectory of closed-loop target system with the initial condition $(X_0, u_0) \triangleq (f(\theta), \varphi(\cdot,\theta)) \in W$ by $(X(t; X_0), u(x,t; u_0))$, the domain of attraction of the closed-loop target system is then the set

$$\mathcal{S} = \left\{(X_0, u_0) \in W : \lim_{t\to\infty} \|(X(t; X_0), u(x,t; u_0))\|_{\mathcal{H}} = 0\right\}.$$

We will obtain an estimate $\mathcal{X}_\beta \subset \mathcal{S}$ on the domain of attraction, where

$$\mathcal{X}_\beta = \left\{(X_0, u_0) \in W : \max_{[-h,0]} |X_0|^2 + \max_{[-h,0]} \|u_0\|^2 \leq \beta^{-1}\right\},$$

$\beta > 0$ is a scalar that will be minimized in the sequel.

We design the state feedback controller in the form:

$$U_{sat}(t) = \mathrm{sat}(U(t), \bar{u}), \tag{2.110}$$

where $U(t)$ is given by (2.75).

Applying the latter control law (2.110), we represent the saturated closed-loop target system as the system (2.76) with the following boundary condition:

$$u(1,t) = \mathrm{sat}(U(t), \bar{u}) - U(t). \tag{2.111}$$

From (2.75), $U(t)$ admits the following representation:

$$\begin{aligned}U(t) &= \int_0^1 n(1,y)w(y,t)dy + \psi(1)X(t) + \int_0^1 l(1,y)u(y,t)dy \\ &= \int_0^1 n(1,y)\left[u(y,t) + \int_0^y l(y,s)u(s,t)ds\right]dy + \psi(1)X(t) + \int_0^1 l(1,y)u(y,t)dy,\end{aligned}$$

provided saturation is avoided. Denote

$$\begin{aligned}c_1 = |\psi(1)|,\ c_2 = &\max_{0\le y\le 1} |n(1,y)| \left(1 + \max_{0\le y\le x\le 1} |l(x,y)|\right) \\ &+ \max_{0\le y\le 1} |l(1,y)|.\end{aligned}$$

Due to (2.70) and (2.74), $n(x,y)$ and $l(x,y)$ are continuous functions bounded on any compact. Then Jensen's inequality implies

$$|U(t)| \le c_1|X| + c_2\|u\|.$$

Applying Young's inequality, we obtain

$$|U(t)|^2 \le 2c_1^2|X|^2 + 2c_2^2\|u\|^2. \tag{2.112}$$

Given $\bar{u} > 0$, we define the following set:

$$\mathcal{L}(c_1, c_2, \bar{u}) = \left\{(X,u) \in \mathcal{H} : c_1^2|X|^2 + c_2^2\|u\|^2 \le \frac{\bar{u}^2}{2}\right\}.$$

From the inequality (2.112) and the definition above, we can obtain the following implication: if $(X,u) \in \mathcal{L}(c_1, c_2, \bar{u})$, then $|U(t)| \le \bar{u}$, and the saturation is avoided. Thus, the system (2.76) subject to (2.111) admits the linear representation (2.76) subject to (2.77).

From Proposition 2.3, we find that if there exist $0 < \delta_1 = \delta_0$ such that the strict LMIs (2.84), (2.85) are feasible, then the following inequality holds

$$\begin{aligned}X^\top(t)PX(t) + p_1\int_0^1 u^2(x,t)dx = V(t) \le \sup_{-h\le\theta\le 0} V(\theta) \\ \le \lambda_{\max}(P)\max_{[-h,0]}|X_0|^2 + p_1\max_{[-h,0]}\|u_0\|^2,\ \forall t \ge 0.\end{aligned}$$

Hence, the following inequalities:

$$P \leq \beta I, \ p_1 \leq \beta \tag{2.113}$$

guarantee that the trajectories $(X(t; X_0), u(x, t; u_0))$ starting from initial function $(X_0, u_0) \in \mathcal{X}_\beta$ remain within $\mathcal{X}_u$, where

$$\mathcal{X}_u = \left\{(X, u) \in \mathcal{H} : X^\top(t)PX(t) + p_1 \int_0^1 u^2(x, t)dx \leq 1\right\}.$$

The "ellipsoid" $\mathcal{X}_u$ is contained in $\mathcal{L}(c_1, c_2, \bar{u})$, if the following implication holds

$$X^\top(t)PX(t) + p_1 \int_0^1 u^2(x, t)dx \leq 1 \Longrightarrow c_1^2|X(t)|^2 + c_2^2\|u(x, t)\|^2 \leq \frac{\bar{u}^2}{2}$$

for all $(X(t), u(x, t))$, i.e. if

$$c_1^2|X(t)|^2 + c_2^2\|u(x, t)\|^2 \leq \frac{\bar{u}^2}{2}\left[X^\top(t)PX(t) + p_1 \int_0^1 u^2(x, t)dx\right].$$

The latter inequality is guaranteed if

$$P\frac{\bar{u}^2}{2} - c_1^2 I \geq 0, \ p_1\frac{\bar{u}^2}{2} - c_2^2 \geq 0. \tag{2.114}$$

Therefore, the inequalities (2.114) guarantee the saturation avoidance, and together with Proposition 2.3 and condition (2.113) imply that

$$\lim_{t\to\infty} \|(X(t; X_0), u(x, t; u_0))\|_{\mathcal{H}} = 0.$$

Returning to the original system by the transformation (2.56) and (2.61), we have

$$\|u\| \leq \left[1 + \max_{0\leq y\leq x\leq 1} |q(x, y)|\right] \|w\|, \tag{2.115}$$

$$\|w\| \leq \left[1 + \max_{0\leq y\leq x\leq 1} |k(x, y)|\right] \|z\| + \left[\max_{0\leq x\leq 1} |\gamma(x)|\right] |X|. \tag{2.116}$$

Hence,

$$|X|^2 + \|u\|^2 \leq M_1|X|^2 + M_2\|z\|^2, \tag{2.117}$$

where

$$M_1 = 1 + 2\left[\max_{0\le x\le 1} |\gamma(x)|\left(1 + \max_{0\le y\le x\le 1} |q(x,y)|\right)\right]^2,$$
$$M_2 = 2\left[1 + \max_{0\le y\le x\le 1} |k(x,y)|\right]^2\left[1 + \max_{0\le y\le x\le 1} |q(x,y)|\right]^2.$$

Denote

$$\mathcal{X}_z = \left\{(X_0, z_0) \in W_1 : M_1 \max_{[-h,0]} |X_0|^2 + M_2 \max_{[-h,0]} \|z_0\|^2 \le \beta^{-1}\right\}.$$

It follows from the inequality (2.117) that if the initial function of (2.52) with the Dirichlet boundary actuation (2.111) satisfies $(X_0, z_0) \in \mathcal{X}_z$, then by backstepping transformation, the initial function of (2.76) subject to (2.111) satisfies $(X_0, u_0) \in \mathcal{X}_\beta$. The following is thus obtained:

Theorem 2.3 *Given gains K and c, and tuning parameters $r > 0$, $0 < \delta_1 = \delta_0$, let there exist an $n \times n$ matrix $P > 0$ and scalars $p_1 > 0$, $0 \le \lambda \le 2p_1$ that satisfy LMIs (2.84), (2.85) with notations given by (2.86)–(2.89) and LMIs (2.113), (2.114). Then for all $h_0 > 0$ and $h > 0$, the classical solutions of (2.52) with Dirichlet boundary actuation (2.111) starting from initial functions $(X_0, z_0) \in \mathcal{X}_z$ converge to zero for all delays τ subject to (2.55), i.e.*

$$\lim_{t\to\infty} \|(X(t; X_0), z(x, t; z_0))\|_{\mathcal{H}} = 0.$$

Example 2.1 Consider (2.52) with Dirichlet actuation, and the scalar $x(t) \in \mathbb{R}$ with $A = 1, B = 1, A_1 = 0.4, a_2 = 0.1, a = 0.2$, and $\bar{u} = 20$. For the target system (2.76), we choose $K = -2$, $c = 0.8$. In order to enlarge the volume of the ellipsoid inside of the domain of attraction, we would like to minimize β. By Proposition 2.4, with $\delta_0 = \delta_1 = 0.3$, $c_1 = 0.91$, $c_2 = 2.93$, $r = 1$, we obtain that $\min \beta = 0.0739$, and the largest obtained ellipsoid inside of domain of attraction is given by

$$\mathcal{X}_\beta = \left\{(X_0, u_0) \in W : \max_{[-h,0]} |X_0|^2 + \max_{[-h,0]} \|u_0\|^2 \le 13.53\right\}.$$

By Theorem 2.3, with $M_1 = 18.15$, $M_2 = 30.31$, we obtain

$$\mathcal{X}_z = \left\{(X_0, z_0) \in W_1 : 1.34 \max_{[-h,0]} |X_0|^2 + 2.24 \max_{[-h,0]} \|z_0\|^2 \le 1\right\}.$$

Next, a finite difference method is applied to compute the state of the closed-loop system. The steps of space and time are taken as 0.04 and 0.0002, respectively. We choose the delay $\tau(t) \equiv h = 0.4$. Figure 2.1 shows the state of the system starting from the initial condition $X(\theta) \equiv 0.82$, $z(x, \theta) \equiv 0.29\cos(\pi x)$ $(\theta \in [-0.4, 0])$, which is inside the ellipsoid $\mathcal{X}_z$. Here

$$1.34 \max_{[-h,0]} |X_0|^2 + 2.24 \max_{[-h,0]} \|z_0\|^2 = 0.99 < 1.$$

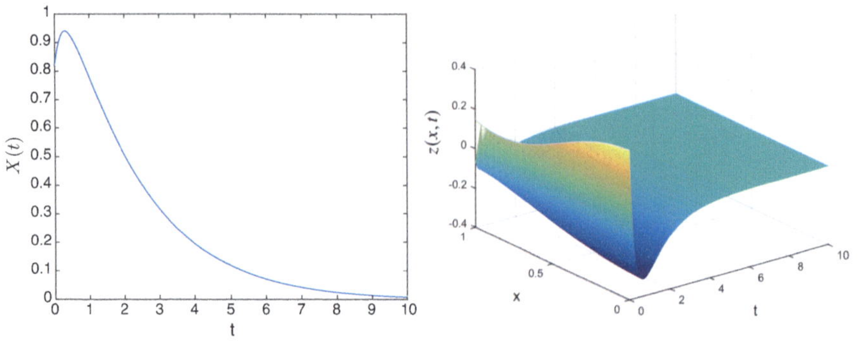

Fig. 2.3 State with initial condition inside $\mathcal{X}_z$

It is seen from Fig. 2.3 that the state converges. Simulations of the state starting outside the ellipsoid $\mathcal{X}_z$ from $X(\theta) \equiv 5$ and $z(x,\theta) \equiv 4\cos(\pi x)$ $(\theta \in [-0.4, 0])$ illustrate that this state is unbounded. See the corresponding plots in [18].

2.2.4.2 Neumann Control Under Saturation

For the case of Neumann actuation, the domain of attraction of the closed-loop target system is the set

$$\mathcal{S} = \left\{ (X_0, u_0) \in W : \lim_{t\to\infty} \|(X(t; X_0), u(x, t; u_0))\|_{\mathcal{H}_1} = 0 \right\}.$$

We will obtain an estimate $\mathcal{X}_\beta \subset \mathcal{S}$ of the domain of attraction, where

$$\mathcal{X}_\beta = \Bigg\{ (X_0, u_0) \in W : \max_{[-h,0]} |X_0|^2 + \max_{[-h,0]} \|u_0\|^2 + \max_{[-h,0]} \|u_0'\|^2 \le \beta^{-1} \Bigg\},$$

$\beta > 0$ is a scalar that will be minimized in the sequel.

Then we design the state feedback controller in the following form

$$U_{sat}(t) = \mathrm{sat}(U(t), \bar{u}), \tag{2.118}$$

where $U(t)$ is given by (2.78).

Applying the latter control law (2.118), we represent the saturated closed-loop target system into the system (2.76) with the following boundary condition:

$$u_x(1, t) = \mathrm{sat}(U(t), \bar{u}) - U(t). \tag{2.119}$$

In this case, from (2.78), $U(t)$ admits the representation:

$$\begin{aligned} U(t) &= \int_0^1 n_x(1,y)w(y,t)dy + \psi'(1)X(t) \\ &+ l(1,1)u(1,t) + \int_0^1 l_x(1,y)u(y,t)dy \\ &= \int_0^1 n_x(1,y)\left[u(y,t) + \int_0^y l(y,s)u(s,t)ds\right]dy \\ &+ \psi'(1)X(t) + l(1,1)u(1,t) + \int_0^1 l_x(1,y)u(y,t)dy. \end{aligned}$$

Here we use the fact that $n(1,1) = 0$.

Denote $c_1 = |\psi'(1)|,\ c_2 = \sqrt{2}|l(1,1)| + \xi,\ c_3 = |l(1,1)|, \xi \triangleq \max_{0\le y\le 1} |n_x(1,y)|$ $(1 + \max_{0\le x\le y\le 1} |l(x,y)|) + \max_{0\le y\le 1} |l_x(1,y)|$. Applying Jensen's and Young's inequalities, we obtain

$$|U(t)| \le |l(1,1)||u(1,t)| + |\psi'(1)||X(t)| + \xi \|u(x,t)\|.$$

By using Agmon's inequality, we have

$$|u(1,t)|^2 \le 2\|u(x,t)\|^2 + \|u_x(x,t)\|^2.$$

Then, $|U(t)|^2 \le 3\left[c_1^2|X|^2 + c_2^2\|u\|^2 + c_3^2\|u_x\|^2\right]$.

Given $\bar{u} > 0$, we define the following set:

$$\begin{aligned} \mathcal{L}(c_1, c_2, c_3, \bar{u}) = \Big\{(X,u) \in \mathcal{H}_1 : c_1^2|X|^2 + c_2^2\|u\|^2 \\ + c_3^2\|u_x\|^2 \le \frac{\bar{u}^2}{3}\Big\}. \end{aligned} \tag{2.120}$$

From the definition (2.120), we can obtain: if $(X,u) \in \mathcal{L}(c_1, c_2, c_3, \bar{u})$, then $|U(t)| \le \bar{u}$, and the saturation is avoided. Thus, the system (2.76) subject to (2.119) admits the linear representation (2.76) subject to (2.79).

From Proposition 2.4, we find that if there exist $0 < \delta_1 = \delta_0$ such that LMIs (2.100), (2.101) and (2.102) are feasible, then for all $t \ge 0$, the following inequality holds:

$$\begin{aligned} &X^\top PX + p_1\|u\|^2 + p_2\|u_x\|^2 \\ &\le \lambda_{\max}(P) \max_{[-h,0]} |X_0|^2 + p_1 \max_{[-h,0]} \|u_0\|^2 + p_2 \max_{[-h,0]} \|u_0'\|^2. \end{aligned}$$

Hence, the following inequalities:

$$P \le \beta I,\ p_1 \le \beta,\ p_2 \le \beta \tag{2.121}$$

guarantee that the trajectories $(X(t;X_0), u(x,t;u_0))$ starting from initial function $(X_0, u_0) \in \mathcal{X}_\beta$ remain within $\mathcal{X}_u$, where

$$\mathcal{X}_u = \left\{(X,u) \in \mathcal{H}_1 : X^\top PX + p_1\|u\|^2 + p_2\|u_x\|^2 \le 1\right\}.$$

Note that the ellipsoid $\mathcal{X}_u$ is contained in $\mathcal{L}(c_1, c_2, c_3, \bar{u})$, if the following implication holds

$$X^\top PX + p_1\|u\|^2 + p_2\|u_x\|^2 \le 1 \Longrightarrow c_1^2|X|^2 + c_2^2\|u\|^2 + c_3^2\|u_x\|^2 \le \frac{\bar{u}^2}{3}$$

for all $(X(t), u(x,t))$, i.e. if

$$c_1^2|X|^2 + c_2^2\|u\|^2 + c_3^2\|u_x\|^2 \le \frac{\bar{u}^2}{3}\left[X^\top PX + p_1\|u\|^2 + p_2\|u_x\|^2\right].$$

The latter inequality is guaranteed if

$$P\frac{\bar{u}^2}{3} - c_1^2 I \ge 0,\ p_1\frac{\bar{u}^2}{3} - c_2^2 \ge 0,\ p_2\frac{\bar{u}^2}{3} - c_3^2 \ge 0. \tag{2.122}$$

Therefore, the inequalities (2.122) guarantee the saturation avoidance, and together with Proposition 2.4 and the condition (2.121) imply that

$$\lim_{t\to\infty} \|(X(t; X_0), u(x,t; u_0))\|_{\mathcal{H}_1} = 0.$$

Returning to the original system by the transformation (2.56) and (2.61), we obtain that

$$\begin{aligned}\|u_x\| \le \|z_x\| &+ \max_{0\le y\le x\le 1}|k_x(x,y)|\|z\| + \max_{0\le x\le 1}|\gamma'(x)||X| \\ &+ \left[\max_{0\le x\le 1}|q(x,x)| + \max_{0\le y\le x\le 1}|q_x(x,y)|\right]\|w\|.\end{aligned}$$

It follows from (2.115) and (2.116) that

$$|X|^2 + \|u\|^2 + \|u_x\|^2 \le M_1|X|^2 + M_2\|z\|^2 + 4\|z_x\|^2,$$

where

$$\begin{aligned}M_1 = &\left\{8\left[\max_{0\le x\le 1}|q(x,x)| + \max_{0\le y\le x\le 1}|q_x(x,y)|\right]^2\right. \\ &\left.+2\left[1+\max_{0\le y\le x\le 1}|q(x,y)|\right]^2\right\}\left[\max_{0\le x\le 1}|\gamma(x)|\right]^2 + 4\max_{0\le x\le 1}|\gamma'(x)|^2 + 1,\end{aligned}$$

$$\begin{aligned}M_2 = &\,8\left[\max_{0\le x\le 1}|q(x,x)| + \max_{0\le y\le x\le 1}|q_x(x,y)|\right]^2\left[1+\max_{0\le y\le x\le 1}|k(x,y)|\right]^2 \\ &+4\max_{0\le y\le x\le 1}|k_x(x,y)|^2 + 2\left[1+\max_{0\le y\le x\le 1}|k(x,y)|\right]^2\left[1+\max_{0\le y\le x\le 1}|q(x,y)|\right]^2.\end{aligned}$$

Denote $\mathcal{X}_z = \left\{(X_0, z_0) \in W_1 : M_1 \max\limits_{[-h,0]} |X_0|^2 + M_2 \max\limits_{[-h,0]} \|z_0\|^2 + 4 \max\limits_{[-h,0]} \|z_0'\|^2 \leq \beta^{-1}\right\}$. Then, we obtain the following result:

Theorem 2.4 *Given gains K and c, and tuning parameters $r > 0$, $0 < r_1 < 2$, $0 < \delta_1 = \delta_0$, let there exist an $n \times n$ matrix $P > 0$, and scalars $p_1 > 0$, $p_2 > 0$, $\lambda > 0$ and $\lambda_1 \geq 0$ that satisfy LMIs (2.100)–(2.102) with notations given by (2.86), (2.87) and LMIs (2.121), (2.122). Then for all $h_0 > 0$ and $h > 0$, the classical solutions of (2.52) with Neumann boundary actuation (2.118) starting from initial functions $(X_0, z_0) \in \mathcal{X}_z$ converge to zero for all delays τ subject to (2.55), i.e.*

$$\lim_{t\to\infty} \|(X(t; X_0), z(x, t; z_0))\|_{\mathcal{H}_1} = 0.$$

Example 2.2 Consider the system (2.52) with Neumann actuation, and the scalar $x(t) \in \mathbb{R}$ with $A = 1$, $B = 1$, $A_1 = 0.4$, $a_2 = 0.1$, $a = 0.2$, and $\bar{u} = 50$. For (2.76), we choose $K = -4$, $c = 1.8$. By Proposition 2.4, with $\delta_0 = \delta_1 = 0.5$, $c_1 = 6.98$, $c_2 = 9.9$, $c_3 = r = r_1 = 1$, we obtain $\min \beta = 0.1176$, and the largest obtained ball inside of domain of attraction is given by

$$\mathcal{X}_\beta = \{(X_0, u_0) \in W : \max_{[-h,0]} |X_0|^2 + \max_{[-h,0]} \|u_0\|^2 + \max_{[-h,0]} \|u_0'\|^2 \leq 8.50\}.$$

By Theorem 2.4, with $M_1 = 118.7$, $M_2 = 141.8$, we obtain

$$\begin{aligned}\mathcal{X}_z = \{(X_0, z_0) \in W_1 : 13.96 \max_{[-h,0]} |X_0|^2 + 16.67 \max_{[-h,0]} \|z_0\|^2 \\ +0.47 \max_{[-h,0]} \|z_0'\|^2 \leq 1\}.\end{aligned}$$

Simulations of the solutions confirm the theoretical results. Thus, the solution starting inside the ellipsoid from the initial conditions $X(\theta) \equiv 0.26$ and $z(x, \theta) \equiv 0.05 \cos(\pi x)$ ($\theta \in [-0.4, 0]$) converges to zero. However, the solution starting outside the ellipsoid from $X(\theta) \equiv 3$ and $z(x, \theta) \equiv 0.05 \cos(\pi x)$ ($\theta \in [-0.4, 0]$) is unbounded.

2.3 Notes

Constrained control was recently suggested for various PDE systems, such as KdV equation [19], Euler-Bernoulli beam [20] and wave equation [21]. Constrained control can also be considered for systems with time-varying constraints [22], input/output constraints [23], et al. Observer-based boundary control of nonlinear PDEs may be a topic for future research.

References

1. Kang, W., Fridman, E.: Boundary constrained control of delayed nonlinear schrodinger equation. IEEE Trans. Automat. Contr. **63**(11), 3873–3880 (2018)
2. Zhao, Z.H., Ge, W.G.: Traveling wave solutions for Schrödinger equation with distributed delay. Appl. Math. Modell. **35**(2), 675–687 (2011)
3. Chen, C.J., Menyuk, C.R., Islam, M.N., Stolen, R.H.: Numerical study of the Raman effect and its impact on soliton-dragging logic gates. Opt. Lett. **21**(16), 1647–1649 (1991)
4. Agrawal, G.P.: Nonlinear Fiber Optics. Academic (2007)
5. Krstic, M., Guo, B.Z., Smyshlyaev, A.: Boundary controllers and observers for the linearized Schrödinger equation. SIAM J. Control Optim. **49**(4), 1479–1497 (2011)
6. Fridman, E.: Introduction to Time-delay Systems: Analysis and Control. Birkhäuser, Basel (2014)
7. Pazy, A.: Semigroups of Linear Operators and Applications to Partial Differential Equations. Springer, New York (1983)
8. Krasnosel'skii, M.A., Zabreyko, P.P., Pustylnik, E.I., Sobolevski, P.E.: Integral Operators in Spaces of Summable Functions. Springer, Dordrecht (2011)
9. Hashimoto, T., Krstic, M.: Stabilization of reaction diffusion equations with state delay using boundary control input. IEEE Trans. Automat. Control **61**(12), 4041–4046 (2016)
10. Susto, G.A., Krstic, M.: Control of PDE-ODE cascades with Neumann interconnections. J. Franklin Inst. **347**, 284–314 (2010)
11. Krstic, M.: Compensating actuator and sensor dynamics governed by diffusion PDEs. Syst. & Control Lett. **58**, 372–377 (2009)
12. Krstic, M., Smyshlyaev, A.: Boundary Control of PDEs: A Course on Backstepping Designs. SIAM, Philadelphia, PA (2008)
13. Wang, J.M., Liu, J.J., Ren, B.B., Chen, J.H.: Sliding mode control to stabilization of cascaded heat PDE-ODE systems subject to boundary control matched disturbance. Automatica **52**, 23–34 (2015)
14. Tucsnak, M., Weiss, G.: Observation and Control for Operator Semigroups. Birkhäuser Advanced Texts: Basler Lehrbücher. Birkhäuser Verlag, Basel (2009)
15. Curtain, R., Zwart, H.: An Introduction to Infinite-dimensional Linear Systems. Springer (1995)
16. Fridman, E., Orlov, Y.: Exponential stability of linear distributed parameter systems with time-varying delays. Automatica **45**(1), 194–201 (2009)
17. Fridman, E., Blighovsky, A.: Robust sampled-data control of a class of semilinear parabolic systems. Automatica **48**, 826–836 (2012)
18. Kang, W., Fridman, E.: Boundary control of delayed ODE-heat cascade under actuator saturation. Automatica **83**, 252–261 (2017)
19. Marx, S., Cerpa, E., Prieur, C., Andrieu, V.: Global stabilization of a Korteweg-de Vries equation with saturating distributed control. SIAM J. Control Opt. **55**(3), 1452–1480 (2017)
20. He, W., Meng, T., Huang, D., Li, X.: Adaptive boundary iterative learning control for an Euler-Bernoulli beam system with input constraint. IEEE Trans. Neural Netw. Learn. Syst. **29**(5), 1539–1549 (2017)
21. Prieur, C., Tarbouriech, S., da Silva Jr, J.M.G.: Well-posedness and stability of 1D wave equation with saturating distributed input. In: Proceeding, IEEE Conference on Decision and Control, pp. 2846–2851. Los Angeles, California, USA (2014)
22. Liu, Y.J., Gong, M., Liu, L., Tong, S., Chen, C.L.P.: Fuzzy observer constraint based on adaptive control for uncertain nonlinear MIMO systems with time-varying state constraints. IEEE Trans. Cybern. **51**(3), 1380–1389 (2019)
23. Li, R., Chen, M., Wu, Q.: Adaptive neural tracking control for uncertain nonlinear systems with input and output constraints using disturbance observer. Neurocomputing **235**, 27–37 (2017)

Chapter 3
Sampled-Data and Delayed Control of PDE Systems

After discussing boundary control for delay systems in Chap. 2, we now focus on sampled-data control, crucial for real-world systems. This chapter explores distributed sampled-data control of nonlinear PDEs, particularly the KSE. We design locally stabilizing controllers for 1D and 2D systems using spatially distributed measurements. We employ the spatial decomposition (or sampling in space) method introduced in [1, 2] for the heat equation. We also introduce a switched sampled-data approach to reduce energy consumption and stabilize the system by using mobile actuators. We apply both continuous-time and sampled-data feedback to stabilize a PDE-PDE cascade system, demonstrating the versatility of our control strategies. Finally we present distributed stabilization of KdVB equation in the presence of constant delay [3].

3.1 Distributed Sampled-Data Control of KSE

This section is devoted to distributed sampled-data control of nonlinear PDE system governed by 1D KSE. It is assumed that N sensors provide sampled in time spatially distributed (either point or averaged) measurements of the state over N sampling spatial intervals. Locally stabilizing sampled-data controllers are designed that are applied through distributed in space shape functions and zero-order hold devices. Given upper bounds on the sampling intervals in time and in space, sufficient conditions ensuring regional exponential stability of the closed-loop system are established in terms of Linear Matrix Inequalities (LMIs) by using the time-delay approach to sampled-data control and Lyapunov-Krasovskii method. As it happened in the case of diffusion equation, the descriptor method appeared to be an efficient tool for the stability analysis of the sampled-data KSE. An estimate on the domain of attraction is also given. A numerical example demonstrates the efficiency of the results.

W. Kang and E. Fridman, *Delay-Robust Control of Distributed Parameter Systems*, Advances in Delays and Dynamics 13, https://doi.org/10.1007/978-981-96-6344-6_3

3.1.1 Problem Formulation and Useful Lemmas

We consider 1-D Kuramoto-Sivashinsky equation

$$\begin{aligned} &z_t(x,t) + z_{xx}(x,t) + \nu z_{xxxx}(x,t) + z(x,t)z_x(x,t) \\ &= \sum_{j=1}^{N} b_j(x)U_j(t),\ 0 < x < L,\ t \geq 0, \end{aligned} \tag{3.1}$$

subject to Dirichlet

$$z(0,t) = z(L,t) = 0, z_x(0,t) = z_x(L,t) = 0 \tag{3.2}$$

or to periodic

$$\frac{\partial^m z}{\partial x^m}(0,t) = \frac{\partial^m z}{\partial x^m}(L,t),\ m = 0, 1, 2, 3 \tag{3.3}$$

boundary conditions. Here ν is a positive constant, $z(x,t)$ is the state of KSE, and $U_j(t) \in \mathbb{R}$, $j = 1, 2, \ldots, N$ are the control inputs. Dirichlet boundary conditions were considered in [4], whereas the periodic ones were studied in [5–7]. The open-loop system (3.1) (subject to $U_j(t) \equiv 0$) may become unstable if ν is small enough. Thus, for $L = 2\pi$ if $\nu < 1$ the open-loop system is unstable (see the example below).

As in [1, 2, 7], consider the points

$$0 = x_0 < x_1 < \cdots < x_N = L$$

that divide $[0, L]$ into N sampling intervals $\Omega_j = [x_{j-1}, x_j)$. Let

$$0 = t_0 < t_1 < \cdots < t_k \cdots,\quad \lim_{k\to\infty} t_k = \infty$$

be sampling time instants. The sampling intervals in time and in space may be variable but bounded,

$$0 \leq t_{k+1} - t_k \leq h,\ 0 < x_j - x_{j-1} = \Delta_j \leq \Delta,$$

where h and Δ are the corresponding upper bounds. The control inputs $U_j(t)$ enter (3.1) through the shape functions

$$\begin{cases} b_j(x) = 1,\ x \in \Omega_j, \\ b_j(x) = 0,\ x \notin \Omega_j, \end{cases} j = 1, \ldots, N. \tag{3.4}$$

Sensors provide either point

$$y_{jk} = z(\bar{x}_j, t_k), \bar{x}_j = \frac{x_{j-1} + x_j}{2},\ j = 1, \ldots, N,\ k = 0, 1, 2\ldots \tag{3.5}$$

or averaged

$$y_{jk} = \frac{\int_{x_{j-1}}^{x_j} z(x, t_k)dx}{\Delta_j}, \quad j = 1, \ldots, N, \quad k = 0, 1, 2\ldots \tag{3.6}$$

measurements of the state. Our main objective is to design for (3.1) an exponentially stabilizing sampled-data controller that can be implemented by zero-order hold devices:

$$U_j(t) = -\mu y_{jk}, \quad j = 1, \ldots, N, \quad t \in [t_k, t_{k+1}), \quad k = 0, 1, \ldots, \tag{3.7}$$

where μ is a positive controller gain and y_{jk} is given by (3.5) or (3.6).

3.1.2 Continuous-Time Global Stabilization

We will start with continuous in time results, where global stabilization can be achieved. Here the stability analysis is similar to [7], but differently from [7] we give a bound on the decay rate. Sampled-data controllers under the point/averaged measurements leading to regional stability will be presented in Sect. 3.1.3.

In this chapter, we consider the stabilization of Eq. (3.1) with the Dirichlet or periodic boundary conditions under the continuous-time measurements

$$y_j(t) = z(\bar{x}_j, t), \quad \bar{x}_j = \frac{x_{j-1} + x_j}{2}, \quad j = 1, \ldots, N, \tag{3.8}$$

or

$$y_j(t) = \frac{\int_{x_{j-1}}^{x_j} z(x, t)dx}{\Delta_j} \tag{3.9}$$

via a continuous-time control law

$$U_j(t) = -\mu y_j(t). \tag{3.10}$$

The control law (3.10) can be presented as

$$U_j(t) = -\mu[z(x, t) - f_j(x, t)],$$

where for (3.8)

$$f_j(x, t) = \int_{\bar{x}_j}^{x} z_\xi(\xi, t)d\xi, \tag{3.11}$$

and for (3.9)

$$f_j(x, t) = \frac{\int_{x_{j-1}}^{x_j} [z(x, t) - z(\zeta, t)]d\zeta}{\Delta_j}. \tag{3.12}$$

Then the closed-loop system (3.1), (3.10) has a form

$$\begin{aligned} &z_t(x,t) + z_{xx}(x,t) + \nu z_{xxxx}(x,t) + z(x,t)z_x(x,t) \\ &= -\sum_{j=1}^{N} \mu b_j(x)[z(x,t) - f_j(x,t)], \quad t \geq 0. \end{aligned} \tag{3.13}$$

We aim to find μ that enlarges Δ (leading to smaller number of sensors).

3.1.2.1 Well-Posedness and Stability of (3.13) Subject to (3.2)

Let $H = L^2(0, L)$ be a Hilbert space with the inner product $\langle \cdot, \cdot \rangle$ and induced norm $\| \cdot \|_{L^2}$. We define an unbounded linear operator $A : D(A) \subset H \to H$ as follows:

$$\begin{cases} Af = -\nu f'''', \ \forall f \in D(A), \\ D(A) = H^4(0, L) \cap H_0^2(0, L). \end{cases} \tag{3.14}$$

It is well-known that A is a dissipative operator, and A generates an analytic semigroup $T(t)$.

The domain $H_1 = D(A) = A^{-1}H$ forms another Hilbert space with the graph inner product $\langle x, y \rangle_1 = \langle Ax, Ay \rangle$, $\forall x, y \in H_1$. The domain $D(A)$ is dense in H.

Operator $-A$ is positive implying that its square root $(-A)^{\frac{1}{2}}$ is also positive. We define the Hilbert space:

$$H_{\frac{1}{2}} = D((-A)^{\frac{1}{2}}) = H_0^2(0, L).$$

The norm of $H_{\frac{1}{2}}$ is endowed by the induced inner product:

$$\langle f, g \rangle_{\frac{1}{2}} = \langle (-A)^{\frac{1}{2}} f, (-A)^{\frac{1}{2}} g \rangle \quad \forall f, g \in H_{\frac{1}{2}},$$

$$\|f\|_{H_{\frac{1}{2}}} = \nu^{\frac{1}{2}} \left[\int_0^1 |f''(x)|^2 dx \right]^{\frac{1}{2}} \quad \forall f \in H_{\frac{1}{2}}. \tag{3.15}$$

Note that $H_{\frac{1}{2}}$ norm is equivalent to the inherent norm $\| \cdot \|_{H^2}$ of Sobolev space $H^2(0, L)$. We define the spaces $H_{-\frac{1}{2}}$ and H_{-1} as the dual spaces of $H_{\frac{1}{2}}$ and H_1 respectively, with respect to the pivot space H. Therefore, $H_{-\frac{1}{2}} = H^{-2}(0, L)$. Then we have $H_1 \subset H_{\frac{1}{2}} \subset H \subset H_{-\frac{1}{2}} \subset H_{-1}$, densely and with continuous embedding. All relevant material on fractional operator degrees can be found in [8] (see pp. 81–83).

The nonlinear term $F : H^2(0, L) \to L^2(0, L)$ is defined on functions $z(\cdot, t)$ according to

$$F(z(\cdot,t)) = -z(x,t)z_x(x,t) - z_{xx}(x,t) \\ - \sum_{j=1}^{N} \mu b_j(x)[z(x,t) - f_j(x,t)],\ t \geq 0.$$

With the operator A at hand, the system (3.13) subject to (3.2) can be written as an evolution equation:

$$\begin{cases} \dfrac{d}{dt}z(\cdot,t) = Az(\cdot,t) + F(z(\cdot,t)), \\ z(\cdot,0) = z_0(\cdot). \end{cases} \tag{3.16}$$

Note that the nonlinear term F is locally Lipschitz continuous, that is, there exists a positive constant $l(K)$ such that the following inequality

$$\|F(z_1) - F(z_2)\|_{L^2} \leq l(K)\|(z_1 - z_2)\|_{H_0^2(0,L)}$$

holds for $z_1, z_2 \in H_0^2(0,L)$ with $\|z_1\|_{H_0^2(0,L)} \leq K$, $\|z_2\|_{H_0^2(0,L)} \leq K$. Thus, by Theorem 6.3.1 of [9], we obtain that for any initial condition $z_0 \in H_0^2(0,L)$, there exists a unique local classical solution $z \in C([0,T), L^2(0,L)) \cap C^1((0,T), L^2(0,L))$ of (3.13) subject to (3.2), where $T = T(z_0) > 0$.

The following proposition provides conditions that guarantee the existence of solution of the closed-loop system (3.13) subject to (3.2) for all $t \geq 0$ as well as the global stability of this system.

Proposition 3.1 *Consider the closed-loop system (3.13) subject to (3.11) or (3.12) under the Dirichlet boundary conditions* (3.2). *Given a scalar $\alpha > 0$ and tunning parameter $\Delta > 0$, let there exit scalars $\mu > 0$, $\lambda \geq 0$ and $\lambda_1 > 0$ that satisfy the following LMI*

$$W = \begin{bmatrix} W_{11} & -1 - \dfrac{\lambda_1}{2} & \mu \\ * & -2\nu + \lambda & 0 \\ * & * & -\lambda_1 \dfrac{\pi^2}{\Delta^2} \end{bmatrix} < 0, \tag{3.17}$$

where

$$W_{11} = -2\mu + 2\alpha - \lambda\frac{\pi^4}{L^4}. \tag{3.18}$$

Then a unique classical solution of the Dirichlet boundary value problem (3.13), (3.2) subject to (3.11) or (3.12) initialized by $z_0 \in H_0^2(0,L)$ exists for all $t \geq 0$ and the system is exponentially stable with a decay rate α, i.e. its classical solutions satisfy

$$\int_0^L z^2(x,t)dx \leq e^{-2\alpha t}\int_0^L z^2(x,0)dx,\ \forall t \geq 0. \tag{3.19}$$

If the LMI (3.17) *is feasible with* $\alpha = 0$, *then the closed-loop system is exponentially stable with a small enough decay rate* $\alpha_0 > 0$. *Particularly, if* $\frac{\Delta^2}{\pi^2} < \nu$ *and* $\mu = \frac{2\pi^2}{\Delta^2}$, *then* (3.17) *is feasible with* $\lambda = \alpha = 0$ *and* $\lambda_1 = 2$, *meaning that the closed-loop system is exponentially stable.*

Proof The proof of this proposition consists of several steps.

Step 1: It has been shown that the local classical solution of (3.13) under the Dirichlet boundary conditions (3.2) exists in the sense that $z \in C([0, T), L^2(0, L)) \cap C^1((0, T), L^2(0, L))$, where $T = T(z_0) > 0$. From Theorem 6.23.5 of [10], it follows that if this solution admits a priori estimate, then the solution exists for any $T > 0$. We will prove in Step 3 that if the LMI (3.17) is feasible, then the solution of (3.13) subject to (3.2) admits a priori bound, which guarantees the existence of the solution of (3.13) subject to (3.2) for all $t \geq 0$.

Step 2: Assume formally that solution of (3.13) subject to (3.2) exists for all $t \geq 0$ and choose the Lyapunov function of the form:

$$V(t) = \int_0^L z^2(x, t)dx. \tag{3.20}$$

Differentiating V along (3.13) subject to (3.2) and integrating by parts, we find

$$\begin{aligned} \dot{V}(t) &= 2\int_0^L z(x, t)z_t(x, t)dx \\ &= -2\int_0^L z(x, t)z_{xx}(x, t)dx - 2\nu\int_0^L z_{xx}^2(x, t)dx \\ &+2\mu\sum_{j=1}^N\int_{x_{j-1}}^{x_j} z(x, t)f_j(x, t)dx \\ &-2\mu\int_0^L z^2(x, t)dx. \end{aligned} \tag{3.21}$$

(a) For the case of the distributed control under the point measurements, $f_j(x, t)$ is given by (3.11).

Applying Wirtinger's inequality, we obtain

$$\begin{aligned} \int_{x_{j-1}}^{x_j} f_j^2(x, t)dx &= \int_{x_{j-1}}^{x_j}\left[\int_{\bar{x}_j}^x z_\xi(\xi, t)d\xi\right]^2 dx \\ &\leq \frac{\Delta^2}{\pi^2}\int_{x_{j-1}}^{x_j} z_x^2(x, t)dx. \end{aligned} \tag{3.22}$$

(b) For the case of the distributed control under the averaged measurements, $f_j(x, t)$ is given by (3.12).

Since

$$\int_{x_{j-1}}^{x_j} \left[z(x,t) - \frac{\int_{x_{j-1}}^{x_j} z(\zeta,t)d\zeta}{\Delta_j} \right] dx = 0,$$

the Poincaré inequality and Jensen inequality allow to obtain

$$\begin{aligned} \int_{x_{j-1}}^{x_j} f_j^2(x,t)dx &= \int_{x_{j-1}}^{x_j} \left[z(x,t) - \frac{\int_{x_{j-1}}^{x_j} z(\zeta,t)d\zeta}{\Delta_j} \right]^2 dx \\ &\leq \frac{\Delta^2}{\pi^2} \int_{x_{j-1}}^{x_j} z_x^2(x,t)dx. \end{aligned} \tag{3.23}$$

Hence, for both cases (a) and (b), from (3.22) and (3.23) the following inequality

$$\int_{x_{j-1}}^{x_j} f_j^2(x,t)dx \leq \frac{\Delta^2}{\pi^2} \int_{x_{j-1}}^{x_j} z_x^2(x,t)dx \tag{3.24}$$

holds.

Note that integration by parts of $-zz_{xx}$ yields

$$-\int_0^L z(x,t)z_{xx}(x,t)dx = \int_0^L z_x^2(x,t)dx. \tag{3.25}$$

Multiplying the inequality (3.24) by some constant $\lambda_1 > 0$, summing and using (3.25) we obtain

$$\begin{aligned} &\sum_{j=1}^N \lambda_1 \left[\int_{x_{j-1}}^{x_j} z_x^2(x,t)dx - \frac{\pi^2}{\Delta^2} \int_{x_{j-1}}^{x_j} f_j^2(x,t)dx \right] \\ &= -\lambda_1 \int_0^L z(x,t)z_{xx}(x,t)dx \quad - \sum_{j=1}^N \frac{\lambda_1 \pi^2}{\Delta^2} \int_{x_{j-1}}^{x_j} f_j^2(x,t)dx \geq 0. \end{aligned} \tag{3.26}$$

From Lemma 1.2, the Wirtinger inequality implies

$$\lambda \int_0^L \left[z_{xx}^2(x,t) - (\frac{\pi^2}{L^2})^2 z^2(x,t) \right] dx \geq 0, \tag{3.27}$$

where $\lambda \geq 0$.

Applying S-procedure, we add to $\dot{V}(t) + 2\alpha V(t)$ the left-hand sides of (3.26) and (3.27). This leads to

$$\dot{V}(t) + 2\alpha V(t) \leq \sum_{j=1}^N \int_{x_{j-1}}^{x_j} \varphi^T W \varphi dx < 0 \tag{3.28}$$

if $W < 0$ holds, where

$$\varphi = \text{col}\{z(x,t), z_{xx}(x,t), f_j(x,t)\}$$

and W is given by (3.17).

Note that the feasibility of the LMI (3.17) with $\alpha = 0$ implies its feasibility with a small enough $\alpha_0 > 0$. Therefore, if the LMI (3.17) holds for $\alpha = 0$, then the system (3.13), (3.2) subject to (3.11) or (3.12) is exponentially stable with a small decay rate $\alpha_0 > 0$.

For $\lambda = \alpha = 0$, by application of Schur complement theorem, the strict LMI (3.17) is feasible iff the following inequalities hold:

$$\mu^2(\lambda_1 \frac{\pi^2}{\Delta^2})^{-1} - 2\mu + (1 + \frac{\lambda_1}{2})^2 (2\nu)^{-1} < 0 \tag{3.29}$$

Minimizing the left-hand side of the latter inequality in μ, we find that for $\mu = \lambda_1 \frac{\pi^2}{\Delta^2}$ this inequality is reduced to $\frac{\Delta^2}{\pi^2} < \frac{2\lambda_1}{(1+\frac{\lambda_1}{2})^2}\nu$. Note that $\max_{\lambda_1} \frac{2\lambda_1}{(1+\frac{\lambda_1}{2})^2} = 1$ (this corresponds to $\lambda_1 = 2$). Therefore, the system (3.13), (3.2) subject to (3.11) or (3.12) is exponentially stable if $\frac{\Delta^2}{\pi^2} < \nu$ and $\mu = \frac{2\pi^2}{\Delta^2}$.

Step 3: We obtain that if the LMI condition (3.17) is satisfied, then the solution (3.13), (3.2) subject to (3.200) or (3.213) initialized with $z_0 \in H_0^2(0, L)$ on $[0, T)$ admits a priori estimate

$$V(t) \leq e^{-2\alpha t} V(0).$$

Thus, continuation of this solution of (3.13), (3.2) subject to (3.200) or (3.213) under a priori bound to entire interval $[0, \infty)$ follows from Theorem 6.23.5 of [10]. Furthermore, the inequality (3.28) implies (3.19) for all $t \geq 0$, which completes the proof. □

3.1.2.2 Extension to Periodic Boundary Conditions

The well-posedness under the periodic boundary conditions can be established similar to the case of Dirichlet boundary conditions. Here instead of $H_0^2(0, L)$ we consider a Hilbert space $H_{per}^2 = \{f \in H^2(0, L) : f^{(m)}(0) = f^{(m)}(L),\ m = 0, 1\}$ with respect to the H^2 norm. By arguments of Proposition 3.1, where in (3.27) $\lambda = 0$, because Writinger's inequality is not applicable here, we arrive at the following result:

If the LMI condition of Proposition 3.1 holds with $\lambda = 0$, then there exists a unique solution of the periodic boundary value problem (3.13), (3.3) subject to (3.11) or (3.12) initialized by $z(\cdot, 0) \in H_{per}^2$ in the sense that $z \in C([0, \infty), L^2(0, L)) \cap$

$C^1((0,\infty), L^2(0,L))$. Moreover, the solution satisfies (3.19) meaning that the closed-loop system (3.13), (3.3) subject to (3.11) or (3.12) is exponentially stable with a decay rate α.

Remark 3.1 In [7] stabilization of (3.1) was studied subject to the periodic boundary conditions (3.3) under the assumption $\int_0^L z_0(x)dx = 0$ (that implies also $\int_0^L z(x,t)dx = 0$). In this case Poincaré inequality allows to use (3.27) with $\lambda \geq 0$ as for the case of Dirichlet boundary conditions leading to LMI (3.17) with $\lambda \geq 0$. The stability conditions of (3.13) subject to (3.3) were given in [7] as $\mu > \frac{4}{\nu}$, $\nu > \mu \frac{\Delta^4}{\pi^4}$ (cf. (15) in [7]). The LMI condition (3.17) of Proposition 3.1 is less conservative (due to the use of $\lambda_1 > 0$ and $\lambda \geq 0$ instead of conservative use of Young inequality with a specific constant), and gives also a bound on the decay rate. The advantages of LMI condition (3.17) are illustrated in the example below (see Sect. 3.1.4).

3.1.3 Sampled-Data Regional Stabilization

3.1.3.1 Sampled-Data Control Under Point Measurements

By selecting the controller (3.10) subject to (3.5), we arrive at the closed-loop system

$$\begin{aligned} &z_t(x,t) + z_{xx}(x,t) + \nu z_{xxxx}(x,t) + z(x,t)z_x(x,t) \\ &= -\mu \sum_{j=1}^{N} b_j(x) z(\bar{x}_j, t_k), \quad t \in [t_k, t_{k+1}), \quad k = 0, 1, \ldots \end{aligned} \tag{3.30}$$

subject to (3.2) or (3.3).

3.1.3.2 Well-Posedness and Stability of (3.30) Subject to (3.2)

We start with the well-posedness of the sampled-data closed-loop system (3.30) under the Dirichlet boundary conditions (3.2) initialized with $z_0(x) = z(x,0)$. We will use the step method for solution of time-delay systems [11, 12].

For $t \in [t_0, t_1]$, we consider the following equation:

$$\begin{cases} z_t(x,t) + z_{xx}(x,t) + \nu z_{xxxx}(x,t) + z(x,t)z_x(x,t) \\ = -\mu \sum_{j=1}^{N} b_j(x) z(\bar{x}_j, t_0), \\ z(0,t) = z(L,t) = z_x(0,t) = z_x(L,t) = 0. \end{cases} \tag{3.31}$$

We can present the system (3.31) as an evolution Eq. (3.16) with A defined by (3.14). Here the nonlinear term $F : H^2(0,L) \to L^2(0,L)$ is defined on functions $z(\cdot,t)$ according to

$$
\begin{aligned}
F(z(\cdot,t)) = & -z(x,t)z_x(x,t) - z_{xx}(x,t) \\
& - \sum_{j=1}^{N} \mu b_j(x) z_0(\bar{x}_j), \ t \in [t_0, t_1).
\end{aligned} \tag{3.32}
$$

A function $z \in C([0,T]; H_0^2(0,L)) \cap L^2([0,T]; D(A))$ such that $\dot{z} \in L^2([0,T]; L^2(0,L))$ is called a strong solution of (3.16) with the nonlinearity F given by (3.32) if (3.16) holds almost everywhere on $[0,T]$.

From (3.32) it follows that the nonlinear term F is locally Lipschitz continuous. Thus, Theorem 3.3.3 of [13] is applicable to (3.16). Given any initial condition $z_0 \in H_0^2(0,L)$, there exists a unique local strong solution $z(\cdot,t) \in H_0^2(0,L)$ of (3.31) on some interval $[0,T] \subset [0,t_1]$, where $T = T(z_0) > 0$. From Theorem 6.23.5 of [10], it follows that if this solution admits a priori estimate, then the solution exists on the entire interval $[0,t_1]$. The a priori estimate on the solutions starting from the domain of attraction will be guaranteed by the stability conditions that we will provide (see Theorem 3.1 below).

For the stability analysis, we present (3.30) as

$$
\begin{aligned}
& z_t(x,t) + z_{xx}(x,t) + \nu z_{xxxx}(x,t) + z(x,t)z_x(x,t) \\
& = -\mu z(x,t_k) + \mu \sum_{j=1}^{N} b_j(x) \int_{\bar{x}_j}^{x} z_\xi(\xi,t_k)d\xi, \\
& \qquad t \in [t_k, t_{k+1}), \ k = 0, 1, \ldots
\end{aligned} \tag{3.33}
$$

We will use an input delay approach to sampled-data control ([14, 15]), where the sampling time t_k is presented as delayed time $t - \tau(t)$ with $\tau(t) = t - t_k$ for $t \in [t_k, t_{k+1})$.

In order to derive the stability conditions for (3.33) we employ the following Lyapunov-Krasovskii functional

$$
\begin{aligned}
V_1(t) = & \, p_1 \int_0^L z^2(x,t)dx + p_3 \nu \int_0^L z_{xx}^2(x,t)dx \\
& + r(t_{k+1} - t) \int_0^L \int_{t_k}^{t} e^{2\delta(s-t)} z_s^2(x,s)dsdx, \\
& t \in [t_k, t_{k+1}), \ p_1 > 0, \ p_3 > 0, \ r > 0.
\end{aligned} \tag{3.34}
$$

Here p_1 and p_3-terms are extensions of the corresponding terms of [2] to KSE, whereas r-term treats sampled-data control as introduced in [16], and $\delta > 0$ stands for the decay rate. In the time-derivative of r-term we have a positive term $r(t_{k+1} - t) \int_0^L z_t^2(x,t)dx$ (see (2.97) below). To compensate such a term in $\dot{V}_1$, we choose the p_3-term that guarantees convergence in H^2-norm (and not in H^1-norm like in [2] for the case of diffusion-reaction equation).

For convenience we define

$$
\|z(\cdot,t)\|_V^2 = p_1 \int_0^L z^2(x,t)dx + p_3 \nu \int_0^L z_{xx}^2(x,t)dx, \tag{3.35}
$$

where p_1 and p_3 are positive constants, and $z(\cdot, t) \in H_0^2(0, L)$. The choice of such norm is motivated by the Lyapunov-Krasovskii functional (3.34). By using Lyapunov-Krasovskii functional (3.34), in Theorem 3.1 below we provide LMI conditions for regional exponential stability of (3.30) and for a bound on the domain of attraction.

Remark 3.2 To find a bound the domain of attraction for system (3.33) subject to (3.2), we use positive invariance principle in Theorem 3.1: we derive stability conditions in terms of matrix inequalities that guarantee $V_1(t) \leq V_1(0)$ for all $t \geq 0$. These matrix inequalities ($\Theta_1 < 0$ and $\Theta_2 < 0$ with Θ_1 and Θ_2 defined by (3.39) and (3.40)) are affine in $z_x(x, t)$. Our objective is to guarantee that $\max_{x\in[0,L]} |z_x(x, t)|^2 < C^2$ for all $t \geq 0$. This allows to verify the matrix inequalities in the vertices $z_x = \pm C$ (see (3.37)). Therefore, if the initial condition satisfies $\|z_0\|_V < \sqrt{\frac{p_3\nu}{L}}C$, then from the Sobolev inequality we obtain the desired bound on z_x:

$$\begin{aligned} &\max_{x\in[0,L]} |z_x(x, t)|^2 \leq L\|z_{xx}(\cdot, t)\|_{L^2}^2 \leq \frac{L}{p_3\nu} V_1(t) \\ &\leq \frac{L}{p_3\nu} V_1(0) = \frac{L}{p_3\nu} \|z_0\|_V^2 < C^2. \end{aligned} \tag{3.36}$$

Now we are in a position to formulate our first main result.

Theorem 3.1 *Consider the closed-loop system* (3.33) *under the Dirichlet boundary conditions* (3.2)*. Given positive scalars C, R, h, μ, Δ and $\delta_1 < 2\delta$, let there exist scalars $r > 0$, $\lambda \geq 0$, $p_i > 0$ (i=1,2,3) that satisfy the LMIs:*

$$\Theta_i|_{z_x=C} < 0, \ \Theta_i|_{z_x=-C} < 0, \ i = 1, 2, \tag{3.37}$$

and

$$\bar{\Theta} = \begin{bmatrix} -\delta_1 p_1 & \frac{\mu}{2}\frac{\Delta}{\pi} R^{-1}(p_2 + p_3) \\ * & -\delta_1 p_3 \nu \end{bmatrix} < 0, \tag{3.38}$$

where

$$\Theta_1 = \begin{bmatrix} \theta_{11} & \theta_{12} & \theta_{13} \\ * & rh - 2p_3 + \frac{\Delta}{\pi}\mu R p_3 & \theta_{23} \\ * & * & \theta_{33} \end{bmatrix}, \tag{3.39}$$

$$\Theta_2 = \begin{bmatrix} \theta_{11} & \theta_{12} & \theta_{13} & \mu p_2 h \\ * & -2p_3 + \frac{\Delta}{\pi}\mu R p_3 & \theta_{23} & \mu p_3 h \\ * & * & \theta_{33} & 0 \\ * & * & * & -re^{-2\delta h} h \end{bmatrix}, \tag{3.40}$$

$$\theta_{11} = 2\delta p_1 + \frac{\Delta}{\pi}\mu R p_2 - 2\mu p_2 - \lambda\frac{\pi^4}{L^4},$$
$$\theta_{12} = p_1 - p_2 - \mu p_3 - p_3 z_x,$$
$$\theta_{13} = -p_2, \ \theta_{23} = -p_3, \ \theta_{33} = 2\delta p_3 \nu - 2p_2\nu + \lambda.$$

Then for any initial function $z(\cdot, 0) \in H_0^2(0, L)$ *subject to* $\|z(\cdot, 0)\|_V < \sqrt{\frac{p_3\nu}{L}}C$, *a unique strong solution of* (3.33), (3.2) *exists and satisfies*

$$\begin{aligned} &p_1 \int_0^L z^2(x,t)dx + p_3\nu \int_0^L z_{xx}^2(x,t)dx \\ &\le e^{-2\alpha t}\left[p_1 \int_0^L z^2(x,0)dx + p_3\nu \int_0^L z_{xx}^2(x,0)dx\right] \end{aligned} \tag{3.41}$$

for all $t \geq 0$, *where* α *is a unique positive solution of (1.43). Furthermore, if the strong equalities (3.37) and* (3.38) *are feasible with* $\delta = \frac{\delta_1}{2} > 0$, *then the strong solutions of (3.33),* (3.2) *initialized with* $z(\cdot, 0) \in H_0^2(0, L)$ *subject to* $\|z(\cdot, 0)\|_V < \sqrt{\frac{p_3\nu}{L}}C$ *are exponentially converging with a small enough decay rate.*

Proof Step 1: It has been shown that a unique local strong solution of (3.31) exists on some interval $[0, T] \subset [0, t_1]$. From Theorem 6.23.5 of [10], it follows that if this solution admits a priori estimate, then the solution exists on the entire interval $[0, t_1]$. We will prove in Step 3 that if the LMIs (3.37), (3.38) are feasible, then the solution of (3.31) starting from $\|z(\cdot, 0)\|_V < \sqrt{\frac{p_3\nu}{L}}C$ admits a priori bound. The latter guarantees the existence of the strong solution of (3.31) on the entire interval $[0, t_1]$. Then, by applying the same arguments step-by-step for $[t_k, t_{k+1}]$, $k = 1, 2, \ldots$ we conclude that the strong solution exists for all $t \geq 0$.

Step 2. Assume formally that strong solutions of (3.33) subject to (3.2) starting from $\|z_0\|_V < \sqrt{\frac{p_3\nu}{L}}C$ exist for all $t \geq 0$. Differentiating V_1 along (3.33) subject to (3.2), we have

$$\begin{aligned} &\dot{V}_1(t) + 2\delta V_1(t) = 2p_1 \int_0^L z(x,t)z_t(x,t)dx + 2p_3\nu \int_0^L z_{xx}(x,t)z_{xxt}(x,t)dx \\ &+2\delta p_1 \int_0^L z^2(x,t)dx + 2\delta p_3\nu \int_0^L z_{xx}^2(x,t)dx - r\int_0^L \int_{t_k}^t e^{2\delta(s-t)}z_s^2(x,s)dsdx \\ &+r(t_{k+1}-t)\int_0^L z_t^2(x,t)dx, \end{aligned} \tag{3.42}$$

where $t \in [t_k, t_{k+1})$.

Denote

$$\rho(x,t) \triangleq \frac{1}{t-t_k}\int_{t_k}^{t} z_s(x,s)ds. \tag{3.43}$$

Here we understand $\lim_{t\to t_k^+}\rho(x,t) = z_t(x,t_k)$ and obtain

$$z(x,t) = z(x,t_k) + (t-t_k)\rho(x,t). \tag{3.44}$$

Jensen's inequality [17] yields

$$\begin{aligned}&-r\int_0^L\int_{t_k}^t e^{2\delta(s-t)}z_s^2(x,s)dsdx \le -re^{-2\delta h}\int_0^L \frac{1}{t-t_k}\left[\int_{t_k}^t z_s(x,s)ds\right]^2 dx\\&= -re^{-2\delta h}(t-t_k)\int_0^L \rho^2(x,t)dx.\end{aligned} \tag{3.45}$$

We apply the descriptor method ([11, 16, 18]) by adding to $\dot V_1 + 2\delta V_1$ the left-hand sides of the following equations:

$$\begin{aligned}&2\int_0^L [p_2 z(x,t) + p_3 z_t(x,t)][-z_t(x,t) - z_{xx}(x,t)\\&-\nu z_{xxxx}(x,t) - z(x,t)z_x(x,t) - \mu z(x,t_k)]dx\\&+2\mu\sum_{j=1}^N\int_{x_{j-1}}^{x_j}[p_2 z(x,t) + p_3 z_t(x,t)]\int_{\bar x_j}^x z_\xi(\xi,t_k)d\xi dx = 0,\end{aligned} \tag{3.46}$$

where $p_2 > 0$ is some scalar. This avoids substitution of z_t from (3.33) into the right-hand side of (3.42). Integration by parts and substitution of the Dirichlet boundary conditions (3.2) lead to

$$-2p_2\int_0^L z(x,t)[z(x,t)z_x(x,t)]dx = 0, \tag{3.47}$$

$$-2p_2\int_0^L z(x,t)[\nu z_{xxxx}(x,t)]dx = -2p_2\nu\int_0^L z_{xx}^2(x,t)dx, \tag{3.48}$$

and

$$-2p_3\int_0^L z_t(x,t)[\nu z_{xxxx}(x,t)]dx = -2p_3\nu\int_0^L z_{xx}(x,t)z_{xxt}(x,t)dx. \tag{3.49}$$

By adding to $\dot V_1 + 2\delta V_1$ the equality (3.46), and using (3.44), (3.45), (3.47), (3.48), (3.49), we obtain

$$
\begin{aligned}
&\dot{V}_1(t) + 2\delta V_1(t) = 2p_1 \int_0^L z(x,t)z_t(x,t)dx - 2p_2\nu \int_0^L z_{xx}^2(x,t)dx \\
&+2\delta p_1 \int_0^L z^2(x,t)dx + 2\delta p_3 \nu \int_0^L z_{xx}^2(x,t)dx \\
&-re^{-2\delta h}(t-t_k) \int_0^L \rho^2(x,t)dx + r(t_{k+1}-t) \int_0^L z_t^2(x,t)dx \\
&+2 \int_0^L [p_2 z(x,t) + p_3 z_t(x,t)][-z_t(x,t) - z_{xx}(x,t) - \mu z(x,t) + \mu(t-t_k)\rho(x,t)]dx \\
&-2p_3 \int_0^L z_t(x,t)z(x,t)z_x(x,t)dx \\
&+2\mu \sum_{j=1}^N \int_{x_{j-1}}^{x_j} [p_2 z(x,t) + p_3 z_t(x,t)] \int_{\bar{x}_j}^x z_\xi(\xi,t_k)d\xi dx.
\end{aligned}
\tag{3.50}
$$

From Young's and Wirtinger's inequalities, we have

$$
\begin{aligned}
&2\mu \sum_{j=1}^N \int_{x_{j-1}}^{x_j} [p_2 z(x,t) + p_3 z_t(x,t)] \int_{\bar{x}_j}^x z_\xi(\xi,t_k)d\xi dx \\
&\le \mu \bar{R} \int_0^L [p_2 z^2(x,t) + p_3 z_t^2(x,t)]dx + \mu \bar{R}^{-1}(p_2+p_3)\frac{\Delta^2}{\pi^2} \int_0^L z_x^2(x,t_k)dx, \ \forall \bar{R} > 0.
\end{aligned}
\tag{3.51}
$$

Substituting (3.51) into the right-hand side of (3.50), adding (3.27) to $\dot{V}_1 + 2\delta V_1$, using Halanay's inequality and employing the inequality

$$
\|z_x(\cdot,t_k)\|_{L^2}^2 \le \|z(\cdot,t_k)\|_{L^2} \|z_{xx}(\cdot,t_k)\|_{L^2},
$$

we obtain

$$
\begin{aligned}
&\dot{V}_1(t) + 2\delta V_1(t) - \delta_1 \sup_{\theta \in [-h,0]} V_1(t+\theta) \le \dot{V}_1(t) + 2\delta V_1(t) - \delta_1 V_1(t_k) \\
&\le (2p_1 - 2p_2 - 2\mu p_3) \int_0^L z(x,t)z_t(x,t)dx \\
&+(2\delta p_3 \nu - 2p_2 \nu + \lambda) \int_0^L z_{xx}^2(x,t)dx \\
&-2 \int_0^L [p_2 z(x,t) + p_3 z_t(x,t)] z_{xx}(x,t)dx \\
&+ \left[2\delta p_1 + \mu \bar{R} p_2 - 2\mu p_2 - \lambda \frac{\pi^4}{L^4} \right] \int_0^L z^2(x,t)dx \\
&+ \left[r(t_{k+1}-t) - 2p_3 + \mu \bar{R} p_3 \right] \int_0^L z_t^2(x,t)dx \\
&-2p_3 \int_0^L z_t(x,t)z(x,t)z_x(x,t)dx - re^{-2\delta h}(t-t_k) \int_0^L \rho^2(x,t)dx \\
&+\mu(p_2+p_3)\frac{\Delta^2}{\pi^2} \bar{R}^{-1} \|z(\cdot,t_k)\|_{L^2} \|z_{xx}(\cdot,t_k)\|_{L^2} \\
&+2\mu \int_0^L [p_2 z(x,t) + p_3 z_t(x,t)][(t-t_k)\rho(x,t)]dx \\
&-\delta_1 p_1 \int_0^L z^2(x,t_k)dx - \delta_1 p_3 \nu \int_0^L z_{xx}^2(x,t_k)dx.
\end{aligned}
\tag{3.52}
$$

Set

$$\begin{aligned}\eta &= \mathrm{col}\{z(x,t), z_t(x,t), z_{xx}(x,t), \rho(x,t)\},\\ \eta_0 &= \mathrm{col}\{z(x,t), z_t(x,t), z_{xx}(x,t)\},\\ \bar{\eta} &= \mathrm{col}\{\|z(\cdot,t_k)\|_{L^2}, \|z_{xx}(\cdot,t_k)\|_{L^2}\},\end{aligned} \tag{3.53}$$

and choose $R = \frac{\pi}{\Delta}\bar{R}$. Since $0 \le t_{k+1} - t_k \le h$, from (3.52) it follows that

$$\begin{aligned}&\dot{V}_1(t) + 2\delta V_1(t) - \delta_1 \sup_{\theta\in[-h,0]} V_1(t+\theta)\\ &\le \int_0^L \frac{h-t+t_k}{h}\eta_0^T\Theta_1\eta_0 + \frac{t-t_k}{h}\eta^T\Theta_2\eta + \bar{\eta}^T\bar{\Theta}\bar{\eta}dx,\\ &\qquad\qquad \forall t\in[t_k, t_{k+1}),\end{aligned} \tag{3.54}$$

where $\bar{\Theta}$, Θ_1 and Θ_2 are given by (3.38), (3.39), (3.40) respectively.

We first assume that

$$\max_{x\in[0,L]} |z_x(x,t)| < C,\ \forall t \ge 0. \tag{3.55}$$

Under the assumption (3.55), from (3.54) we obtain

$$\dot{V}_1(t) + 2\delta V_1(t) - \delta_1 \sup_{\theta\in[-h,0]} V_1(t+\theta) \le 0 \tag{3.56}$$

if $\Theta_1 < 0,\ \Theta_2 < 0,\ \bar{\Theta} < 0$ hold for all $z_x \in (-C, C)$.

Matrices Θ_1 and Θ_2 given by (3.39), (3.40) are affine in z_x. Hence, $\Theta_1 < 0$ and $\Theta_2 < 0$ for all $z_x \in (-C, C)$ if these inequalities hold in the vertices $z_x = \pm C$ hold, i.e. if LMIs (3.37) are feasible.

We prove next that (3.55) holds. From (3.36) it follows that it is sufficient to show that $V_1(t) < \frac{p_3\nu}{L}C^2$. The initial condition $V_1(0) = \|z_0\|_V^2 < \frac{p_3\nu}{L}C^2$ implies $\max_{x\in[0,L]} |z_x(x,0)|^2 < C^2$. Let $t^* \in (0,\infty)$ be the smallest time instance such that $V_1(t^*) \ge \frac{p_3\nu}{L}C^2$. Since V_1 is continuous in time, we have $V_1(t^*) = \frac{p_3\nu}{L}C^2$ and $V_1(t) < \frac{p_3\nu}{L}C^2$ for $t\in[0,t^*)$. Together with (3.36) this implies $\max_{x\in[0,L]} |z_x(x,t)|^2 < C^2$ for $t \in [0,t^*)$ and, therefore, the feasibility of (3.37) and (3.38) guarantees that (3.56) is true for $t\in[0,t^*)$. Hence, $V_1(t) \le e^{-2\alpha t}V_1(0) < \frac{p_3\nu}{L}C^2$ holds for $t\in[0,t^*]$, which contradicts to the definition of t^*. Thus, for all $t \ge 0$, $\|z_0\|_V < \sqrt{\frac{p_3\nu}{L}}C \Rightarrow (3.55) \Rightarrow (3.41)$.

Step 3: Now we continue to prove the well-posedness. When $k = 0$, we obtain that if the LMIs conditions (3.37) and (3.38) are satisfied, then any strong solution of (3.31) initialized with $z_0 \in H_0^2(0,L)$ subject to $\|z_0\|_V < \sqrt{\frac{p_3\nu}{L}}C$ admits a priori estimate

$$V_1(t) \leq e^{-2\alpha t} V_1(0), \tag{3.57}$$

where α is the solution of (1.45). The latter bound guarantees the existence of these strong solutions for all $t \in [0, t_1]$. Then, by step method, the strong solution exists for all $t \geq 0$. Furthermore, Halanay's inequality implies (3.41) for all $t \geq 0$.

Note that the feasibility of strong inequalities (3.37) and (3.38) with $\delta = \frac{\delta_1}{2} > 0$ implies their feasibility with a slightly larger $\bar{\delta} = \delta + \alpha_0 > 0$, where $\alpha_0 > 0$ is small. Therefore, if (3.37) and (3.38) hold with $\delta = \frac{\delta_1}{2} > 0$, then the system (3.33), (3.2) is regionally exponentially stable with a small decay rate. □

3.1.3.3 Sampled-Data Control Under the Averaged Measurements

Under the controller (3.10) subject to (3.6), the closed-loop system of (3.1) becomes:

$$\begin{aligned} &z_t(x,t) + z_{xx}(x,t) + \nu z_{xxxx}(x,t) + z(x,t)z_x(x,t) \\ &= -\mu \sum_{j=1}^{N} \left[b_j(x) \frac{\int_{x_{j-1}}^{x_j} z(x,t_k)dx}{\Delta_j} \right], \ t \in [t_k, t_{k+1}), \ k = 0, 1, \ldots \end{aligned} \tag{3.58}$$

subject to the Dirichlet or periodic boundary conditions.

For $j = 1, \ldots, N$, $k = 0, 1, \ldots$ we consider the quantities

$$f_j(x,t) = z(x,t) - \frac{\int_{x_{j-1}}^{x_j} z(\zeta,t)d\zeta}{\Delta_j}, \ k_j(t) = \frac{1}{t - t_k} \frac{\int_{x_{j-1}}^{x_j} \int_{t_k}^{t} z_\xi(x,\xi)d\xi dx}{\Delta_j},$$

where by $k_j|_{t=t_k}$ we understand the following:

$$\lim_{t \to t_k^+} k_j(t) = \frac{\int_{x_{j-1}}^{x_j} z_t(x,t_k)dx}{\Delta_j}.$$

Similar to [19], we use the following presentation of $U_j(t)$:

$$U_j(t) = -\mu[z(x,t) - f_j(x,t) - (t - t_k)k_j(t)]. \tag{3.59}$$

Theorem 3.2 *Consider the closed-loop system* (3.58) *under the Dirichlet boundary conditions* (3.2). *Given positive scalars* C, h, μ, Δ *and* δ, *let there exist scalars* $r > 0$, $\lambda \geq 0$, $\lambda_1 > 0$, $p_i > 0$ *(*$i = 1, 2, 3$*) satisfy the linear matrix inequalities:*

$$\Phi_i|_{z_x = C} < 0, \ \Phi_i|_{z_x = -C} < 0, \ i = 1, 2 \tag{3.60}$$

where

$$\Phi_1 = \begin{bmatrix} \phi_{11} & \phi_{12} & \phi_{13} & \phi_{14} \\ * & rh - 2p_3 & \phi_{23} & \phi_{24} \\ * & * & \phi_{33} & \phi_{34} \\ * & * & * & \phi_{44} \end{bmatrix}, \tag{3.61}$$

$$\Phi_2 = \begin{bmatrix} \phi_{11} & \phi_{12} & \phi_{13} & \phi_{14} & \mu p_2 h \\ * & -2p_3 & \phi_{23} & \phi_{24} & \mu p_3 h \\ * & * & \phi_{33} & \phi_{34} & 0 \\ * & * & * & \phi_{44} & 0 \\ * & * & * & * & -re^{-2\delta h}h \end{bmatrix}, \tag{3.62}$$

$$\begin{aligned} &\phi_{11} = -2p_2\mu + 2\delta p_1 - \lambda\frac{\pi^4}{L^4},\ \phi_{12} = p_1 - p_2 - p_3\mu - p_3 z_x, \\ &\phi_{13} = -p_2 - \frac{\lambda_1}{2},\ \phi_{14} = \mu p_2,\ \phi_{23} = -p_3,\ \phi_{24} = \mu p_3, \\ &\phi_{33} = -2p_2\nu + 2\delta p_3\nu + \lambda,\ \phi_{34} = 0,\ \phi_{44} = -\frac{\lambda_1\pi^2}{\Delta^2}. \end{aligned}$$

Then for any initial function $z(\cdot, 0) \in H_0^2(0, L)$ *satisfying* $\|z(\cdot, 0)\|_V < \sqrt{\frac{p_3\nu}{L}}C$*, a unique strong solution of* (3.58), (3.2) *exists and satisfies*

$$\begin{aligned} &p_1\int_0^L z^2(x,t)dx + p_3\nu\int_0^L z_{xx}^2(x,t)dx \\ &\le e^{-2\delta t}\left[p_1\int_0^L z^2(x,0)dx + p_3\nu\int_0^L z_{xx}^2(x,0)dx\right] \end{aligned} \tag{3.63}$$

for all $t \ge 0$*. Furthermore, if the strong inequalities* (3.60) *are feasible with* $\delta = 0$*, then the strong solutions of (3.58),* (3.2) *initialized with* $z(\cdot, 0) \in H_0^2(0, L)$ *subject to* $\|z(\cdot, 0)\|_V < \sqrt{\frac{p_3\nu}{L}}C$ *are exponentially converging with a small enough decay rate.*

Proof For the case of sampled-data controller under the averaged measurements, by arguments of Theorem 3.1, the well-posedness of (3.58) subject to (3.2) can be established via the step method.

Step 1: For $[t_0, t_1]$, we consider the following equation:

$$\begin{cases} z_t(x,t) + z_{xx}(x,t) + \nu z_{xxxx}(x,t) + z(x,t)z_x(x,t) \\ = -\mu\sum_{j=1}^{N}\left[b_j(x)\frac{\int_{x_{j-1}}^{x_j} z(\zeta, t_0)d\zeta}{\Delta_j}\right], \\ z(0,t) = z(L,t) = z_x(0,t) = z_x(L,t) = 0. \end{cases} \tag{3.64}$$

We define the nonlinear term $F : H^2(0, L) \to L^2(0, L)$ according to

$$F(z(\cdot, t)) = -z(x, t)z_x(x, t) - z_{xx}(x, t) \\ -\mu \sum_{j=1}^{N} \left[b_j(x) \frac{\int_{x_{j-1}}^{x_j} z_0(\zeta)d\zeta}{\Delta_j} \right], \ t \in [t_0, t_1].$$

The nonlinear term F is locally Lipschitz continuous. Thus, there exists a unique local strong solution $z(\cdot, t) \in H_0^2(0, L)$ of (3.58) subject to (3.2) on some interval $[0, T] \subset [0, t_1]$, where $T = T(z_0) > 0$. By Theorem 6.23.5 of [10], we only need to prove that if the LMIs (3.60) are feasible, the solution of (3.58) subject to (3.2) starting from $\|z(\cdot, 0)\|_V < \sqrt{p_3\nu}C$ admits a priori bound, which guarantees the existence of a strong solution of (3.58) subject to (3.2) on the entire interval $[0, t_1]$. Then, by applying the same arguments step-by-step for $[t_k, t_{k+1}], k = 1, 2, \ldots$ we conclude that the strong solution exists for all $t \geq 0$.

Step 2: Assume formally that strong solution of (3.58) subject to (3.2) starting from $\|z_0\|_V < \sqrt{\frac{p_3\nu}{L}}C$ exists for all $t \geq 0$ and choose the same Lyapunov function V_1 as in (3.34). Differentiating V_1 along (3.58) subject to (3.2), we obtain the inequalities (3.42) and (3.45). Taking into account (3.59), we further apply the descriptor method to (3.58), where we add to $\dot{V}_1 + 2\delta V_1$ the left-hand side of the following equation

$$\begin{aligned} &2\int_0^L [p_2 z(x, t) + p_3 z_t(x, t)][-z_t(x, t) - z_{xx}(x, t) \\ &-\nu z_{xxxx}(x, t) - z(x, t)z_x(x, t) - \mu z(x, t)]dx \\ &+2\mu \sum_{j=1}^{N} \int_{x_{j-1}}^{x_j} [p_2 z(x, t) + p_3 z_t(x, t)][f_j(x, t) + (t - t_k)k_j(t)]dx = 0. \end{aligned} \tag{3.65}$$

Integration by parts and substitution of the dir1ichlet boundary conditions (3.2) lead to (3.47), (3.48) and (3.49). Adding the left-hand side of the expression (3.65) into the right-hand side of (3.42), we obtain

$$\begin{aligned} &\dot{V}_1(t) + 2\delta V_1(t) \\ &= 2p_1 \int_0^L z(x, t)z_t(x, t)dx - 2p_2\nu \int_0^L z_{xx}^2(x, t)dx \\ &-re^{-2\delta h} \int_0^L \frac{1}{t - t_k} \left[\int_{t_k}^t z_s(x, s)ds \right]^2 dx \\ &+r(t_{k+1} - t) \int_0^L z_t^2(x, t)dx + 2\delta p_1 \int_0^L z^2(x, t)dx \\ &+2\delta p_3\nu \int_0^L z_{xx}^2(x, t)dx \\ &+2\int_0^L [p_2 z(x, t) + p_3 z_t(x, t)][-z_t(x, t) - z_{xx}(x, t) \\ &-\mu z(x, t)]dx - 2p_3 \int_0^L z_t(x, t)z(x, t)z_x(x, t)dx \\ &+2\mu \sum_{j=1}^{N} \int_{x_{j-1}}^{x_j} [p_2 z(x, t) + p_3 z_t(x, t)][f_j(x, t) + (t - t_k)k_j(t)]dx. \end{aligned} \tag{3.66}$$

From Jensen's inequality, we have

$$
\begin{aligned}
&-re^{-2\delta h}\int_0^L \frac{1}{t-t_k}\left[\int_{t_k}^t z_s(x,s)ds\right]^2 dx \\
&= -\frac{r}{t-t_k}e^{-2\delta h}\sum_{j=1}^N \int_{x_{j-1}}^{x_j}\left[\int_{t_k}^t z_s(x,s)ds\right]^2 dx \\
&\le -r(t-t_k)e^{-2\delta h}\sum_{j=1}^N \int_{x_{j-1}}^{x_j} k_j^2 dx.
\end{aligned}
\tag{3.67}
$$

To bound the term "$\int_0^L z_t(x,t)z(x,t)z_x(x,t)dx$" in (3.66), we need to bound $z_x(x,t)$ by some constant $C>0$ for all $t\ge 0$ and $x\in[0,L]$. Similar to [20], we first assume that (3.55) is satisfied.

Set

$$
\begin{aligned}
\eta_1 &= \mathrm{col}\{z(x,t), z_t(x,t), z_{xx}(x,t), f_j(x,t)\},\\
\eta_2 &= \mathrm{col}\{z(x,t), z_t(x,t), z_{xx}(x,t), f_j(x,t), k_j\}.
\end{aligned}
\tag{3.68}
$$

Applying S-procedure, we add the left-hand sides of (3.26) and (3.27) to $\dot{V}_1(t) + 2\delta V_1(t)$. Substituting (3.67) into (3.66), we obtain

$$
\begin{aligned}
&\dot{V}_1(t) + 2\delta V_1(t)\\
&\le \sum_{j=1}^N \int_{x_{j-1}}^{x_j} \frac{h-t+t_k}{h}\eta_1^T\Phi_1\eta_1 + \frac{t-t_k}{h}\eta_2^T\Phi_2\eta_2 dx < 0
\end{aligned}
$$

if $\Phi_1<0$ and $\Phi_2<0$ hold for all $z_x\in(-C,C)$, where Φ_1 and Φ_2 are given by (3.61), (3.62) respectively.

Similarly to Theorem 3.1, LMIs (3.60) yield $\Phi_1<0$ and $\Phi_2<0$ for all $z_x\in(-C,C)$, and

$$
V_1(t)\le e^{-2\delta t}V_1(0)\quad \forall t\ge 0
$$

for initial conditions $z(\cdot,0)\in H_0^2(0,L)$ satisfying $\|z(\cdot,0)\|_V < \sqrt{\frac{p_3\nu}{L}}C$. Moreover, (3.55) is satisfied.

Step 3: By the arguments of the proof of Step 3 in Theorem 3.1, if the LMIs (3.60) are satisefied, then there exists a strong solution for $t\in[0,t_1]$ initialized with $\|z(\cdot,0)\|_V < \sqrt{\frac{p_3\nu}{L}}C$. Then, by applying the same arguments step-by-step for $[t_k,t_{k+1}]$, $k=1,2,\ldots$ we conclude that the strong solution exists for all $t\ge 0$. □

3.1.3.4 Extension to Periodic Boundary Conditions

Finding a bound on the domain of attraction of the closed-loop system (3.30) or (3.58) under the periodic boundary conditions (3.3) will be based on the following useful Lemma.

Lemma 3.1 *Let $w(x) \in H^1(0, L)$, then*

$$\max_{x\in[0,L]} |w(x)|^2 \leq (1+\frac{1}{L})\|w(\cdot,t)\|^2_{L^2} + \|w_x(\cdot,t)\|^2_{L^2}.$$

Proof By Proposition 5.22 of [21], $w(\cdot) \in H^1(0, L)$ implies $w(\cdot) \in C[0, L]$. Then by mean value theorem, there exists $c \in (0, L)$ such that

$$w(c) = \frac{1}{L}\int_0^L w(x)dx.$$

Then, by integration by parts and further application of Jensen's and Young's inequalities, for all $x_1 \in [0, L]$ we have

$$\begin{aligned} w^2(x_1) &= w^2(c) + 2\int_c^{x_1} w(x)w_x(x)dx = \left[\frac{1}{L}\int_0^L w(x)dx\right]^2 + 2\int_c^{x_1} w(x)w_x(x)dx \\ &\leq \frac{1}{L}\int_0^L w^2(x)dx + \int_0^L w^2(x)dx + \int_0^L w_x^2(x)dx \\ &\leq (1+\frac{1}{L})\int_0^L w^2(x)dx + \int_0^L w_x^2(x)dx. \end{aligned}$$

□

Denote

$$M = \max\{\frac{1}{2p_1}, \frac{3}{2p_3\nu}\}. \tag{3.69}$$

Similar to Sect. 3.2, Theorems 3.1 and 3.2, can be easily extended to periodic boundary conditions:

(i) If the conditions of Theorem 3.1 with $\lambda = 0$ hold, then for the initial function $z(\cdot, 0) \in H^2_{per}$ satisfying $\|z(\cdot,0)\|_V < \sqrt{\frac{L}{(L+1)M}}C$, a unique strong solution of (3.30) under the periodic boundary conditions (3.3) exists and satisfies (3.41).

(ii) If the conditions of Theorem 3.2 with $\lambda = 0$ hold, then for the initial function $z(\cdot, 0) \in H^2_{per}$ satisfying $\|z(\cdot,0)\|_V < \sqrt{\frac{L}{(L+1)M}}C$, a unique strong solution of (3.58) under the periodic boundary conditions (3.3) exists and satisfies (3.63).

Proof of (i) The proof is similar to the case of Dirichlet boundary conditions. Due to the periodic boundary conditions, the Writinger inequality is not applicable (cf. (3.27)). Therefore, by arguments of Theorem 3.1, we obtain that if the conditions of Theorem 3.1 with $\lambda = 0$ hold, then (3.56), (3.57) are satisfied (and, thus, (3.41) hold) provided that (3.55) holds for all $t \geq 0$.

By Lemma 3.1 and Young's inequality, using the inequality $\|z_x(\cdot,t)\|^2_{L^2} \le \|z(\cdot,t)\|_{L^2}\|z_{xx}(\cdot,t)\|_{L^2}$ we have

$$\begin{aligned}
&\max_{x\in[0,L]} |z_x(x,t)|^2 \le (1+\frac{1}{L})\|z_x(\cdot,t)\|^2_{L^2} + \|z_{xx}(\cdot,t)\|^2_{L^2} \\
&\le (1+\frac{1}{L})\|z(\cdot,t)\|_{L^2}\|z_{xx}(\cdot,t)\|_{L^2} + \|z_{xx}(\cdot,t)\|^2_{L^2} \\
&\le (1+\frac{1}{L})\left[\frac{1}{2}\|z(\cdot,t)\|^2_{L^2} + (\frac{1}{2}+1)\|z_{xx}(\cdot,t)\|^2_{L^2}\right] \\
&\le (1+\frac{1}{L})M\left[p_1\|z(\cdot,t)\|^2_{L^2} + p_3\nu\|z_{xx}(\cdot,t)\|^2_{L^2}\right] \\
&= (1+\frac{1}{L})M\|z(\cdot,t)\|^2_V \le (1+\frac{1}{L})MV_1(t),
\end{aligned} \tag{3.70}$$

where M is given by (3.69). Therefore, if the initial function $z(\cdot,0)\in H^2_{per}$ satisfies $\|z(\cdot,0)\|_V < \sqrt{\frac{L}{(L+1)M}}C$, then

$$\max_{x\in[0,L]} |z_x(x,0)|^2 \le (1+\frac{1}{L})M\|z(\cdot,0)\|^2_V < C^2.$$

Due to (3.70), the LMI conditions of Theorem 3.1 with $\lambda = 0$ guarantee that

$$\max_{x\in[0,L]} |z_x(x,t)|^2 \le (1+\frac{1}{L})MV_1(t) \le (1+\frac{1}{L})MV_1(0) = (1+\frac{1}{L})M\|z(\cdot,0)\|^2_V < C^2$$

meaning that (3.55) holds.

Proof of (ii) follows arguments of (i).

Remark 3.3 The presented results can be extended to the case of time-varying input delay. Similar to [2, 19], LMI stability conditions for KSE with time-varying delay can be derived. However, a special attention should be paid to the bound on the domain of attraction. This is because of the initial time interval, where the system is in the open-loop (see [22]).

3.1.4 Example

Consider KSE (3.1) under the Dirichlet boundary conditions (3.2) with $L = 2\pi$ and $\nu = 0.5 < 1$:

$$\begin{aligned}
&z_t(x,t) + z_{xx}(x,t) + 0.5z_{xxxx}(x,t) + z(x,t)z_x(x,t) \\
&= \sum_{j=1}^{N} b_j(x)U_j(t),\ \ 0 < x < 2\pi,\ \ t \ge 0.
\end{aligned}$$

Figure 3.1 demonstrates the time evolution of the L^2–norm $\|z(\cdot,t)\|^2_{L^2}$ for the open-loop system initialized by $z(x,0) = (1-\cos x)\sin x$ $(0 \le x \le 2\pi)$. It is seen that the open-loop system is unstable.

(i) We start with the continuous in time controller under the point and averaged measurements. Verifying the LMI condition of Proposition 3.1 with $\alpha = 0$ by Yalmip Toolbox of Matlab, we find that the closed-loop system preserves the exponential stability till $\Delta \le 2.2$ (this corresponds to $\mu = 3.0853$, $\lambda_1 = 1.5131$ and $\lambda = 2.2905 * 10^{-7}$). Note that for $\Delta \le 2.2$, the condition $\frac{\Delta^2}{\pi^2} \le 0.49 < \nu$ of particular case of Proposition 3.1 is satisfied, and the corresponding bound for $\mu = \frac{2\pi^2}{\Delta^2}$ is essentially larger: $\mu = 4.07 > 3.0853$. For both values of μ, the condition of [7] $\mu > 8$ is not satisfied.

Using Proposition 3.1 with $\alpha = 0$, $\mu > 4/\nu = 8$ that satisfies the condition of [7], we find that the closed-loop system is exponentially stable till $\Delta \le 2.1$ (allowing 3 sensors under the point measurements). This corresponds to $\mu = 8.0004$, $\lambda_1 = 2.7795$ and $\lambda = 3.3738 * 10^{-5}$. Here we added the restriction $\mu > 4/\nu$ (that is used to compare with [7]). The condition $\mu\frac{\Delta^4}{\pi^4} < \nu$ of [7] with the same $\mu = 8.0004$ leads to essentially smaller $\Delta < 1.56$ (that requires 5 point sensors) meaning that the LMIs conditions of Proposition 3.1 are less restrictive.

(ii) Consider next the closed-loop system (3.58) subject to (3.2) under the sampled-data control law (3.7) with the averaged measurements. To enlarge the sampling time interval, we choose $\mu = 3.0853$ (the smallest μ found from the LMI of Proposition 3.1). We verify LMI conditions of Theorem 3.2 with $\delta = 0.0002$ and $C = h = 0$ and find that the LMIs are feasible till $\Delta \le 2.2$ (as in the continuous-time case). For $\delta =$

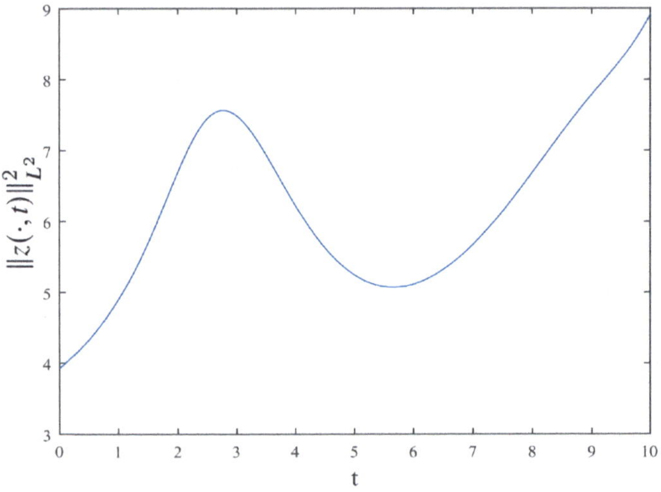

Fig. 3.1 Open-loop system (without control input)

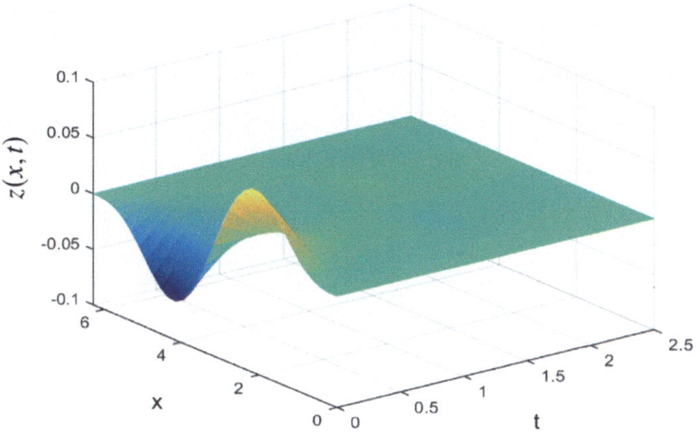

Fig. 3.2 Closed-loop system with $x_j - x_{j-1} = \frac{\pi}{8}$, $t_{k+1} - t_k = 0.28$ under the averaged measurements

$0.2, C = 1$ and $\Delta = \frac{\pi}{8}$, we find that the closed-loop system preserves the exponential stability within a given domain of initial conditions $\|z(\cdot, 0)\|_V < \sqrt{\frac{p_3 \nu}{L}} C = 0.58$ for $t_{k+1} - t_k \le h = 0.28$.

Next, a finite difference method is applied to compute the displacement of the closed-loop system (3.58) under the Dirichlet boundary conditions (3.2). We choose initial condition $z(x, 0) = 0.05(1 - \cos x) \sin x, 0 \le x \le 2\pi$ satisfying $\|z(\cdot, 0)\|_V < 0.58$. The steps of space and time are taken as $\frac{\pi}{16}$ and 0.0001, respectively. Simulations of solutions under the sampled-data in time and in space controller $U_j(t) = -3.0853 \frac{\int_{x_{j-1}}^{x_j} z(x, t_k) dx}{\Delta_j}$ with $x_j - x_{j-1} = \Delta_j = \frac{\pi}{8}, j = 1, \ldots, 16$, $t_{k+1} - t_k = 0.28$, where the spatial domain is divided into sixteen sub-domains, show that the closed-loop system is exponentially stable (see Fig. 3.2). By enlarging the sampling period, we find that the system becomes unstable for $t_{k+1} - t_k \ge 1$ (see Fig. 3.3).

(iii) Consider now the sampled-data controller under the point measurements. Here we choose the same value of μ as in (ii). For sampled-data control law (3.7) with the point measurements, we verify LMI conditions of Theorem 3.1 with $\delta_1 = 0.5$, $\delta = 0.26$, $R = 1$, $C = 1$ and $\Delta = \pi/8$. We find that the closed-loop system is exponentially stable for $t_{k+1} - t_k \le h = 0.23$ for any initial values satisfying $\|z(\cdot, 0)\|_V < \sqrt{\frac{p_3 \nu}{L}} C = 0.5$. For $h = 0.23$, the above controller locally exponentially stabilizes the closed-loop system with a decay rate $\alpha = 0.0089$. Table 3.1 shows the maximum values of h that preserve the exponential stability. The corresponding

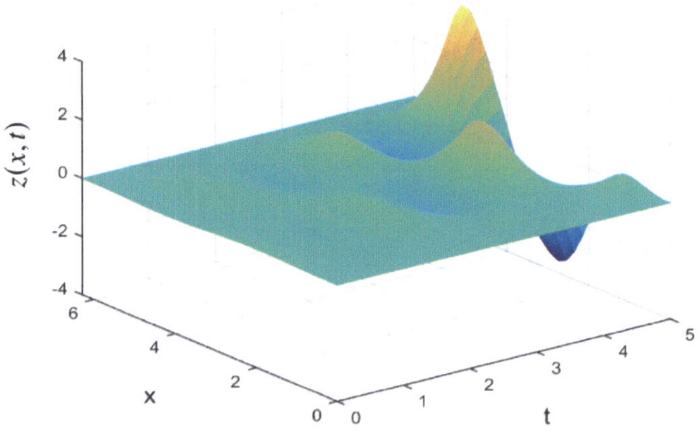

Fig. 3.3 Closed-loop system with $x_j - x_{j-1} = \frac{\pi}{8}$, $t_{k+1} - t_k = 1$ under the averaged measurements

Table 3.1 Max. values of h preserving the stability of the closed-loop system in Example 1

δ	0.4	0.3	0.26
h	0.2	0.22	0.23

value of δ_1 is given by $\delta_1 = 0.5$, whereas the values of $\delta > \frac{\delta_1}{2}$ are chosen to be close to $\frac{\delta_1}{2}$. The latter leads to a small decay rate α but enlarges the sampling intervals.

We proceed further with the numerical simulations of the solutions of the closed-loop system (3.30) subject to (3.2) under the sampled in time and in space controller $U_j(t) = -3.0853z(\bar{x}_j, t_k)$ with $x_j - x_{j-1} = \frac{\pi}{8}$, $j = 1, \ldots, 16$, $t_{k+1} - t_k = 0.23$. Here we choose the same initial conditions and the same value of μ as in (ii). The simulations show that the state of KSE converges to zero (see Fig. 3.4). Simulations of the solutions confirm the theoretical results that follow from LMIs.

As expected (because averaged measurements use more information on the state), the averaged measurements allow larger sampling intervals than point measurements.

3.2 Sampled-Data Control of 2D KSE

This chapter addresses sámpled-data control of 2D KSE over a rectangular domain Ω. We suggest to divide the 2D rectangular Ω into N sub-domains, where sensors provide spatially averaged or point state measurements to be transmitted through communication network to the controller. Note that differently from 2D heat equation, here we manage with sampled-data control under point measurements. We design

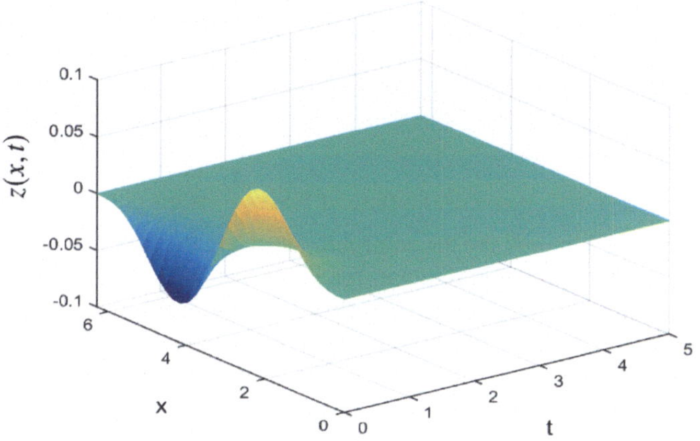

Fig. 3.4 Closed-loop system with $x_j - x_{j-1} = \frac{\pi}{8}$, $t_{k+1} - t_k = 0.23$ under the point measurements

a regionally stabilizing controller applied through distributed in space characteristic functions. Sufficient conditions ensuring regional stability of the closed-loop system are established in terms of LMIs. By solving these LMIs, we find an estimate on the set of initial conditions starting from which the state trajectories of the system are exponentially converging to zero. A numerical example demonstrates the efficiency of the results. This section follows [23].

3.2.1 Problem Formulation and Useful Lemmas

Denote by Ω the two dimensional (2-D) unit square

$$\Omega = [0, 1] \times [0, 1] \subset \mathbb{R}^2.$$

Consider the biharmonic operator:

$$\Delta^2 = \frac{\partial^4}{\partial x_1^4} + 2\frac{\partial^4}{\partial x_1^2 \partial x_2^2} + \frac{\partial^4}{\partial x_2^4}.$$

As in [24], we consider the following 2D KSE over Ω under the Dirichlet boundary conditions:

$$\begin{cases} z_t + zz_{x_1} + (1-\kappa)z_{x_1x_1} - \kappa z_{x_2x_2} + \Delta^2 z = \sum_{j=1}^{N} \chi_j(x)U_j(t), \quad (x,t) \in \Omega \times [0,\infty), \\ z|_{\partial\Omega} = 0, \ \frac{\partial z}{\partial n}|_{\partial\Omega} = 0, \\ z(x,0) = z_0(x), \end{cases} \tag{3.71}$$

where $x = (x_1, x_2) \in \Omega$, $z \in \mathbb{R}$ is the state of KSE, $\frac{\partial z}{\partial n}$ is the normal derivative, and $U_j(t) \in \mathbb{R}$, $j = 1, 2, \ldots, N$ are the control inputs. Here the parameter κ denotes the angle of the substrate to the horizontal: for $\kappa > 0$ we have overlying film flows, a vertical film flow for $\kappa = 0$, and hanging flows when $\kappa < 0$.

Motivated by [1, 2, 7, 19, 25] we suggest to divide Ω into N square sub-domains Ω_j covering the whole region $\cup_{j=1}^{N}\Omega_j = \Omega$ (see Fig. 3.5) with an actuator and a sensor placed in each Ω_j. Here

$$\Omega_j = \{x = (x_1, x_2)^T \in \Omega | x_i \in [x_i^{\min}(j), x_i^{\max}(j)], i = 1, 2\}, \ \ j = 1, 2, \ldots, N.$$

The measure of their intersections is zero. Let $0 = t_0 < t_1 < \cdots < t_k \cdots$, $\lim_{k\to\infty} t_k = \infty$ be sampling time instants. The sampling sub-domains in time and in space may be bounded,

$$0 \le t_{k+1} - t_k \le h, \ 0 < x_i^{\max}(j) - x_i^{\min}(j) = \Delta_j \le \bar{\Delta}, \quad i = 1, 2; \ \ j = 1, 2, \ldots, N,$$

where h and $\bar{\Delta}$ are the corresponding upper bounds.

Remark 3.4 For simplicity, we consider each sub-domain Ω_j is a square (i.e. $x_1^{\max}(j) - x_1^{\min}(j)) = x_2^{\max}(j) - x_2^{\min}(j)$, $j = 1, 2, \ldots, N$). Indeed, Ω_j can be a rectangular. For the case that $x_1^{\max}(j) - x_1^{\min}(j)) \neq x_2^{\max}(j) - x_2^{\min}(j)$ for some j (see Fig. 1 of [26]), Δ_j can be chosen as follows

$$\Delta_j = \max\{x_1^{\max}(j) - x_1^{\min}(j), x_2^{\max}(j) - x_2^{\min}(j)\}.$$

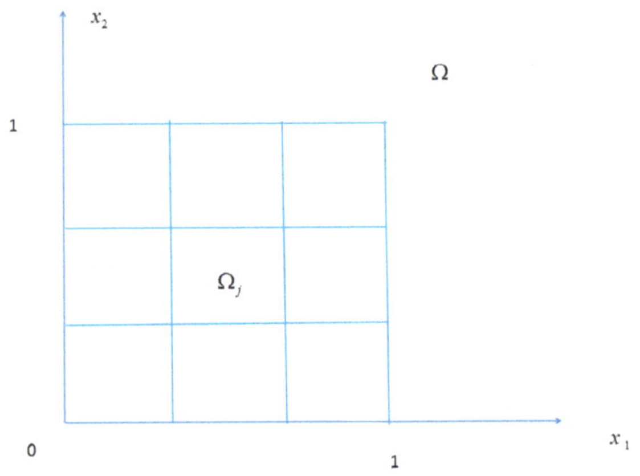

Fig. 3.5 Unit square Ω and sub-domains Ω_j

Thus, the results of this work are applicable to the case of nonsquare sub-domains.

The spatial characteristic functions are taken as

$$\begin{cases} \chi_j(x) = 1, \ x \in \Omega_j, \\ \chi_j(x) = 0, \ x \notin \Omega_j, \end{cases} \quad j = 1, \ldots, N. \tag{3.72}$$

We assume that sensors provide the following averaged measurements

$$y_{jk} = \frac{\int_{\Omega_j} z(x, t_k) dx}{|\Omega_j|}, \quad j = 1, \ldots, N; \ k = 0, 1, 2 \ldots \tag{3.73}$$

or point measurements

$$y_{jk} = z(\bar{x}_j, t_k), \quad j = 1, \ldots, N; \ k = 0, 1, 2 \ldots \tag{3.74}$$

where $\bar{x}_j$ locates in the center of the square sub-domain Ω_j, and $|\Omega_j|$ stands for the Lebesgue measure of the domain Ω_j.

We aim to design for (3.71) an exponentially stabilizing sampled-data controller that can be implemented by zero-order hold devices:

$$U_j(t) = -\mu y_{jk}, \quad j = 1, \ldots, N; \ t \in [t_k, t_{k+1}), \ k = 0, 1, 2 \ldots, \tag{3.75}$$

where μ is a positive controller gain and y_{jk} is given by (3.73) or (3.74).

3.2.2 Global Stabilization: Continuous Static Output-Feedback

In this section, we will establish the well-posedness and stability analysis for the system (3.71) under the continuous-time averaged measurements

$$y_j(t) = \frac{\int_{\Omega_j} z(x, t) dx}{|\Omega_j|}, \quad j = 1, \ldots, N \tag{3.76}$$

or point measurements

$$y_j(t) = z(\bar{x}_j, t), \quad j = 1, \ldots, N \tag{3.77}$$

via a controller

$$U_j(t) = -\mu y_j(t), \quad j = 1, \ldots, N. \tag{3.78}$$

The closed-loop system can be represented in the following form:

$$\begin{cases} z_t + zz_{x_1} + (1-\kappa)z_{x_1x_1} - \kappa z_{x_2x_2} + \Delta^2 z = -\mu \sum_{j=1}^{N} \chi_j(x)[z - f_j], \\ \qquad\qquad (x,t) \in \Omega\times > [0,\infty), \\ z|_{\partial\Omega} = 0, \ \frac{\partial z}{\partial n}|_{\partial\Omega} = 0, \\ z(x,0) = z_0(x), \end{cases} \tag{3.79}$$

where for (3.76)

$$f_j(x,t) = z(x,t) - \frac{\int_{\Omega_j} z(\zeta,t)d\zeta}{|\Omega_j|}, \tag{3.80}$$

for (3.77)

$$f_j(x,t) = z(x,t) - z(\bar{x}_j,t). \tag{3.81}$$

Now we establish the well-posedness of the system (3.79) subject to (3.80) or (3.81). Define the spatial differential operator $A : D(A) \subset L^2(\Omega) \to L^2(\Omega)$ as follows:

$$\begin{cases} Af = -\Delta^2 f, \ \forall f \in D(A), \\ D(A) = \{f \in H^4(\Omega) : f|_{\partial\Omega} = 0, \ \frac{\partial f}{\partial n}|_{\partial\Omega} = 0\}. \end{cases} \tag{3.82}$$

Note that $A^* = A$ and $\mathrm{Re}\langle Af, f\rangle \le 0, \ \forall f \in D(A)$. Thus, the operator A is self-adjoint and dissipative. Moreover, the inverse A^{-1} is bounded, and hence $0 \in \rho(A)$. By the Lumer-Phillips theorem [9], A generates a C_0-semigroup. Since the resolvent of A is compact on $L^2(\Omega)$, the spectrum of A consists of isolated eigenvalues only, and a sequence of corresponding eigenfunctions of A forms an orthonormal basis of $L^2(\Omega)$. Let $\{\lambda_n\}$ be the eigenvalues of A and let $\{\phi_n\}$ be the corresponding eigenfunctions, i.e. $A\phi_n = \lambda_n\phi_n$. Since A is negative, all the eigenvalues are located on the negative real axis, i.e. $\lambda_n < 0$. For any $x_0 \in L^2(\Omega)$, it can be presented in the following form: $x_0 = \sum_{n=1}^{\infty} a_n\phi_n$ with $\sum_{n=1}^{\infty} |a_n|^2 = \|x_0\|^2_{L^2(\Omega)}$. Therefore, for any $\tau \in \mathbb{R}$,

$$\begin{aligned} \|(i\tau I - A)^{-1}x_0\|^2_{L^2(\Omega)} &= \sum_{n=1}^{\infty} \frac{|a_n|^2}{|i\tau - \lambda_n|^2} = \sum_{n=1}^{\infty} \frac{|a_n|^2}{\tau^2 + \lambda_n^2} \\ &\le \sum_{n=1}^{\infty} \frac{|a_n|^2}{\tau^2} = \frac{1}{\tau^2}\|x_0\|^2_{L^2(\Omega)}, \end{aligned}$$

which implies

$$\|(i\tau I - A)^{-1}\| \le \frac{1}{|\tau|}, \ \forall \tau \in \mathbb{R}.$$

Hence,

$$\overline{\lim_{|\tau|\to\infty}} \|\tau(i\tau I - A)^{-1}\| < \infty.$$

Then from Theorem 1.3.3 of [27], it follows that A generates an analytic semigroup. Since $-A$ is positive, $(-A)^{\frac{1}{2}}$ is also positive and

$$D((-A)^{\frac{1}{2}}) = \{f \in H^2(\Omega) : f|_{\partial\Omega} = 0, \ \frac{\partial f}{\partial n}|_{\partial\Omega} = 0\}.$$

The norm of $D((-A)^{\frac{1}{2}})$ is given by

$$\|f\|^2_{D((-A)^{\frac{1}{2}})} = \int_\Omega |\Delta f|^2 dx.$$

Throughout the paper, we assume that $z_0 \in D((-A)^{\frac{1}{2}})$. We can rewrite the system (3.79) subject to (3.80) or (3.81) as the evolution equation:

$$\begin{cases} \frac{d}{dt} z(\cdot, t) = Az(\cdot, t) + F(z(\cdot, t)), \\ z(\cdot, 0) = z_0(\cdot) \end{cases} \tag{3.83}$$

subject to

$$F(z(\cdot, t)) = -z(x, t)z_{x_1}(x, t) - (1 - \kappa)z_{x_1x_1}(x, t) \\ +\kappa z_{x_2x_2}(x, t) - \mu \sum_{j=1}^{N} \chi_j(x)[z(x, t) - f_j(x, t)].$$

Note that the nonlinear term F is locally Lipschitz continuous, that is, there exists a positive constant $l(C)$ such that the following inequality holds:

$$\|F(z_1) - F(z_2)\|_{L^2(\Omega)} \le l(C)\|z_1 - z_2\|_{D((-A)^{\frac{1}{2}})}$$

for any z_1, $z_2 \in D((-A)^{\frac{1}{2}})$ with $\|z_1\|_{D((-A)^{\frac{1}{2}})} \le C$, $\|z_2\|_{D((-A)^{\frac{1}{2}})} \le C$. Here we prove the nonlinear term F is locally Lipschitz continuous for the case of point measurements. From the expression of $F(z(\cdot, t))$, using Minkowskii's inequality we have

$$\begin{aligned} \|F(z_1) - F(z_2)\|_{L^2(\Omega)} &\le \|z_1 z_{1x_1} - z_2 z_{2x_1}\|_{L^2(\Omega)} \\ &+|1 - \kappa| \cdot \|z_{1x_1x_1} - z_{2x_1x_1}\|_{L^2(\Omega)} \\ &+|\kappa| \cdot \|z_{1x_2x_2} - z_{2x_2x_2}\|_{L^2(\Omega)} + \mu\|z_1 - z_2\|_{L^2(\Omega)} \\ &+\mu \sum_{j=1}^{N} \| \int_{\bar{x}_j}^{x} z_{1\xi}(\xi, t) - z_{2\xi}(\xi, t)d\xi \|_{L^2(\Omega_j)}. \end{aligned} \tag{$*$}$$

Note that $D((-A)^{\frac{1}{2}}) = \{f \in H^2(\Omega) : f|_{\partial\Omega} = 0, \ \frac{\partial f}{\partial n}|_{\partial\Omega} = 0\}$. Since z_1, $z_2 \in D((-A)^{\frac{1}{2}})$ with $\|z_1\|_{D((-A)^{\frac{1}{2}})} \le C$, $\|z_2\|_{D((-A)^{\frac{1}{2}})} \le C$, we obtain that $z_1 - z_2 \in D((-A)^{\frac{1}{2}})$, and there exist some positive constants M_1 and M_2 such that

$$\begin{cases} \|z_1\|_{L^2(\Omega)} \le M_1\left[\|z_{1x_1}\|_{L^2(\Omega)} + \|z_{1x_2}\|_{L^2(\Omega)}\right], \\ \|z_{1x_1}\|_{L^2(\Omega)} + \|z_{1x_2}\|_{L^2(\Omega)} \le M_2\|z_1\|_{D((-A)^{\frac{1}{2}})}, \\ |z_2\|_{L^2(\Omega)} \le M_1\left[\|z_{2x_1}\|_{L^2(\Omega)} + \|z_{2x_2}\|_{L^2(\Omega)}\right], \\ \|z_{2x_1}\|_{L^2(\Omega)} + \|z_{2x_2}\|_{L^2(\Omega)} \le M_2\|z_2\|_{D((-A)^{\frac{1}{2}})}, \\ \|z_1 - z_2\|_{L^2(\Omega)} \le M_1\left[\|z_{1x_1} - z_{2x_1}\|_{L^2(\Omega)} + \|z_{1x_2} - z_{2x_2}\|_{L^2(\Omega)}\right], \\ \|z_{1x_1} - z_{2x_1}\|_{L^2(\Omega)} + \|z_{1x_2} - z_{2x_2}\|_{L^2(\Omega)} \le M_2\|z_1 - z_2\|_{D((-A)^{\frac{1}{2}})}. \end{cases}$$

Hence, the following holds:

$$\begin{aligned} &\|z_1 z_{1x_1} - z_2 z_{2x_1}\|_{L^2(\Omega)} = \|z_1 z_{1x_1} - z_2 z_{1x_1} + z_2 z_{1x_1} - z_2 z_{2x_1}\|_{L^2(\Omega)} \\ &\le \|z_1 - z_2\|_{L^2(\Omega)}\|z_{1x_1}\|_{L^2(\Omega)} + \|z_2\|_{L^2(\Omega)}\|z_{1x_1} - z_{2x_1}\|_{L^2(\Omega)} \\ &\le M_1 M_2\|z_1 - z_2\|_{D((-A)^{\frac{1}{2}}} M_2\|z_1\|_{D((-A)^{\frac{1}{2}}} \\ &\qquad + M_1 M_2\|z_2\|_{D((-A)^{\frac{1}{2}}} M_2\|z_1 - z_2\|_{D((-A)^{\frac{1}{2}}} \\ &\le 2M_1 M_2^2 C\|z_1 - z_2\|_{D((-A)^{\frac{1}{2}}}, \end{aligned}$$

$$\begin{aligned} &|1-\kappa| \cdot \|z_{1x_1x_1} - z_{2x_1x_1}\|_{L^2(\Omega)} + |\kappa| \cdot \|z_{1x_2x_2} - z_{2x_2x_2}\|_{L^2(\Omega)} \\ &\qquad \le \max\{|1-\kappa|, |\kappa|\}\|z_1 - z_2\|_{D((-A)^{\frac{1}{2}}}, \end{aligned}$$

$$\mu\|z_1 - z_2\|_{L^2(\Omega)} \le \mu M_1 M_2\|z_1 - z_2\|_{D((-A)^{\frac{1}{2}}}.$$

Moreover, Lemma 1.5 implies that there exist some positive constants M_3, M_4 and M_5 such that

$$\begin{aligned} &\mu\sum_{j=1}^{N}\|\int_{\bar{x}_j}^{x} z_{1\xi}(\xi,t) - z_{2\xi}(\xi,t)d\xi\|_{L^2(\Omega_j)} \\ &\le \mu\left[M_3\|z_{1x_1} - z_{2x_1}\|_{L^2(\Omega)} + M_4\|z_{1x_2} - z_{2x_2}\|_{L^2(\Omega)} + M_5\|z_{1x_1x_2} - z_{2x_1x_2}\|_{L^2(\Omega)}\right] \\ &\le \mu M_3 M_2\|z_1 - z_2\|_{D((-A)^{\frac{1}{2}}} + \mu M_4 M_2\|z_1 - z_2\|_{D((-A)^{\frac{1}{2}}} + \mu M_5\|z_1 - z_2\|_{D((-A)^{\frac{1}{2}}}. \end{aligned}$$

Substitution of the above inequalities into the right-hand side of (*) yields

$$\|F(z_1) - F(z_2)\|_{L^2(\Omega)} \le l(C)\|z_1 - z_2\|_{D(-A)^{\frac{1}{2}}},$$

where

$$l(C) = 2M_1 M_2^2 C + \max\{|1-\kappa|, |\kappa|\} + \mu(M_1 M_2 + M_3 M_2 + M_4 M_2 + M_5).$$

From Theorem 6.3.1 of [9], it follows that the system (3.79) subject to (3.80) or (3.81) has a unique local classical solution $z \in C([0,T), L^2(\Omega)) \cap C^1((0,T), L^2(\Omega))$ for any initial function $z_0 \in D((-A)^{\frac{1}{2}})$.

Remark 3.5 The above mentioned well-posedness result is dependent on the initial condition $z_0 \in D((-A)^{\frac{1}{2}})$. If the initial function $z_0 \in L^2(\Omega)$, the solution of the system (3.79) subject to (3.80) or (3.81) may become a mild solution or weak solution.

3.2.2.1 Distributed Controller Under Averaged Measurements

Proposition 3.2 *Consider the closed-loop system* (3.79) *subject to* (3.80)*. Given positive scalars* $\bar{\Delta}$*, if there exist* $\delta > 0$*,* $\mu > 0$ *and* $\lambda_i \geq 0$ $(i = 1, 2)$ *such that the following LMI holds:*

$$\Upsilon \triangleq \begin{bmatrix} -2\mu + 2\delta - \lambda_1 \frac{\pi^2}{2} & \Upsilon_{12} & \mu \\ * & -2 & 0 \\ * & * & -\lambda_2 \end{bmatrix} \leq 0 \tag{3.84}$$

where

$$\Upsilon_{12} = -\frac{\lambda_1}{2} - \lambda_2 \frac{\bar{\Delta}^2}{\pi^2} - (1 - \kappa),$$

then the closed-loop system is globally exponentially stable in the L^2*-sense:*

$$\int_\Omega z^2(x, t)dx \leq e^{-2\delta t} \int_\Omega z^2(x, 0)dx, \ \forall t \geq 0. \tag{3.85}$$

Furthermore, if the strict LMI (3.84) *is feasible for* $\delta = 0$*, then the closed-loop system is exponentially stable with a small enough decay rate.*

Proof The proof is divided into three parts.

Step 1: We have shown that there exists a local clasical solution to (3.79) subject to (3.80), where $T = T(z_0)$. By Theorem 6.23.5 of [10], we obtain that the solution exists for any $T > 0$ if this solution admits a priori estimate. In Step 3, it will be shown that the feasibility of LMI (3.84) guarantees that the solution of (3.79) subject to (3.80) admits a priori bound, which can further guarantee the existence of the solution for all $t \geq 0$.

Step 2: Assume formally that there exists a classical solution of (3.79) subject to (3.80) for all $t \geq 0$. We consider the following Lyapunov-Krasovskii functional:

$$V(t) = \|z(\cdot, t)\|^2_{L^2(\Omega)}. \tag{3.86}$$

Since $z|_{\partial\Omega} = 0$ and $\frac{\partial z}{\partial n}|_{\partial\Omega} = 0$, integration by parts leads to

$$\int_\Omega z^2 z_{x_1} dx = 0, \tag{3.87}$$

$$\textstyle\int_\Omega z^2_{x_1x_2} dx = \int_\Omega z_{x_1x_1} z_{x_2x_2} dx. \tag{3.88}$$

Furthermore, from (3.88) it follows that

$$-2\int_\Omega [z_{x_1x_1}^2 + z_{x_2x_2}^2 + 2z_{x_1x_2}^2]dx = -2\int_\Omega [z_{x_1x_1} + z_{x_2x_2}]^2 dx = -2\int_\Omega |\Delta z|^2 dx. \tag{3.89}$$

Differentiating (3.86) along (3.79), integrating by part and using (3.87), (3.89) we obtain

$$\begin{array}{l} \dot V(t) + 2\delta V(t) = 2\int_\Omega z z_t dx + 2\delta \int_\Omega z^2 dx \\ = 2(1-\kappa)\int_\Omega z_{x_1}^2 dx - 2\kappa \int_\Omega z_{x_2}^2 dx - 2\int_\Omega |\Delta z|^2 dx \\ -(2\mu - 2\delta)\int_\Omega z^2 dx + 2\mu \sum\limits_{j=1}^N \int_{\Omega_j} z f_j dx. \end{array} \tag{3.90}$$

From Lemma 1.3, the Wirtinger's inequality yields

$$\lambda_1 \left[\int_\Omega \nabla z^T \nabla z dx - \tfrac{\pi^2}{2}\int_\Omega z^2 dx\right] \ge 0, \tag{3.91}$$

where $\lambda_1 \ge 0$.

For the case of averaged measurements, f_j is given by (3.80). Since $\int_{\Omega_j} f_j(x,t)dx = 0$, from Lemma 1.3, the Poincaré inequality leads to

$$\int_{\Omega_j} f_j^2 dx \le \tfrac{2\bar\Delta^2}{\pi^2}\int_{\Omega_j} \nabla z^T \nabla z dx.$$

Hence,

$$\lambda_2 \sum_{j=1}^N \left[\tfrac{2\bar\Delta^2}{\pi^2}\int_{\Omega_j} \nabla z^T \nabla z dx - \int_{\Omega_j} f_j^2 dx\right] \ge 0, \tag{3.92}$$

where $\lambda_2 \ge 0$.

Integration by parts yields

$$-\int_\Omega \nabla z^T \nabla z dx = \int_\Omega z \Delta z dx. \tag{3.93}$$

Applying S-procedure [28], we add to $\dot V(t) + 2\delta V(t)$ the left-hand side of (3.91), (3.92) and use (3.93). Then it follows that

$$\begin{array}{l} \dot V(t) + 2\delta V(t) \le \dot V(t) + 2\delta V(t) + \lambda_1 \left[\int_\Omega \nabla z^T \nabla z dx - \tfrac{\pi^2}{2}\int_\Omega z^2 dx\right] \\ +\lambda_2 \left[\tfrac{2\bar\Delta^2}{\pi^2}\int_\Omega \nabla z^T \nabla z dx - \sum\limits_{j=1}^N \int_{\Omega_j} f_j^2 dx\right] \\ \le 2(1-\kappa)\int_\Omega z_{x_1}^2 dx - 2\kappa \int_\Omega z_{x_2}^2 dx - 2\int_\Omega |\Delta z|^2 dx - (2\mu - 2\delta + \lambda_1 \tfrac{\pi^2}{2})\int_\Omega z^2 dx \\ +2\mu \sum\limits_{j=1}^N > \int_{\Omega_j} z f_j dx - (\lambda_1 + \lambda_2 \tfrac{2\bar\Delta^2}{\pi^2})\int_\Omega z \Delta z dx - \lambda_2 \sum\limits_{j=1}^N \int_{\Omega_j} f_j^2 dx. \end{array}$$

Note that

$$\begin{aligned}&2(1-\kappa)\int_{\Omega} z_{x_1}^2 dx - 2\kappa \int_{\Omega} z_{x_2}^2 dx = 2(1-\kappa)\int_{\Omega} \nabla z^T \nabla z dx - 2\int_{\Omega} z_{x_2}^2 dx\\ &\leq 2(1-\kappa)\int_{\Omega} \nabla z^T \nabla z dx.\end{aligned} \tag{3.94}$$

Then substitution of (3.93) into (3.94) yields

$$2(1-\kappa)\int_{\Omega} z_{x_1}^2 dx - 2\kappa \int_{\Omega} z_{x_2}^2 dx \leq -2(1-\kappa)\int_{\Omega} z\Delta z dx. \tag{3.95}$$

Hence,

$$\dot{V}(t) + 2\delta V(t) \leq \sum_{j=1}^{N} \int_{\Omega_j} \begin{bmatrix} z & \Delta z & f_j \end{bmatrix} \Upsilon \begin{bmatrix} z \\ \Delta z \\ f_j \end{bmatrix} dx \leq 0$$

if $\Upsilon \leq 0$ holds. Therefore,

$$V(t) \leq e^{-2\delta t} V(0), \ \forall t \geq 0.$$

Note that the feasibility of the strict LMI (3.84) with $\delta = 0$ implies its feasibility with a small enough $\delta_0 > 0$. Therefore, if the strict LMI (3.84) holds for $\delta = 0$, then the closed-loop system is exponentially stable with a small decay rate $\delta_0 > 0$.

Step 3: The feasibility of LMI (3.84) yields that the solution of (3.79) subject to (3.80) admits a priori estimate $V(t) \leq e^{-2\delta t} V(0)$. By Theorem 6.23.5 of [10], continuation of this solution under a priori bound to entire interval $[0, \infty)$. □

3.2.2.2 Distributed Controller Under Point Measurements

Proposition 3.3 *Consider the closed-loop system* (3.79) *subject to* (3.81). *Given positive scalars* $\bar{\Delta}$, *if there exist* $\delta > 0$, $\mu > 0$, $\eta > 0$, $\lambda_1 \geq 0$, $\lambda_2 \in \mathbb{R}$ *and* $\beta_i > 0$ $(i = 1, 2, 3)$ *such that (1.6) is satisfied and the following LMIs hold:*

$$2(1-\kappa) + \beta_1 \left(\frac{\bar{\Delta}}{\pi}\right)^2 + \lambda_1 - \lambda_2 \leq 0, \tag{3.96}$$

$$-2\kappa + \beta_2 \left(\frac{\bar{\Delta}}{\pi}\right)^2 + \lambda_1 - \lambda_2 \leq 0, \tag{3.97}$$

$$\Lambda = \begin{bmatrix} -2\mu + 2\delta - \lambda_1 \frac{\pi^2}{2} & -\frac{\lambda_2}{2} & -\frac{\lambda_2}{2} & \mu \\ * & -2 & -2 + \frac{\beta_3}{2}\left(\frac{\bar{\Delta}}{\pi}\right)^4 & 0 \\ * & * & -2 & 0 \\ * & * & * & -\eta \end{bmatrix} \leq 0, \tag{3.98}$$

the closed-loop system is globally exponentially stable satisfying (3.85). *Furthermore, if the strict LMI* (3.98) *is feasible for* $\delta = 0$, *then the closed-loop system is exponentially stable with a small enough decay rate.*

Proof Step 1: We have shown that there exists a local clasical solution to (3.79) subject to (3.81), where $T = T(z_0)$. By Theorem 6.23.5 of [10], we obtain that the solution exists for any $T > 0$ if this solution admits a priori estimate. In Step 3, it will be shown that the feasibility of LMIs (3.96)–(3.98) guarantees that the solution of (3.79) subject to (3.81) admits a priori bound, which can further guarantee the existence of the solution for all $t \geq 0$.

Step 2: Assume formally that there exists a classical solution of (3.79) subject to (3.81) for all $t \geq 0$. Consider V given by (3.86). Differentiating V along (3.79) and integrating by parts, we have (3.90).

For the case of point measurements, f_j is given by (3.81). From Lemma 1.5, we have

$$\eta\|f_j\|^2_{L^2(\Omega_j)} \leq \beta_1\left(\tfrac{\bar{\Delta}}{\pi}\right)^2\|z_{x_1}\|^2_{L^2(\Omega_j)} + \beta_2\left(\tfrac{\bar{\Delta}}{\pi}\right)^2\|z_{x_2}\|^2_{L^2(\Omega_j)} + \beta_3\left(\tfrac{\bar{\Delta}}{\pi}\right)^4\|z_{x_1x_2}\|^2_{L^2(\Omega_j)}$$

for any scalars $\beta_1, \beta_2, \beta_3$ such that (1.6) holds.

Hence,

$$\begin{aligned}\sum_{j=1}^{N}\Big[&\beta_1\left(\tfrac{\bar{\Delta}}{\pi}\right)^2\|z_{x_1}\|^2_{L^2(\Omega_j)} + \beta_2\left(\tfrac{\bar{\Delta}}{\pi}\right)^2\|z_{x_2}\|^2_{L^2(\Omega_j)}\\ &+\beta_3\left(\tfrac{\bar{\Delta}}{\pi}\right)^4\|z_{x_1x_2}\|^2_{L^2(\Omega_j)} - \eta\|f_j\|^2_{L^2(\Omega_j)}\Big] \geq 0.\end{aligned} \tag{3.99}$$

From (3.93), for any $\lambda_2 \in \mathbb{R}$ we have

$$\lambda_2\left[\int_\Omega \nabla z^T\nabla z dx + \int_\Omega z\Delta z dx\right] = 0. \tag{3.100}$$

Similarly, we add to $\dot{V}(t) + 2\delta V(t)$ the left-hand side of (3.91) and (3.99). Then by taking into account (3.90), we obtain

$$\begin{aligned}&\dot{V}(t) + 2\delta V(t)\\ &\leq \dot{V}(t) + 2\delta V(t) + \lambda_1\left[\textstyle\int_\Omega \nabla z^T\nabla z dx - \frac{\pi^2}{2}\int_\Omega z^2 dx\right] + \lambda_2\left[-\textstyle\int_\Omega \nabla z^T\nabla z dx - \int_\Omega z\Delta z dx\right]\\ &+\left[\beta_1\left(\tfrac{\bar{\Delta}}{\pi}\right)^2\|z_{x_1}\|^2_{L^2(\Omega_j)} + \beta_2\left(\tfrac{\bar{\Delta}}{\pi}\right)^2\|z_{x_2}\|^2_{L^2(\Omega_j)} + \beta_3\left(\tfrac{\bar{\Delta}}{\pi}\right)^4\|z_{x_1x_2}\|^2_{L^2(\Omega_j)} - \eta\|f_j\|^2_{L^2(\Omega_j)}\right]\\ &\leq \left[2(1-\kappa) + \beta_1\left(\tfrac{\bar{\Delta}}{\pi}\right)^2 + \lambda_1 - \lambda_2\right]\textstyle\int_\Omega z^2_{x_1}dx + \left[-2\kappa + \beta_2\left(\tfrac{\bar{\Delta}}{\pi}\right)^2 + \lambda_1 - \lambda_2\right]\int_\Omega z^2_{x_2}dx\\ &+\beta_3\left(\tfrac{\bar{\Delta}}{\pi}\right)^4\textstyle\int_\Omega z^2_{x_1x_2}dx - 2\int_\Omega|\Delta z|^2 dx - (2\mu - 2\delta + \lambda_1\frac{\pi^2}{2})\int_\Omega z^2 dx + 2\mu\sum_{j=1}^{N}\int_{\Omega_j} zf_j dx\\ &-\lambda_2\textstyle\int_\Omega z\Delta z dx - \eta\sum_{j=1}^{N}\int_{\Omega_j} f_j^2 dx.\end{aligned}$$

By using (3.88) and (3.89), the following inequality holds for all $t \geq 0$

$$\dot{V}(t) + 2\delta V(t) \leq \sum_{j=1}^{N} \int_{\Omega_j} \psi^T(x,t)\Lambda\psi(x,t)dx \tag{3.101}$$

where

$$\psi(x,t) = \mathrm{col}\{z, z_{x_1x_1}, z_{x_2x_2}, f_j\}.$$

Therefore, the LMIs (3.96)–(3.98) yield (3.85).

Step 3: The feasibility of LMIs (3.96)–(3.98) yields that the solution of (3.79) subject to (3.81) admits a priori estimate $V(t) \leq e^{-2\delta t}V(0)$. By Theorem 6.23.5 of [10], continuation of this solution under a priori bound to entire interval $[0, \infty)$. □

3.2.3 Sampled-Data Regional Stabilization

3.2.3.1 Sampled-Data Control Under Averaged Measurements

For $j = 1, \ldots, N; k = 0, 1, \ldots$ we consider the quantities

$$f_j(x,t) = z(x,t) - \frac{\int_{\Omega_j} z(\zeta,t)d\zeta}{|\Omega_j|}, \tag{3.102}$$

$$g_j(t) = \frac{1}{t - t_k} \frac{\int_{\Omega_j}\int_{t_k}^{t} z_s(\zeta,s)dsd\zeta}{|\Omega_j|}. \tag{3.103}$$

Then the controller (3.75) subject to (3.73) leads to the closed-loop system

$$\begin{cases} z_t + zz_{x_1} + (1-\kappa)z_{x_1x_1} - \kappa z_{x_2x_2} + \Delta^2 z = -\mu \sum_{j=1}^{N} \chi_j(x)[z - f_j - (t-t_k)g_j], \\ \qquad (x,t) \in \Omega \times [t_k, t_{k+1}), \\ z|_{\partial\Omega} = 0, \quad \frac{\partial z}{\partial n}|_{\partial\Omega} = 0. \end{cases} \tag{3.104}$$

Now we use the step method (see e.g. [11, 12]) to establish the proof of the well-posedness for system (3.104). For $t \in [t_0, t_1]$, we consider the following equation:

$$\begin{cases} z_t + zz_{x_1} + (1-\kappa)z_{x_1x_1} - \kappa z_{x_2x_2} + \Delta^2 z = -\mu \sum_{j=1}^{N} \chi_j(x)\frac{\int_{\Omega_j} z_0(x)dx}{|\Omega_j|}, \\ z|_{\partial\Omega} = 0, \quad \frac{\partial z}{\partial n}|_{\partial\Omega} = 0. \end{cases} \tag{3.105}$$

Then system (3.105) can be represented as an evolution Eq. (3.83) subject to

$$F(z(\cdot,t)) = -z(x,t)z_{x_1}(x,t) - (1-\kappa)z_{x_1x_1}(x,t) + \kappa z_{x_2x_2}(x,t)$$
$$-\mu \sum_{j=1}^{N} \chi_j(x) \frac{\int_{\Omega_j} z_0(x)dx}{|\Omega_j|}.$$

Note that the nonlinearity $F(z(\cdot,t))$ is locally Lipschitz continuous. From Theorem 3.3.3 of [13], it follows that there exists a unique local strong solution $z(\cdot,t) \in C([0,T]; D((-A)^{\frac{1}{2}})) \cap C^1((0,T]; D(A))$ of (3.105) initialized with $z_0 \in D((-A)^{\frac{1}{2}})$ on some interval $[0,T] \subset [0,t_1]$, where $T = T(z_0) > 0$. By Theorem 6.23.5 of [10], we obtain that if this solution admits a priori estimate, then the solution exists on the entire $[0,t_1]$. The priori estimate on the solutions starting from the domain of attraction will be guaranteed by the stability conditions that we will provide (see Theorem 3.3). Then we apply the same line of reasoning step-by-step to the time segments $[t_1,t_2]$, $[t_2,t_3]$, ... Following this procedure, we find that the strong solution exists for all $t \geq 0$.

In order to derive the stability conditions for (3.104) we employ the following Lyapunov-Krasovskii functional

$$V_1(t) = p_1\|z\|^2_{L^2(\Omega)} + p_2\|\Delta z\|^2_{L^2(\Omega)} + r(t_{k+1}-t)\int_\Omega \int_{t_k}^t e^{2\delta(s-t)} z_s^2(x,s)dsdx,$$
$$t \in [t_k, t_{k+1}),\ p_1 > 0,\ p_2 > 0,\ r > 0. \tag{3.106}$$

Remark 3.6 Note that without delay/sampling behavior, the energy norm is usually used. In the present work, due to the sampling terms, we need to use Lyapunov-Krasovskii functionals (see e.g. [2, 29]). Therefore, additionally to the energy norm $p_1\|z\|^2_{L^2(\Omega)} + p_2\|\Delta z\|^2_{L^2(\Omega)}$, we employ the term $r(t_{k+1} - t)\int_\Omega \int_{t_k}^t e^{2\delta(s-t)} z_s^2(x,s)dsdx$ to deal with the sampling.

For convenience we define $V = D((-A)^{\frac{1}{2}})$ with the norm $\|z\|_V^2 = p_1\|z\|^2_{L^2(\Omega)} + p_2\|\Delta z\|^2_{L^2(\Omega)}$. Here p_1 and p_2 are positive constants that are related to the Lyapunov-Krasovskii functional (3.106). By using Lyapunov-Krasovskii functional (3.106), in Theorem 3.3 we provide LMI conditions for regional exponential stability of (3.104) and for a bound on the domain of attraction.

Theorem 3.3 *Consider the closed-loop system* (3.104). *Given positive scalars C, h, μ, $\bar{\Delta}$ and δ, let there exist scalars $r > 0$, $\Gamma > 0$, $p_1 > 0$, $p_2 > 0$, $\lambda_i \geq 0$ $(i = 1,2)$ and $\lambda_3 \in \mathbb{R}$ satisfy the linear matrix inequalities:*

$$\Xi_i|_{z=C} < 0,\ \Xi_i|_{z=-C} < 0,\ i = 1,2 \tag{3.107}$$

$$\begin{bmatrix} p_2 - (1+\Gamma)\frac{1}{\pi^2} & \sqrt{\frac{1}{2}} \\ * & \Gamma \end{bmatrix} > 0, \tag{3.108}$$

where

$$\Xi_1 = \left[\begin{array}{c:c} \Phi_1 & \begin{array}{c} -rhz \\ 0 \\ -(1-\kappa)rh \\ \kappa rh \\ -rh \\ -\mu rh \\ \mu rh \end{array} \\ \hdashline * & -rh \end{array}\right], \tag{3.109}$$

$$\Xi_2 = \left[\begin{array}{c:c} \Phi_2 & \begin{array}{c} -rhz \\ 0 \\ -(1-\kappa)rh \\ \kappa rh \\ -rh \\ -\mu rh \\ \mu rh \\ \mu rh^2 \end{array} \\ \hdashline * & -rh \end{array}\right], \tag{3.110}$$

$\Phi_1 = \{\phi_{ij}\}$ *is a symmetric matrix composed from*

$$\phi_{11} = 2p_1(1-\kappa) + \lambda_1 + \frac{2\bar{\Delta}^2}{\pi^2}\lambda_2 - \lambda_3,\ \phi_{15} = -p_2 z,$$
$$\phi_{22} = -2p_1\kappa + \lambda_1 + \frac{2\bar{\Delta}^2}{\pi^2}\lambda_2 - \lambda_3,\ \phi_{33} = -2p_1 + 2\delta p_2,\ \phi_{34} = 2\delta p_2,$$
$$\phi_{35} = -p_2(1-\kappa), \phi_{36} = -\frac{\lambda_3}{2},\ \phi_{44} = -2p_1 + 2\delta p_2,\ \phi_{45} = p_2\kappa,\ \phi_{46} = -\frac{\lambda_3}{2},$$
$$\phi_{55} = -2p_2,\ \phi_{56} = -p_2\mu,\ \phi_{57} = p_2\mu,\ \phi_{66} = -2p_1\mu + 2\delta p_1 - \frac{\pi^2}{2}\lambda_1,$$
$$\phi_{67} = p_1\mu,\ \phi_{77} = -\lambda_2,$$

$$\Phi_2 = \left[\begin{array}{cccccccc:c} & & & & \Phi_1 & & & & * \\ \hdashline 0 & 0 & 0 & 0 & p_2\mu h & p_1\mu h & 0 & & -rhe^{-2\delta h} \end{array}\right]. \tag{3.111}$$

Then for any initial state $z_0 \in V$ *satisfying* $\|z_0\|_V^2 < C^2$, *a unique solution of* (3.104) *exists and satisfies*

$$p_1\|z\|^2_{L^2(\Omega)} + p_2\|\Delta z\|^2_{L^2(\Omega)} \le e^{-2\delta t}[p_1\|z_0\|^2_{L^2(\Omega)} + p_2\|\Delta z_0\|^2_{L^2(\Omega)}],\ t \ge 0.$$

Furthermore, if the strict LMI (3.107) *is feasible for* $\delta = 0$, *then the closed-loop system is exponentially stable with a small enough decay rate.*

Proof The proof is divided into three parts.

Step 1: We have shown that there exists a unique local strong solution to (3.104) on some interval $[0, T] \subset [0, t_1]$. By Theorem 6.23.5 of [10], we obtain that the solution exists on the entire interval $[0, t_1]$ if this solution admits a priori estimate. In Step 3, it will be shown that the feasibility of LMIs (3.107), (3.108) guarantees that the solution of (3.104) admits a priori bound, which can further guarantee the existence of the solution for all $t \geq 0$.

Step 2: Assume formally that there exists a solution of (3.104) for all $t \geq 0$. Differentiating V_1 along (3.104), we have

$$\begin{aligned}\dot{V}_1(t) + 2\delta V_1(t) = 2p_1 \int_\Omega z z_t dx + 2p_2 \int_\Omega \Delta z \Delta z_t dx - r \int_\Omega \int_{t_k}^t e^{2\delta(s-t)} z_s^2(x,s) ds dx \\ + r(t_{k+1} - t) \int_\Omega z_t^2(x,t) dx + 2\delta p_1 \int_\Omega z^2 dx + 2\delta p_2 \int_\Omega |\Delta z|^2 dx.\end{aligned} \tag{3.112}$$

Substitution of z_t from (3.104) leads to

$$\begin{aligned}& r(t_{k+1} - t) \int_\Omega z_t^2(x,t) dx = r(t_{k+1} - t) \sum_{j=1}^N \int_{\Omega_j} [-z z_{x_1} - (1-\kappa) z_{x_1 x_1} + \kappa z_{x_2 x_2} \\ & -\Delta^2 z - \mu z + \mu f_j + \mu(t - t_k) g_j]^2 dx \\ & \leq rh \sum_{j=1}^N \int_{\Omega_j} [-z z_{x_1} - (1-\kappa) z_{x_1 x_1} + \kappa z_{x_2 x_2} - \Delta^2 z - \mu z + \mu f_j + \mu(t - t_k) g_j]^2 dx.\end{aligned} \tag{3.113}$$

Integration by parts yields

$$\begin{aligned}& 2p_1 \int_\Omega z z_t dx = 2p_1(1-\kappa) \int_\Omega z_{x_1}^2 dx - 2p_1 \kappa \int_\Omega z_{x_2}^2 dx - 2p_1 \int_\Omega |\Delta z|^2 dx \\ & -2p_1 \mu \int_\Omega z^2 dx + 2p_1 \mu \sum_{j=1}^N \int_{\Omega_j} z[f_j + (t - t_k) g_j] dx,\end{aligned} \tag{3.114}$$

and

$$\begin{aligned}& 2p_2 \int_\Omega \Delta z \Delta z_t dx = 2p_2 \int_\Omega \Delta^2 z \cdot z_t dx \\ & = 2p_2 \int_\Omega \Delta^2 z[-z z_{x_1} - (1-\kappa) z_{x_1 x_1} + \kappa z_{x_2 x_2} - \Delta^2 z] dx \\ & -2p_2 \mu \int_\Omega \Delta^2 z[z - f_j - (t - t_k) g_j] dx.\end{aligned} \tag{3.115}$$

The Jensen inequality leads to

$$\begin{aligned}& -r \int_\Omega \int_{t_k}^t e^{2\delta(s-t)} z_s^2(x,s) ds dx \leq -r e^{-2\delta h} \int_\Omega \frac{1}{t - t_k} \left[\int_{t_k}^t z_s(x,s) ds\right]^2 dx \\ & \leq -r(t - t_k) e^{-2\delta h} \sum_{j=1}^N \int_{\Omega_j} g_j^2 dx.\end{aligned} \tag{3.116}$$

From (3.93), we obtain

$$\lambda_3 \left[-\int_\Omega \nabla z^T \nabla z dx - \int_\Omega z \Delta z dx\right] = 0, \tag{3.117}$$

where $\lambda_3 \in \mathbb{R}$.

Set

$$\eta_1 = \mathrm{col}\{z_{x_1}, z_{x_2}, z_{x_1x_1}, z_{x_2x_2}, \Delta^2 z, z, f_j\}, \eta_2 = \mathrm{col}\{\eta_1, g_j\},$$
$$\beta \triangleq \left[-z \; 0 \; -(1-\kappa) \; \kappa \; -1 \; -\mu \; \mu \; \mu(t-t_k) \right].$$

Then

$$[-zz_{x_1} - (1-\kappa)z_{x_1x_1} + \kappa z_{x_2x_2} - \Delta^2 z - \mu z + \mu f_j + \mu(t-t_k)g_j]^2 = \eta_2^T \beta^T \beta \eta_2. \tag{3.118}$$

From (3.113) and (3.118) we have

$$r(t_{k+1}-t)\int_\Omega z_t^2(x,t)dx \le rh\sum_{j=1}^N \int_{\Omega_j} \eta_2^T \beta^T \beta \eta_2 dx. \tag{3.119}$$

Applying S-procedure, we add to $\dot{V}_1(t) + 2\delta V_1(t)$ the left-hand side of (3.91), (3.92), (3.117). Then,

$$\begin{aligned}
&\dot{V}_1(t) + 2\delta V_1(t) \\
&\le \dot{V}_1(t) + 2\delta V_1(t) + \lambda_1 \left[\int_\Omega \nabla z^T \nabla z dx - \tfrac{\pi^2}{2}\int_\Omega z^2 dx \right] \\
&+\lambda_2 \left[\tfrac{2\bar{\Delta}^2}{\pi^2}\int_\Omega \nabla z^T \nabla z dx - \sum_{j=1}^N \int_{\Omega_j} f_j^2 dx \right] + \lambda_3 \left[-\int_\Omega \nabla z^T \nabla z dx - \int_\Omega z\Delta z dx \right] \\
&\le \sum_{j=1}^N \int_{\Omega_j} \frac{h-t+t_k}{h}\eta_1^T \Phi_1 \eta_1 + \frac{t-t_k}{h}\eta_2^T \Phi_2 \eta_2 dx \\
&+rh\sum_{j=1}^N \int_{\Omega_j} \eta_2^T \beta^T \beta \eta_2 dx - (4p_1 - 4\delta p_2)\int_\Omega z_{x_1x_2}^2 dx.
\end{aligned} \tag{3.120}$$

Note that LMIs (3.107) imply that $\phi_{33} < 0$, i.e. $p_1 > \delta p_2$. As in [20], first we assume that

$$\|z(\cdot,t)\|_{C^0(\bar{\Omega})} < C, \; \forall t \ge 0. \tag{3.121}$$

Note that $\dfrac{h-t+t_k}{h} + \dfrac{t-t_k}{h} = 1$ and $t - t_k \le h$. Under the assumption (3.121), applying Schur complement to (3.119), from (3.118)–(3.120) we obtain

$$\begin{aligned}
\dot{V}_1(t) + 2\delta V_1(t) \le \sum_{j=1}^N \int_{\Omega_j} \frac{h-t+t_k}{h}\left[\eta_1^T \; 1 \right] \Xi_1 \begin{bmatrix} \eta_1 \\ 1 \end{bmatrix} dx \\
+\sum_{j=1}^N \int_{\Omega_j} \frac{t-t_k}{h}\left[\eta_2^T \; 1 \right] \Xi_2 \begin{bmatrix} \eta_2 \\ 1 \end{bmatrix} dx \le 0
\end{aligned} \tag{3.122}$$

if $\Xi_1 < 0$, $\Xi_2 < 0$ for all $z \in (-C, C)$.

Matrices Ξ_1 and Ξ_2 given by (3.109), (3.110) are affine in z. Hence, $\Xi_1 < 0$ and $\Xi_2 < 0$ for all $z \in (-C, C)$ if these inequalities hold in the vertices $z = \pm C$ hold, i.e. if LMIs (3.107) are feasible.

We prove next that (3.121) holds. Lemma 1.9 and Schur complement theorem lead to

$$\begin{aligned} \|z\|^2_{C^0(\bar{\Omega})} &\leq \frac{1}{2}(1+\Gamma)\left[\|z_{x_1}\|^2_{L^2(\Omega)} + \|z_{x_2}\|^2_{L^2(\Omega)}\right] + \frac{1}{\Gamma}\|z_{x_1x_2}\|^2_{L^2(\Omega)} \\ &\leq \left[(1+\Gamma)\frac{1}{\pi^2} + \frac{1}{2\Gamma}\right]\|\Delta z\|^2_{L^2(\Omega)} \leq V_1(t). \end{aligned} \tag{3.123}$$

The last inequality in (3.123) follows from (3.108), and for the second inequality in (3.123) we use the Wirtinger's inequality, (3.88) and (3.89). Therefore, it is sufficient to show that

$$V_1(t) < C^2, \ \forall t \geq 0. \tag{3.124}$$

Indeed, for $t = 0$, the inequality (3.124) holds. Let (3.124) be false for some t_1. Then $V_1(t_1) \geq C^2 > V_1(0)$. Since V_1 is continuous in time, there must exist $t^* \in (0, t_1]$ such that

$$V_1(t) < C^2 \ \forall t \in [0, t^*) \text{ and } V_1(t^*) = C^2. \tag{3.125}$$

The first relation of (3.125), together with the feasibility of (3.107), guarantees that $\dot{V}_1(t) + 2\delta V_1(t) \leq 0$ on $[0, t^*)$. Therefore, $V_1(t^*) \leq V_1(0) < C^2$. This contradicts the second relation of (3.125). Thus, (3.124) and consequently, (3.122) is true, which implies provided that $\|z_0\|_V < C$.

Note that the feasibility of LMI (3.107) with $\delta = 0$ implies its feasibility with a small enough $\delta_0 > 0$. Therefore, if LMI (3.107) holds for $\delta = 0$, then the closed-loop system is exponentially stable with a small decay rate.

Step 3: The feasibility of LMIs (3.107), (3.108) yields that the solution of (3.104) admits a priori estimate $V_1(t) \leq e^{-2\delta t} V_1(0)$. By Theorem 6.23.5 of [10], this solution (under a priori bound) can be continued to entire interval $[0, \infty)$. □

3.2.3.2 Sampled-Data Control Under Point Measurements

Under the controller (3.75) subject to (3.74), the closed-loop system becomes

$$\begin{cases} z_t + zz_{x_1} + (1-\kappa)z_{x_1x_1} - \kappa z_{x_2x_2} + \Delta^2 z \\ = -\mu \sum\limits_{j=1}^{N} \chi_j(x) z(\bar{x}_j, t_k), \ (x,t) \in \Omega \times [t_k, t_{k+1}), \\ z|_{\partial\Omega} = 0, \ \dfrac{\partial z}{\partial n}|_{\partial\Omega} = 0. \end{cases} \tag{3.126}$$

Theorem 3.4 *Consider the closed-loop system* (3.126). *Given positive scalars* C, h, μ, $\bar{\Delta}$ *and* $\delta_1 < 2\delta$, *let there exist scalars* $r > 0$, $\Gamma > 0$, $p_1 > 0$, $p_2 > 0$, $\eta > 0$, $\lambda_i \geq 0$ $(i = 1, 2)$, $\lambda_3 \in \mathbb{R}$ *and* $\beta_i > 0$ $(i = 1, 2, 3)$ *such that (1.6) is satisfied and the following LMIs hold:*

$$- 2\delta_1 p_2 + \beta_3 \left(\frac{\bar{\Delta}}{\pi}\right)^4 \leq 0, \tag{3.127}$$

$$\bar{\Theta} = \begin{bmatrix} -\delta_1 p_2 & -\frac{\beta_1}{2}\left(\frac{\bar{\Delta}}{\pi}\right)^2 & -\frac{\beta_2}{2}\left(\frac{\bar{\Delta}}{\pi}\right)^2 \\ * & -\delta_1 p_2 & 0 \\ * & * & -\delta_1 p_2 \end{bmatrix} \leq 0, \tag{3.128}$$

$$\begin{bmatrix} p_2 - (1+\Gamma)\frac{1}{\pi^2} & \sqrt{\frac{1}{2}} \\ * & \Gamma \end{bmatrix} > 0, \tag{3.129}$$

$$\Lambda_i|_{z=C} < 0, \ \Lambda_i|_{z=-C} < 0, \ i = 1, 2 \tag{3.130}$$

where

$$\Lambda_1 = \left[\begin{array}{c:c} \Theta_0 & \Theta_1 \\ \hdashline * & -rh \end{array}\right],$$

$$\Lambda_2 = \left[\begin{array}{c:c} \Theta_2 & \Theta_3 \\ \hdashline * & -rh \end{array}\right],$$

$\Theta_0 = \{\theta_{ij}\}$ *is a symmetric matrix composed from*

$$\begin{aligned}
&\theta_{11} = 2p_1(1-\kappa) + \lambda_1 - \lambda_2, \ \theta_{22} = -2p_1\kappa + \lambda_1 - \lambda_2, \ \theta_{33} = -2p_1 + 2\delta p_2, \\
&\theta_{35} = -p_2(1-\kappa), \ \theta_{36} = -\frac{\lambda_2}{2}, \ \theta_{44} = -2p_1 + 2\delta p_2, \ \theta_{45} = p_2\kappa, \\
&\theta_{46} = -\frac{\lambda_2}{2}, \ \theta_{55} = -2p_2, \ \theta_{56} = -p_2\mu, \ \theta_{57} = p_2\mu, \\
&\theta_{66} = -2p_1\mu + 2\delta p_1 - \frac{\pi^2}{2}\lambda_1, \ \theta_{67} = p_1\mu, \ \theta_{77} = -\eta,
\end{aligned}$$

$$\Theta_1 = [\, -rhz \ 0 \ -(1-\kappa)rh \ \kappa rh \ -rh \ -\mu rh \ \mu rh \,]^T \tag{3.131}$$

$$\Theta_2 = \left[\begin{array}{c:c} \Theta_0 & * \\ \hdashline 0 \ \ 0 \ \ 0 \ \ 0 \ \ p_2\mu h \ \ p_1\mu h \ \ 0 & -rhe^{-2\delta h} \end{array}\right], \tag{3.132}$$

$$\Theta_3 = [\, \Theta_1 \ \mu rh^2 \,]^T \tag{3.133}$$

Then for any initial state $z_0 \in V$ *satisfying* $\|z_0\|_V^2 < C^2$, *a unique solution of* (3.126) *exists and satisfies*

$$p_1\|z\|_{L^2(\Omega)}^2 + p_2\|\Delta z\|_{L^2(\Omega)}^2 \leq e^{-2\sigma t}[p_1\|z_0\|_{L^2(\Omega)}^2 + p_2\|\Delta z_0\|_{L^2(\Omega)}^2], \ t \geq 0,$$

where σ *is a unique positive solution of (1.43).*

Proof Step 1: We have shown that there exists a unique local strong solution to (3.126) on some interval $[0, T] \subset [0, t_1]$. By Theorem 6.23.5 of [10], we obtain that the solution exists on the entire interval $[0, t_1]$ if this solution admits a priori estimate. In Step 3, it will be shown that the feasibility of LMIs (3.107), (3.108) guarantees that the solution of (3.126) admits a priori bound, which can further guarantees the existence of the solution for all $t \geq 0$.

Step 2: Assume formally that there exists a strong solution of (3.126) starting from $\|z_0\|_V < C$ for all $t \geq 0$. Differentiating V_1 along (3.126), we obtain the inequality (3.112). Denote

$$f_j(x,t) = z(x,t) - z(\bar{x}_j, t) = \int_{\bar{x}_j}^{x} z_\xi(\xi, t)d\xi, \tag{3.134}$$

From Lemma 1.5, we have

$$\eta\|f_j\|^2_{L^2(\Omega_j)} \leq \beta_1\left(\frac{\bar{\Delta}}{\pi}\right)^2\|z_{x_1}\|^2_{L^2(\Omega_j)} + \beta_2\left(\frac{\bar{\Delta}}{\pi}\right)^2\|z_{x_2}\|^2_{L^2(\Omega_j)} + \beta_3\left(\frac{\bar{\Delta}}{\pi}\right)^4\|z_{x_1x_2}\|^2_{L^2(\Omega_j)} \tag{3.135}$$

for any scalars $\beta_1, \beta_2, \beta_3$ such that (1.6) holds.

Hence,

$$\begin{aligned}\sum_{j=1}^{N}\Bigg[&\beta_1\left(\frac{\bar{\Delta}}{\pi}\right)^2\|z_{x_1}\|^2_{L^2(\Omega_j)} + \beta_2\left(\frac{\bar{\Delta}}{\pi}\right)^2\|z_{x_2}\|^2_{L^2(\Omega_j)}\\ &+\beta_3\left(\frac{\bar{\Delta}}{\pi}\right)^4\|z_{x_1x_2}\|^2_{L^2(\Omega_j)} - \eta\|f_j\|^2_{L^2(\Omega_j)}\Bigg] \geq 0.\end{aligned} \tag{3.136}$$

Denote

$$\rho(x,t) \triangleq \frac{1}{t-t_k}\int_{t_k}^{t} z_s(x,s)ds.$$

Then, we have

$$z(x,t) = z(x,t_k) + (t-t_k)\rho(x,t).$$

By using Jensen's inequality, we have

$$-r\int_\Omega\int_{t_k}^{t} e^{2\delta(s-t)}z_s^2(x,s)dsdx \leq -re^{-2\delta h}(t-t_k)\int_\Omega \rho^2(x,t)dx. \tag{3.137}$$

Integration by parts leads to

$$\begin{aligned}&2p_1\int_\Omega zz_tdx = 2p_1(1-\kappa)\int_\Omega z_{x_1}^2dx - 2p_1\kappa\int_\Omega z_{x_2}^2dx - 2p_1\int_\Omega|\Delta z|^2dx\\ &+2p_1\mu\sum_{j=1}^{N}\int_{\Omega_j} z\left[(t-t_k)\rho(x,t) + f_j(x,t_k)\right]dx - 2p_1\mu\int_\Omega z^2dx,\end{aligned} \tag{3.138}$$

$$\begin{aligned}&2p_2\int_\Omega \Delta z\Delta z_tdx = 2p_2\int_\Omega \Delta^2 z\cdot z_tdx\\ &= 2p_2\int_\Omega\Delta^2 z[-zz_{x_1} - (1-\kappa)z_{x_1x_1} + \kappa z_{x_2x_2} - \Delta^2 z]dx\\ &-2p_2\mu\sum_{j=1}^{N}\int_{\Omega_j}\Delta^2 z[z(x,t) - (t-t_k)\rho(x,t) - f_j(x,t_k)]dx.\end{aligned} \tag{3.139}$$

Then from (3.137)–(3.139), adding (3.136) into $\dot{V}_1 + 2\delta V_1$ we obtain

$$\begin{aligned}
&\dot{V}_1(t) + 2\delta V_1(t) - \delta_1 \sup_{\theta\in[-h,0]} V_1(t+\theta) \\
&\leq \dot{V}_1(t) + 2\delta V_1(t) - \delta_1 V_1(t_k) + \lambda_1 \left[\int_\Omega \nabla z^T \nabla z dx - \tfrac{\pi^2}{2}\int_\Omega z^2 dx\right] \\
&+\lambda_2\left[-\int_\Omega \nabla z^T\nabla z dx - \int_\Omega z\Delta z dx\right] \\
&+\left[\beta_1\left(\tfrac{\bar{\Delta}}{\pi}\right)^2 \int_\Omega z_{x_1}^2(x,t_k)dx + \beta_2\left(\tfrac{\bar{\Delta}}{\pi}\right)^2\int_\Omega z_{x_2}^2(x,t_k)dx\right. \\
&\left.+\beta_3\left(\tfrac{\bar{\Delta}}{\pi}\right)^4\int_\Omega z_{x_1x_2}^2(x,t_k)dx - \eta\sum_{j=1}^N \int_{\Omega_j} f_j^2(x,t_k)dx\right] \\
&\leq 2p_1(1-\kappa)\int_\Omega z_{x_1}^2 dx - 2p_1\kappa\int_\Omega z_{x_2}^2 dx - 2p_1\int_\Omega |\Delta z|^2 dx \\
&+2p_1\mu\sum_{j=1}^N\int_{\Omega_j} z\left[(t-t_k)\rho + f_j(x,t_k)\right]dx - 2p_1\mu\int_\Omega z^2 dx \\
&+2p_2\int_\Omega \Delta^2 z[-zz_{x_1} - (1-\kappa)z_{x_1x_1} + \kappa z_{x_2x_2} - \Delta^2 z]dx \\
&-2p_2\mu\int_\Omega \Delta^2 z[z - (t-t_k)\rho - f_j(x,t_k)d\xi]dx \\
&-re^{2\delta h}(t-t_k)\int_\Omega \rho^2(x,t)dx + r(t_{k+1}-t)\int_\Omega z_t^2(x,t)dx \\
&+2\delta p_1\int_\Omega z^2 dx + 2\delta p_2\int_\Omega|\Delta z|^2 dx - \delta_1 p_1\int_\Omega z^2(x,t_k)dx \\
&-\delta_1 p_2\int_\Omega|\Delta z(x,t_k)|^2 dx + (\lambda_1-\lambda_2)\int_\Omega[z_{x_1}^2 + z_{x_2}^2]dx \\
&-\lambda_1\tfrac{\pi^2}{2}\int_\Omega z^2 dx - \lambda_2\int_\Omega z\Delta z dx + \beta_1\left(\tfrac{\bar{\Delta}}{\pi}\right)^2\int_\Omega z_{x_1}^2(x,t_k)dx \\
&+\beta_2\left(\tfrac{\bar{\Delta}}{\pi}\right)^2\int_\Omega z_{x_2}^2(x,t_k)dx + \beta_3\left(\tfrac{\bar{\Delta}}{\pi}\right)^4\int_\Omega z_{x_1x_2}^2(x,t_k)dx - \eta\sum_{j=1}^N\int_{\Omega_j} f_j^2(x,t_k)dx.
\end{aligned} \tag{3.140}$$

Using (3.89), we have

$$-2p_1\int_\Omega|\Delta z|^2 dx = -2p_1\int_\Omega[z_{x_1x_1}^2(x,t) + z_{x_2x_2}^2(x,t) + 2z_{x_1x_2}^2(x,t)]dx, \tag{3.141}$$

$$-\delta_1 p_2\int_\Omega|\Delta z(x,t_k)|^2 dx = -\delta_1 p_2\int_\Omega[z_{x_1x_1}^2(x,t_k) + z_{x_2x_2}^2(x,t_k) + 2z_{x_1x_2}^2(x,t_k)]dx. \tag{3.142}$$

Set

$$\begin{aligned}
\eta_0 &= \{z_{x_1}(x,t), z_{x_2}(x,t), z_{x_1x_1}(x,t), z_{x_2x_2}(x,t), \Delta^2 z(x,t), \\
&\quad z(x,t), f_j(x,t_k)\}, \\
\eta_1 &= \{z_{x_1}(x,t), z_{x_2}(x,t), z_{x_1x_1}(x,t), z_{x_2x_2}(x,t), \Delta^2 z(x,t), \\
&\quad z(x,t), f_j(x,t_k), \rho(x,t)\}, \\
\bar{\eta} &= \{z(x,t_k), z_{x_1x_1}(x,t_k), z_{x_2x_2}(x,t_k)\}
\end{aligned}$$

Substituting (3.141), (3.142) into (3.140), we obtain

$$\begin{aligned}
&\dot{V}_1(t) + 2\delta V_1(t) - \delta_1\sup_{\theta\in[-h,0]} V_1(t+\theta) \\
&\leq \sum_{j=1}^N\int_{\Omega_j}\frac{h-t+t_k}{h}\eta_0^T\Theta_0\eta_0 + \frac{t-t_k}{h}\eta_1^T\Theta_1\eta_1 dx + \int_\Omega \bar{\eta}^T\bar{\Theta}\bar{\eta}dx \\
&+r(t_{k+1}-t)\int_\Omega z_t^2(x,t)dx - (4p_1 - 4\delta p_2)\int_\Omega z_{x_1x_2}^2(x,t)dx \\
&-[2\delta_1 p_2 - \beta_3(\tfrac{\bar{\Delta}}{\pi})^4]\int_\Omega z_{x_1x_2}^2(x,t_k)dx.
\end{aligned} \tag{3.143}$$

Substitution of z_t from (3.126) yields

$$\begin{aligned} & r(t_{k+1}-t)\int_\Omega z_t^2(x,t)dx \\ & \leq rh\sum_{j=1}^{N}\int_{\Omega_j}[-zz_{x_1}-(1-\kappa)z_{x_1x_1}+\kappa z_{x_2x_2}-\Delta^2 z \\ & -\mu z+\mu(t-t_k)\rho+\mu f_j(x,t_k)]^2dx. \end{aligned} \tag{3.144}$$

Set $\bar{\psi}=\text{col}\{z_{x_1}(x,t), z_{x_1x_1}(x,t), z_{x_2x_2}(x,t), \Delta^2 z(x,t), z(x,t), f_j(x,t_k), \rho(x,t)\}$ and

$$\beta \triangleq \begin{bmatrix} -z & -(1-\kappa) & \kappa & -1 & -\mu & \mu & \mu(t-t_k) \end{bmatrix}.$$

Then

$$\begin{aligned} & [-zz_{x_1}-(1-\kappa)z_{x_1x_1}+\kappa z_{x_2x_2}-\Delta^2 z-\mu z+\mu f_j(x,t_k)+\mu(t-t_k)\rho]^2 \\ & =\bar{\psi}^T\beta^T\beta\bar{\psi}. \end{aligned} \tag{3.145}$$

Application of Schur complement theorem to (3.145), together with (3.143) and (3.144), implies

$$\dot{V}_1(t)+2\delta V_1(t)-\delta_1\sup_{\theta\in[-h,0]}V_1(t+\theta)\leq 0$$

if (3.127)–(3.129) are satisfied, and $\Lambda_1<0$, $\Lambda_2<0$ hold for all $z\in(-C,C)$. Similar to Theorem 3.3, LMI (3.130) imply $\Lambda_1<0$, $\Lambda_2<0$ for all $z\in(-C,C)$. Thus the result is established via Halanay's inequality.

Step 3: The feasibility of LMIs (3.127)–(3.130) yields that the solution of (3.126) admits a priori estimate $V_1(t)\leq e^{-2\sigma t}\sup_{-h\leq\theta\leq 0}V_1(\theta)$. By Theorem 6.23.5 of [10], continuation of this solution under a priori bound to entire interval $[0,\infty)$. □

3.2.4 Numerical Example

Consider the system (3.71) under the sampled-data control law (3.75) with the averaged measurements (3.73). Here we choose $\mu=0.95$. By verifying LMI conditions of Theorem 3.3 with $\delta=0.1$, $\kappa=-0.5$, $\bar{\Delta}=1/4$, $C=2$, we find that the closed-loop system (3.104) preserves the exponential stability within a given domain of initial conditions $\|z_0\|_V^2<4$ for $t_{k+1}-t_k\leq h\leq 0.39$. Note that $\bar{\Delta}=1/4$ corresponds to $N=16$ square subdomains with the sides length 1/4. The feasible solutions of LMIs with $h=0.35$ are given as follows: $p_1=80.6354$, $p_2=5.145$.

We compute the solution of the closed-loop system (3.104) numerically via finite element method. Let $\xi=0.1$ and $M=1/\xi=10$. Define $(x_1^i,x_2^j)=(i\xi,j\xi), i,j=0,1,2,\ldots,M$. We divide Ω on $M^2=100$ squares R_{ij} defined by

$$R_{ij} \triangleq \{(x_1, x_2) \in \Omega | x_1^i \le x_1 \le x_1^{i+1}, x_2^j \le x_2 \le x_2^{j+1}\}.$$

On the node (x_1^i, x_2^j), four finite element basis functions are selected as

$$N_1^{i,j}(x) = \begin{cases} \left(1 - \dfrac{x_1 - x_1^i}{\xi}\right)\left(\dfrac{x_2 - x_2^j}{\xi}\right), & x \in R_{ij}, \\ 0, & otherwise \end{cases}$$
$$N_2^{i,j}(x) = \begin{cases} \left(1 - \dfrac{x_1 - x_1^i}{\xi}\right)\left(1 - \dfrac{x_2 - x_2^j}{\xi}\right), & x \in R_{ij}, \\ 0, & otherwise \end{cases}$$
$$N_3^{i,j}(x) = \begin{cases} \left(\dfrac{x_1 - x_1^i}{\xi}\right)\left(1 - \dfrac{x_2 - x_2^j}{\xi}\right), & x \in R_{ij}, \\ 0, & otherwise \end{cases}$$
$$N_4^{i,j}(x) = \begin{cases} \left(\dfrac{x_1 - x_1^i}{\xi}\right)\left(\dfrac{x_2 - x_2^j}{\xi}\right), & x \in R_{ij}, \\ 0, & otherwise \end{cases}$$

We consider the Galerkin approximation solution of the closed-loop system in finite dimensional space generated by these basis functions, which takes the form

$$z^M(x, t) = \sum_{k=1}^{4} \sum_{i,j=1}^{M} m_k^{i,j}(t) N_k^{i,j}(x),$$

where $m_k^{i,j}(t)$ are determined by standard finite element Galerkin method to satisfy some ODEs. Figure 3.6 shows snapshots of the state $z(x, t) = z(x_1, x_2, t)$ at different times for the closed-loop system (3.104) with $t_{k+1} - t_k = 0.35$, $N = 16$ and initial condition $z(x_1, x_2, 0) = 0.236 \sin(\pi x_1) \sin(\pi x_2)$, $(x_1, x_2) \in \Omega = (0, 1)^2$. It is seen that the closed-loop system is stable. Figure 3.7 demonstrates the time evolution of $V_1(t)$ via the finite difference method, where the steps of space and time are taken as 1/4 and 0.00025, respectively.

By verifying the LMI conditions of Theorem 3.3, we obtain the maximum value $h = 0.39$ that preserves the exponential stability. By simulation of the solution to the closed-loop system starting from the same initial condition, we find that stability is preserved for essentially larger values of h till approximately $h = 2.45$.

For the sampled-data controller (3.75) under the point measurement (3.74), by choosing $\mu = 0.95$, and using Yalmip we verify LMI conditions of Theorem 3.4 with $\delta = 0.2, \delta_1 = 0.15, \kappa = -0.5, \bar{\Delta} = 1/4, C = 2$. Then we find that the resulting closed-loop system is exponentially stable for $t_{k+1} - t_k \le h \le 0.37$ for any initial values satisfying $\|z_0\|_{L^2(0,1)} < 1$.

Since point measurements use less information on the state, the point measurements allow smaller sampling intervals than averaged measurements. Simulations of the solutions to the closed-loop system under the point measurements confirm the theoretical results.

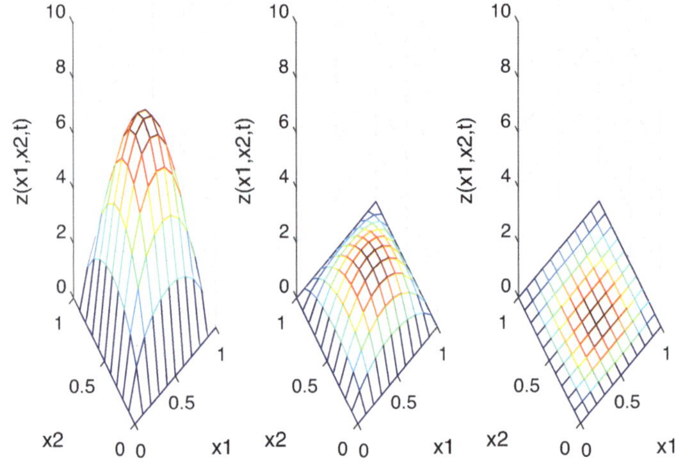

Fig. 3.6 Snapshots of state $z(x_1, x_2, t)$ at different time $t \in \{0, 1.4, 14\}$

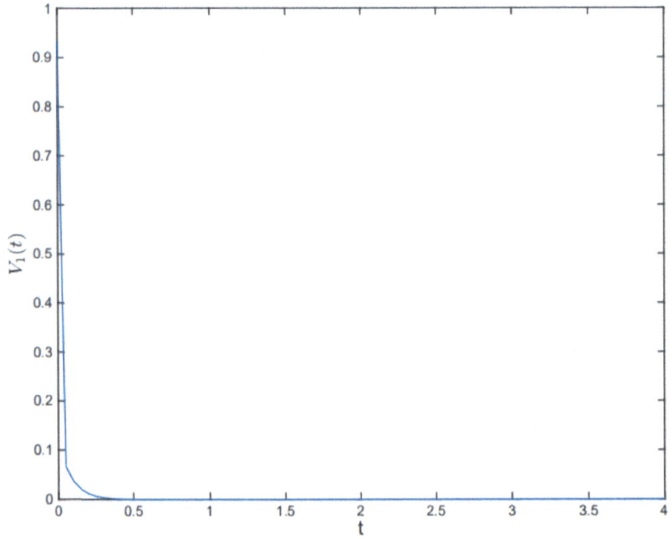

Fig. 3.7 Lyapunov function $V_1(t)$

3.3 Stabilization by Switching of Parabolic PDEs with Spatially Scheduled Actuators and Sensors

This section addresses a switched sampled-data control design for stabilization of KSE under the Dirichlet/periodic boundary conditions with spatially scheduled actuators. It is supposed that discrete-time point-like or averaged measurements are available. The system is known to be stabilizable by static output-feedback employing several distributed in space actuators and sensors, but is not stabilizable by only one of the actuator-sensor pairs. Does there exist a switching stabilizing static output-feedback such that at all times, only one actuator-sensor pair is active? We give a positive answer and find the appropriate switching sampled-data control law. The proposed switching controller can be implemented either by N actuators and sensors placed in each subdomain (here switching control may reduce the energy that the system spends) or by using one actuator-sensor pair that can move to the active subdomain. For implementation of the control law by moving actuators and sensors, we take into account a moving time by treating it as an additional switching between the open-loop system and the closed-loop switched system. The guidance of active (or mobile) actuators and sensors is provided by using output-dependent switching. Constructive conditions are derived to ensure that the resulting closed-loop system is regionally stable by means of the Lyapunov approach. Numerical example illustrates the efficiency of the method.

3.3.1 Problem Formulation

Let $0 = t_0 < t_1 < \cdots < t_k < t_{k+1} < \ldots,\ \lim_{k\to\infty} t_k = \infty$ be a sequence of sampling instants. Consider the following plant governed by KSE:

$$\begin{cases} z_t(x,t) + z_{xx}(x,t) + \nu z_{xxxx}(x,t) + z(x,t)z_x(x,t) \\ = b_{\sigma_k}(x)u_{\sigma_k}(t),\ x \in (0, L),\ t \in [t_k, t_{k+1}), \\ z(x,0) = z_0(x), \end{cases} \tag{3.146}$$

where $k \in \mathbb{Z}_+$, under the Dirichlet boundary conditions:

$$z(0,t) = z(L,t) = 0,\ z_x(0,t) = z_x(L,t) = 0,\ t > 0. \tag{3.147}$$

Here $z(x,t)$ is the state of KSE, $z_0(x)$ is the initial state, and $u_{\sigma_k}(t)$ is the control input. The switching function $\sigma_k : k \in \mathbb{Z}_+ \to \{1, ..., N\}$ selects at each sampling time t_k one of the N available actuators corresponding to the shape function $b_{\sigma_k}(x)$ that will be shortly defined. We make the following assumptions:

- Spatial sampling: Similar to [1, 2, 7, 25], we assume that the points $0 = x_0 < x_1 < \cdots < x_N = L$ divide $[0, L]$ into N equal-length subintervals $\Omega_j = [x_{j-1}, x_j)$ such that $\cup_j \Omega_j = [0, L]$ and $|\Omega_j| = \frac{L}{N}$ meaning that all the subintervals have the same length. The shape functions $b_j(x)$ are chosen to be characteristic functions $b_j(x)$ of Ω_j as follows:

$$\begin{cases} b_j(x) = 0, \ x \notin \Omega_j, \\ b_j(x) = 1, \ otherwise, \end{cases} \quad j = 1, \ldots, N. \tag{3.148}$$

- Time sampling: The length of sampling subintervals in time is supposed to be uniformly bounded:

$$0 < h_0 \leq t_{k+1} - t_k \leq h, \quad \forall k \in \mathbb{Z}_+. \tag{3.149}$$

- Moving time: The moving time $\delta \in (0, h_0)$ for sensors and actuators to the appropriate domain Ω_{σ_k} is taken into account.

We first consider sensors that provide the discrete-time point-like measurements:

$$y_j(t_k) = \int_{\Omega_j} c_j(x) z(x, t_k) dx, \quad k \in \mathbb{Z}_+ \tag{3.150}$$

with

$$0 \leq c_j \in L^2(\Omega_j), \ \int_{\Omega_j} c_j(x) dx = 1, \tag{3.151}$$

$$c_j(x) = \begin{cases} \frac{1}{\varepsilon}, & \text{if } x \in \bar{\Omega}_j, \\ 0, & otherwise, \end{cases} \quad j = 1, \ldots, N, \tag{3.152}$$

where $\bar{\Omega}_j$ is subinterval of Ω_j with the length ε independent of j (see Fig. 3.8, where $\bar{x}_j$ is the midpoint of Ω_j).

We will also consider the averaged state measurements

$$y_j(t_k) = \frac{\int_{\Omega_j} z(x, t_k) dx}{|\Omega_j|} = \frac{N}{L} \int_{\Omega_j} z(x, t_k) dx, \quad j = 1, \ldots, N, k \in \mathbb{Z}_+. \tag{3.153}$$

Differently from point-like measurements, the sensors in the case of averaged measurements cover the whole subdomain. However, our method under the averaged measurements leads to a smaller number of actuators and sensors or allow larger sampling in time (see comparison in Table 3.2). Note that the presented method under the averaged measurements can be extended to N-D PDEs for any N (on the

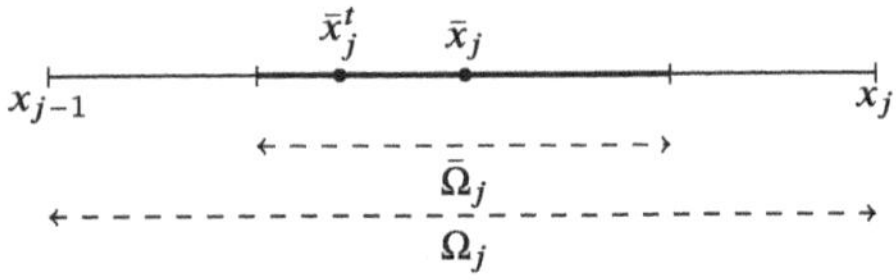

Fig. 3.8 Subdomains $\bar{\Omega}_j$ of point-like measurements

Table 3.2 Maximal values of h, δ under the Dirichlet boundary conditions

	Averaged measurements		Point-like measurements		
N	h	δ	ε	h	δ
10	0.020	0.003	$\pi/90$	0.0033	0.0005
			$\pi/30$	0.0038	0.0006
13	0.027	0.004	$\pi/90$	0.0045	0.0007
			$\pi/30$	0.0054	0.0009
15	0.030	0.005	$\pi/90$	0.0050	0.0008
			$\pi/30$	0.0060	0.0010

basis of static output-feedback without switching presented in [19]), whereas such extension under the point-like measurements is questionable (see [26], where non-switched static output-feedback for heat equation under point-like measurements is confined to $N \leq 2$).

For both measurements, our aim is to find a sampled-data switching law and a sampled-data regionally exponentially stabilizing controller for KSE (3.146) implemented by zero-order hold device. As already mentioned, in this paper we will take into account the moving time δ for sensors and actuators. For the actuators moving time, we consider additional switching between the open-loop system (when the actuator is moving) during the part of the sampling interval and the closed-loop switched system during the remaining part of the interval, where

$$u_{\sigma_k}(t) = \begin{cases} 0, & t \in [t_k, t_k + \delta), \\ -K y_{\sigma_k}(t_k), & t \in [t_k + \delta, t_{k+1}) \end{cases} \tag{3.154}$$

with some $K > 0$. The switching signal σ_k is calculated at time t_k, whereas it takes δ seconds for actuators and sensors to move to the domain Ω_{σ_k}.

Our main objective is to find an appropriate output-depending switching law. Denote the characteristic function of the time interval $[t_k, t_k + \delta]$ by $\chi_{[t_k, t_k+\delta]}(t)$. Consider first the case of the averaged state measurements (3.153), where the closed-loop system (3.146), (3.154) has the form

$$\begin{aligned} &z_t(x,t) + z_{xx}(x,t) + \nu z_{xxxx}(x,t) + z(x,t)z_x(x,t) \\ &= -\tfrac{KN}{L}(1 - \chi_{[t_k,t_k+\delta]}(t))b_{\sigma_k}(x)\int_{\Omega_{\sigma_k}} z(x,t_k)dx, \\ &\qquad x \in (0, L),\ t \in [t_k, t_{k+1}) \end{aligned} \tag{3.155}$$

subject to (3.147). Note that if $b_{\sigma_k}(x)u_{\sigma_k}(t)$ in (3.146) is changed by $\sum\limits_{j=1}^{N} b_j(x)u_j(t)$, then there exists $K > 0$ that regionally exponentially stabilizes the system by $u_j(t) = -K y_j(t)$ [3]. The latter means that the average of systems (3.146) with $b_{\sigma_k}(x)u_{\sigma_k}(t)$ changed by $b_j(x)u_j(t)$ is stabilizable by the static output-feedback (3.154). Similar to state-dependent switching for ODEs in the case of stable convex combination of systems [30], we will define a min-type switching function by using the corresponding Lyapunov function $V(t)$ according to

$$\sigma_k \approx \arg\min \dot{V}(t)$$

for $t \in [t_k + \delta, t_{k+1})$ along the closed-loop system. Thus, for $V(t) = \int_0^L z^2(x,t)dx$ we have

$$\begin{aligned}\dot{V}(t) &= -\tfrac{KN}{L} \int_{\Omega_j} z(x,t)dx \int_{\Omega_j} z(x,t_k)dx \\ &\quad - \int_0^L z(x,t)\left[z_{xx}(x,t) + \nu z_{xxxx}(x,t) + z(x,t)z_x(x,t)\right] dx\end{aligned}$$

that leads (for small enough h) to

$$\begin{aligned}\arg\min \dot{V}(t) &= \arg\min_j \left[- \int_{\Omega_j} z(x,t)dx \int_{\Omega_j} z(x,t_k)dx\right] \\ &\approx \arg\max_j \left[\int_{\Omega_j} z(x,t_k)dx\right]^2\end{aligned}$$

i.e. to the following discrete-time switching law:

$$\sigma_k = \arg\max_j \left[\int_{\Omega_j} z(x,t_k)dx\right]^2. \tag{3.156}$$

Similarly to (3.156) for the point-like measurements we choose

$$\sigma_k = \arg\max_j \left[\int_{\Omega_j} c_j(x)z(x,t_k)dx\right]^2. \tag{3.157}$$

Our sampled-data switching law (3.157) with (3.149) and $\lim_{k=\infty} t_k = \infty$ rules out the possibility of Zeno behaviour. Note that (3.157) is calculated at time t_k. The law (3.157) means that the σ_k-th mode is active if

$$\left[\int_{\Omega_j} c_j(x)z(x,t_k)dx\right]^2 \le \left[\int_{\Omega_{\sigma_k}} c_{\sigma_k}(x)z(x,t_k)dx\right]^2, \quad \forall j = 1, \ldots, N. \tag{3.158}$$

3.3.2 Main Results

In this section, we will analyze the well-posedness and regional exponential stability of the system (3.146) under the static output-feedback (3.154) and the switching law (3.154) (where $c_j = 1$ in the case of averaged measurements).

3.3.2.1 Well-Posedness of the Controlled System

We establish the existence, uniqueness and regularity of the system (3.146) under the switching control law (3.154), (3.157) and Dirichlet boundary conditions (3.147) by using the step method (see e.g. Sect. 1.2 in [11]). For the switching law (3.154), (3.157), we assume that the σ_k-th mode is active. We first consider $t \in [0, \delta]$. Then (3.146), (3.147) becomes

$$\begin{cases} z_t(x,t) + z_{xx}(x,t) + \nu z_{xxxx}(x,t) + z(x,t)z_x(x,t) = 0, \\ \qquad\qquad\qquad\qquad x \in (0, L),\ t \in [0, \delta], \\ z(0,t) = z(L,t) = 0,\ z_x(0,t) = z_x(L,t) = 0, \\ z(x,0) = z_0(x). \end{cases} \tag{3.159}$$

Define the system operator $A : D(A) \subset L^2(0, L) \to L^2(0, L)$ as follows:

$$\begin{cases} Af = -\nu \left[\frac{\partial^4 f}{\partial x^4}\right], \\ D(A) = H^4(0, L) \cap H_0^2(0, L). \end{cases}$$

It is well-known that A is a dissipative operator, and A generates an analytic semigroup. Operator $-A$ is positive implying that its square root $(-A)^{\frac{1}{2}}$ is also positive. Moreover, $D((-A)^{\frac{1}{2}}) = H_0^2(0, L)$ with the norm $\|f\|_{D((-A)^{\frac{1}{2}})} = \nu^{\frac{1}{2}} \|f''\|_{L^2(0,L)}$. Then the system (3.159) can be represented as an evolution equation:

$$\begin{cases} \frac{d}{dt} z(\cdot, t) = Az(\cdot, t) + F(z(\cdot, t)), \\ z(\cdot, 0) = z_0(\cdot), \end{cases} \tag{3.160}$$

where the nonlinear term F is defined on function $z(\cdot, t)$ according to

$$F(z(\cdot, t)) = -z(x,t)z_x(x,t) - z_{xx}(x,t),\ t \in [0, \delta].$$

It should be noticed that the nonlinear term F is locally Lipschitz continuous, that is, there exists a positive constant $l(M)$ such that

$$\|F(z_1) - F(z_2)\|_{L^2(0,L)} \leq l(M)\|z_1 - z_2\|_{H_0^2(0,L)}$$

holds for $z_1, z_2 \in H_0^2(0, L)$ with $\|z_1\|_{H_0^2(0,L)} \leq M$, $\|z_2\|_{H_0^2(0,L)} \leq M$. Therefore, Theorem 3.3.3 of [13] is applicable to (3.160). For any initial condition $z_0 \in H_0^2(0, L)$, there exists a unique local strong solution of (3.160) on some interval $[0, T] \subset [0, \delta]$, where $T = T(z_0) > 0$:

$$\begin{aligned} &z \in C([0, T]; H_0^2(0, L)) \cap L^2([0, T]; D(A)), \\ &\dot{z} \in L^2([0, T]; L^2(0, L)). \end{aligned}$$

From Theorem 6.23.5 of [10], we obtain that if the solution admits a priori estimate (i.e. bounded), then it exists on the entire interval $[0, \delta]$. The a priori estimate on the solutions will be guaranteed by the conditions that we will provide (see Theorem 3.5).

For $t \in [\delta, t_1]$, the system (3.146) under the switching control law (3.154), (3.157) can be also written in the form of (3.160) with the following nonlinearity

$$\begin{array}{l}F(z(\cdot,t)) = -z(x,t)z_x(x,t) - z_{xx}(x,t)\\ -Kb_{\sigma_k}(x)\int_{\Omega_{\sigma_k}} c_{\sigma_k}(x)z_0(x)dx, \quad t \in [\delta, t_1].\end{array}$$

Since F is locally Lipschitz continuous, the same line of reasoning is applied to the time interval $[\delta, t_1]$. The strong solution exists on $[\delta, t_1]$ due to a priori estimate on the solutions starting from the domain of attraction, which is guaranteed by the stability conditions of Theorem 3.5.

3.3.2.2 Stability Analysis of the Switched System

By the mean-value theorem, from (3.151) it follows that there exists $\bar{x}_j^t \in \mathrm{conv}(\mathrm{supp}c_j)$ (see Fig. 3.8) such that

$$\int_{\Omega_j} c_j(x)z(x,t)dx = z(\bar{x}_j^t, t), \quad t \in [t_k, t_{k+1}).$$

Denote

$$f_j(x,t) \triangleq z(x,t) - z(\bar{x}_j^t, t), \quad t \in [t_k, t_{k+1}), \tag{3.161}$$

$$\rho_j(t) \triangleq \int_{\Omega_j} \int_{t_k}^t c_j(x)z_s(x,s)dsdx, \quad t \in [t_k, t_{k+1}). \tag{3.162}$$

Then the switching controller (3.154) can be rewritten as

$$u_{\sigma_k}(t) = \begin{cases} 0, & t \in [t_k, t_k+\delta), \\ -K[z(x,t) - f_{\sigma_k}(x,t) - \rho_{\sigma_k}(t)], & t \in [t_k+\delta, t_{k+1}) \end{cases} \tag{3.163}$$

whereas the switching law chooses σ_k that satisfies

$$\begin{array}{l}\int_{\Omega_j} \left[z(x,t) - f_j(x,t) - \rho_j(t)\right]^2 dx \\ \le \int_{\Omega_{\sigma_k}} \left[z(x,t) - f_{\sigma_k}(x,t) - \rho_{\sigma_k}(t)\right]^2 dx, \quad j = 1, 2, \ldots, N.\end{array} \tag{3.164}$$

Thus, under the controller (3.163), the closed-loop system becomes

$$\begin{array}{l}z_t(x,t) + z_{xx}(x,t) + \nu z_{xxxx}(x,t) + z(x,t)z_x(x,t)\\ = -Kb_{\sigma_k}(x)(1 - \chi_{[t_k,t_k+\delta]}(t))[z(x,t) - f_{\sigma_k}(x,t) - \rho_{\sigma_k}(t)],\\ \qquad\qquad x \in (0,L), \quad t \in [t_k, t_{k+1}),\end{array} \tag{3.165}$$

subject to (3.147), (3.157).

Note that (3.146) may be not stabilizable with a desired decay rate by the non-switched control. The challenge in the stability analysis is to take efficiently into account the switching condition (3.157) in order to derive feasible stability conditions (see (3.182) below and the resulting expression in (3.183)).

Now we focus on the stability of the closed-loop system that switches at times t_k and $t_k + \delta$. Consider the following Lyapunov-Krasovskii functional:

$$V(t) = V_{P_1}(t) + V_{P_2}(t) + V_R(t), \ t \in [t_k, t_{k+1}) \tag{3.166}$$

where

$$\begin{aligned} V_{P_1}(t) &= P_1 \int_0^L z^2(x,t)dx, \\ V_{P_2}(t) &= P_2 \nu \int_0^L z_{xx}^2(x,t)dx, \\ V_R(t) &= R\frac{4h^2}{\pi^2} \sum_{j=1}^{N} \int_{\Omega_j} \int_{t_k}^t e^{-2\alpha(t-s)}[\rho_{js}(s)]^2 ds dx \\ &\quad -Re^{-2\alpha h} \sum_{j=1}^{N} \int_{\Omega_j} \int_{t_k}^t e^{-2\alpha(t-s)}[\rho_j(s)]^2 ds dx \end{aligned}$$

with $P_1 > 0$, $P_2 > 0$, $R > 0$. Here $\rho_{js}(s)$ is the derivative of $\rho_j(s)$ with respect to s. By Wirtinger's inequality, $V_R(t)$ is non-negative (see Lemma 1 in [31]), it does not grow in the switching times t_k and it is continuous in the switching times $t_k + \delta$. Moreover, V_R extends the corresponding terms in [26] to Wirtinger-based Lyapunov functional.

For $z(\cdot, t) \in H_0^2(0, L)$ we define

$$\|z(\cdot,t)\|_V^2 = P_1\|z(\cdot,t)\|_{L^2(0,L)}^2 + P_2\nu\|z_{xx}(\cdot,t)\|_{L^2(0,L)}^2$$

with $P_1 > 0$, $P_2 > 0$.

Remark 3.7 In order to find a bound on the domain of attraction for closed-loop system (3.165) subject to (3.147), we use positive invariance principle in Theorem 3.5 below: we show that if $\Psi_0 < 0$, $\Psi_1 < 0$ and $\Psi_2 < 0$, where Ψ_0, Ψ_1, Ψ_2 are given by (3.170)–(3.172), then $V(t) \le V(0)$ for all $t \ge 0$. Matrices Ψ_0, Ψ_1, Ψ_2 are affine in z_x. Let $C > 0$ be the upper bound of z_x, i.e., $\max_{x\in[0,L]} |z_x(x,t)| \le C$ for all $t \ge 0$. Then it is sufficient to verify the matrix inequalities $\Psi_0 < 0$, $\Psi_1 < 0$ and $\Psi_2 < 0$ in the vertices $z_x = \pm C$ (see (3.167)–(3.169)).

The following result provides sufficient stability conditions in the form of linear matrix inequalities (LMIs) for the closed-loop system (3.165), (3.147), (3.157).

Theorem 3.5 *Consider the closed-loop system* (3.165) *subject to* (3.147) *and the switching law* (3.157). *Given positive scalars h, α, K and tuning parameter $C > 0$, $\alpha_0 > 0$ such that $\alpha h_0 > (\alpha_0 + \alpha)\delta$, let there exist scalars $R > 0$, $P_n > 0$, $\lambda_n \ge 0$* $(n = 1, 2, 3)$ *that satisfy the inequalities:*

$$\Psi_1|_{z_x=\pm C} < 0, \tag{3.167}$$

$$\Psi_2|_{z_x=\pm C} < 0, \tag{3.168}$$

$$\Psi_0|_{z_x=\pm C} < 0, \tag{3.169}$$

where

$$\Psi_1 = \begin{bmatrix} \psi_{11} & \psi_{12} & \psi_{13} & \frac{\lambda_1}{N-1} & \frac{\lambda_1}{N-1} \\ * & \psi_{22} & -P_2 & 0 & 0 \\ * & * & \psi_{33} & 0 & 0 \\ * & * & * & -\lambda_2 - \frac{\lambda_1}{N-1} & -\frac{\lambda_1}{N-1} \\ * & * & * & * & -Re^{-2\alpha h} - \frac{\lambda_1}{N-1} \end{bmatrix}, \tag{3.170}$$

$$\Psi_2 = \begin{bmatrix} \tilde{\psi}_{11} & \tilde{\psi}_{12} & \psi_{13} & P_3K - \lambda_1 & P_3K - \lambda_1 \\ * & \psi_{22} & -P_2 & P_2K & P_2K \\ * & * & \psi_{33} & 0 & 0 \\ * & * & * & \lambda_1 - \lambda_2 & \lambda_1 \\ * & * & * & * & \lambda_1 - Re^{-2\alpha h} \end{bmatrix}, \tag{3.171}$$

$$\Psi_0 = \begin{bmatrix} -2\alpha_0 P_1 & P_1 - P_4 - P_2 z_x & -P_4 \\ * & -2P_2 + R\frac{4h^2}{\pi^2}\frac{L}{N\varepsilon} & -P_2 \\ * & * & -2\alpha_0 P_2\nu - 2P_4\nu \end{bmatrix}, \tag{3.172}$$

$$\begin{aligned} \psi_{11} &= 2\alpha P_1 - \lambda_3\tfrac{\pi^4}{L^4} - \tfrac{\lambda_1}{N-1}, \\ \psi_{12} &= P_1 - P_3 - P_2 z_x, \\ \psi_{13} &= -P_3 - \frac{\lambda_2(\frac{L}{N}+\varepsilon)^2}{2\pi^2}, \\ \psi_{22} &= R\tfrac{4h^2}{\pi^2}\tfrac{L}{N\varepsilon} - 2P_2, \\ \psi_{33} &= 2\alpha P_2\nu - 2P_3\nu + \lambda_3, \\ \tilde{\psi}_{11} &= 2\alpha P_1 - \lambda_3\tfrac{\pi^4}{L^4} - 2P_3K + \lambda_1, \\ \tilde{\psi}_{12} &= P_1 - P_3 - P_2 z_x - P_2 K. \end{aligned}$$

Let α_1 *be subject to*

$$0 < \alpha_1 h_0 \leq \alpha h_0 - (\alpha_0 + \alpha)\delta. \tag{3.173}$$

Then for any initial function $z_0 \in H_0^2(0, L)$ *that satisfies the bound* $\|z_0\|_V < \sqrt{\frac{P_2\nu}{L}}C$, *the closed-loop system* (3.165) *subject to* (3.147) *and* (3.157) *is exponentially stable with a decay rate* α_1, *i.e. the following holds*

$$\|z(\cdot, t)\|_V^2 \leq V(t) \leq e^{-2\alpha_1(t-h)+2\alpha_0\delta} V(0).$$

Proof **Step 1**: Let us only highlight that there exists a unique local strong solution of (3.159) on some interval $[0, T] \subset [0, \delta]$, where $T = T(z_0)$. Due to Theorem 6.23.5 of [10], the solution exists on the entire interval $[0, \delta]$ provided it is bounded. Therefore, by applying the same arguments at $[\delta, t_1]$ and any step $k \in \mathbb{N}$, one can conclude that the strong solution exists for all $t \geq 0$.

Step 2: Assume formally that strong solution of (3.165) subject to (3.147) starting from $\|z_0\|_V < \sqrt{\frac{P_2\nu}{L}C}$ exists for all $t \geq 0$. We first derive sufficient LMI-based conditions to guarantee that $\dot{V} + 2\alpha V \leq 0$ for $[t_k + \delta, t_{k+1})$. Differentiating $V(t)$ along the solution of the closed-loop system and integrating by parts, we obtain

$$\begin{array}{l}\dot{V}(t) + 2\alpha V(t) = 2P_1\int_0^L z(x,t)z_t(x,t)dx + 2\alpha P_1\int_0^L z^2(x,t)dx \\ +2\alpha P_2\nu\int_0^L z_{xx}^2(x,t)dx + 2P_2\nu\int_0^L z_{xx}(x,t)z_{xxt}(x,t)dx \\ +R\frac{4h^2}{\pi^2}\sum_{j=1}^{N}\int_{\Omega_j}[\rho_{jt}(t)]^2dx - Re^{-2\alpha h}\sum_{j=1}^{N}\int_{\Omega_j}[\rho_j(t)]^2dx.\end{array} \tag{3.174}$$

Jensen's inequality leads to

$$\begin{array}{l}\int_{\Omega_j}[\rho_{jt}(t)]^2dx = \frac{L}{N}\left(\int_{\Omega_j}c_j(x)z_t(x,t)dx\right)^2 \\ \leq \frac{L}{N}\int_{\Omega_j}c_j(x)dx\int_{\Omega_j}c_j(x)z_t^2(x,t)dx \leq \frac{L}{N\varepsilon}\int_{\Omega_j}z_t^2(x,t)dx.\end{array} \tag{3.175}$$

Note that $f_j(x,t) \triangleq z(x,t) - z(\bar{x}_j^t, t)$ and $f_{jx}(x,t) = z_x(x,t)$. Then, application of Wirtinger's inequality yields

$$\begin{array}{l}\int_{\Omega_j}f_j^2(x,t)dx = \int_{x_{j-1}}^{\bar{x}_j^t}[z(x,t) - z(\bar{x}_j^t,t)]^2dx + \int_{\bar{x}_j^t}^{x_j}[z(x,t) - z(\bar{x}_j^t,t)]^2dx \\ \leq \frac{(\frac{L}{N}+\varepsilon)^2}{\pi^2}\int_{\Omega_j}z_x^2(x,t)dx.\end{array} \tag{3.176}$$

Moreover, we obtain

$$\|z(\cdot,t)\|_{L^2(0,L)}^2 \leq \left(\frac{L^2}{\pi^2}\right)^2\|z_{xx}(\cdot,t)\|_{L^2(0,L)}^2. \tag{3.177}$$

Therefore, (3.164), (3.176) and (3.177) lead to

$$\begin{array}{l}-\frac{\lambda_1}{N-1}\sum_{j\neq\sigma_k}^{N}\int_{\Omega_j}[z(x,t) - f_j(x,t) - \rho_j(t)]^2dx \\ +\lambda_1\int_{\Omega_{\sigma_k}}[z(x,t) - f_{\sigma_k}(x,t) - \rho_{\sigma_k}(t)]^2dx \geq 0,\end{array} \tag{3.178}$$

$$\begin{array}{l}\lambda_2\left[\frac{(\frac{L}{N}+\varepsilon)^2}{\pi^2}\|z_x(\cdot,t)\|_{L^2(0,L)}^2 - \sum_{j=1}^{N}\|f_j(\cdot,t)\|_{L^2(\Omega_j)}^2\right] \\ = \lambda_2\left[\frac{(\frac{L}{N}+\varepsilon)^2}{\pi^2}\sum_{j\neq\sigma_k}^{N}\int_{\Omega_j}z_x^2(x,t)dx - \sum_{j\neq\sigma_k}^{N}\int_{\Omega_j}f_j^2(x,t)dx\right] \\ +\lambda_2\left[\frac{(\frac{L}{N}+\varepsilon)^2}{\pi^2}\int_{\Omega_{\sigma_k}}z_x^2(x,t)dx - \int_{\Omega_{\sigma_k}}f_{\sigma_k}^2(x,t)dx\right] \geq 0,\end{array} \tag{3.179}$$

$$
\begin{aligned}
&\lambda_3\left[\|z_{xx}(\cdot,t)\|^2_{L^2(0,L)}-\left(\tfrac{\pi^2}{L^2}\right)^2\|z(\cdot,t)\|^2_{L^2(0,L)}\right]\\
&=\lambda_3\left[\sum_{j\neq\sigma_k}^{N}\int_{\Omega_j}z_{xx}^2(x,t)dx-\left(\tfrac{\pi^2}{L^2}\right)^2\sum_{j\neq\sigma_k}^{N}\int_{\Omega_j}z^2(x,t)dx\right]\\
&+\lambda_3\left[\int_{\Omega_{\sigma_k}}z_{xx}^2(x,t)dx-\left(\tfrac{\pi^2}{L^2}\right)^2\int_{\Omega_{\sigma_k}}z^2(x,t)dx\right]\geq 0.
\end{aligned}
\tag{3.180}
$$

We further apply the descriptor method (see Sect. 3.5 in [11]), where the left-hand side of the following equation

$$
\begin{aligned}
&2\int_0^L[P_3z(x,t)+P_2z_t(x,t)]\{-z_t(x,t)-z_{xx}(x,t)\\
&-z(x,t)z_x(x,t)-\nu z_{xxxx}(x,t)\\
&-Kb_{\sigma_k}(x)[z(x,t)-f_{\sigma_k}(x,t)-\rho_{\sigma_k}(t)]\}dx=0
\end{aligned}
\tag{3.181}
$$

with some $P_3>0$ is added to $\dot V$. Then adding the left-hand sides of (3.178)–(3.180) to (3.174) and taking into account (3.175), we obtain

$$
\begin{aligned}
&\dot V(t)+2\alpha V(t)\leq(2P_1-2P_3)\sum_{j\neq\sigma_k}^{N}\int_{\Omega_j}z(x,t)z_t(x,t)dx\\
&+\left(2\alpha P_1-\lambda_3\tfrac{\pi^4}{L^4}\right)\sum_{j\neq\sigma_k}^{N}\int_{\Omega_j}z^2(x,t)dx\\
&-\left[2P_3+\frac{\lambda_2(\frac{L}{N}+\varepsilon)^2}{\pi^2}\right]\sum_{j\neq\sigma_k}^{N}\int_{\Omega_j}z(x,t)z_{xx}(x,t)dx\\
&+(2\alpha P_2\nu-2P_3\nu+\lambda_3)\sum_{j\neq\sigma_k}^{N}\int_{\Omega_j}z_{xx}^2(x,t)dx\\
&+\left(R\tfrac{4h^2}{\pi^2}\tfrac{L}{N\varepsilon}-2P_2\right)\sum_{j\neq\sigma_k}^{N}\int_{\Omega_j}z_t^2(x,t)dx\\
&-2P_2\sum_{j\neq\sigma_k}^{N}\int_{\Omega_j}z_t(x,t)[z(x,t)z_x(x,t)+z_{xx}(x,t)]dx\\
&+(2P_1-2P_3)\int_{\Omega_{\sigma_k}}z(x,t)z_t(x,t)dx\\
&+\left(2\alpha P_1-\lambda_3\tfrac{\pi^4}{L^4}\right)\int_{\Omega_{\sigma_k}}z^2(x,t)dx\\
&-\left[2P_3+\frac{\lambda_2(\frac{L}{N}+\varepsilon)^2}{\pi^2}\right]\int_{\Omega_{\sigma_k}}z(x,t)z_{xx}(x,t)dx\\
&+(2\alpha P_2\nu-2P_3\nu+\lambda_3)\int_{\Omega_{\sigma_k}}z_{xx}^2(x,t)dx\\
&+\left(R\tfrac{4h^2}{\pi^2}\tfrac{L}{N\varepsilon}-2P_2\right)\int_{\Omega_{\sigma_k}}z_t^2(x,t)dx\\
&-2P_2\int_{\Omega_{\sigma_k}}z_t(x,t)[z(x,t)z_x(x,t)+z_{xx}(x,t)]dx\\
&-2P_3K\int_{\Omega_{\sigma_k}}z(x,t)[z(x,t)-f_{\sigma_k}(x,t)-\rho_{\sigma_k}(t)]dx\\
&-2P_2K\int_{\Omega_{\sigma_k}}z_t(x,t)[z(x,t)-f_{\sigma_k}(x,t)-\rho_{\sigma_k}(t)]dx\\
&-Re^{-2\alpha h}\int_{\Omega_{\sigma_k}}\rho_{\sigma_k}^2(t)dx-\lambda_2\int_{\Omega_{\sigma_k}}f_{\sigma_k}^2(x,t)dx\\
&-\tfrac{\lambda_1}{N-1}\sum_{j\neq\sigma_k}^{N}\int_{\Omega_j}[z(x,t)-f_j(x,t)-\rho_j(t)]^2dx\\
&+\lambda_1\int_{\Omega_{\sigma_k}}[z(x,t)-f_{\sigma_k}(x,t)-\rho_{\sigma_k}(t)]^2dx.
\end{aligned}
\tag{3.182}
$$

From (3.182), we have

$$\dot{V}(t) + 2\alpha V(t) \leq \sum_{j \neq \sigma_k}^{N} \int_{\Omega_j} \eta_1^T \Psi_1 \eta_1 dx + \int_{\Omega_{\sigma_k}} \eta_2^T \Psi_2 \eta_2 dx, \quad t \in [t_k + \delta, t_{k+1}), \tag{3.183}$$

where $\eta_1 = \mathrm{col}\{z(x,t), z_t(x,t), z_{xx}(x,t), f_j(x,t), \rho_j(x,t)\}$, $\eta_2 = \mathrm{col}\{z(x,t), z_t(x,t), z_{xx}(x,t), f_{\sigma_k}(x,t), \rho_{\sigma_k}(x,t)\}$, Ψ_i $(i = 1, 2)$ are given by (3.170), (3.171) respectively.

Let us first assume that

$$\max_{x \in [0,L]} |z_x(x,t)| < C, \ \forall t \geq 0. \tag{3.184}$$

Under the assumption (3.184), we have

$$\dot{V}(t) + 2\alpha V(t) \leq 0, \tag{3.185}$$

if $\Psi_1 < 0$ and $\Psi_2 < 0$ hold for all $-C \leq z_x \leq C$.

Matrices Ψ_i $(i = 1, 2)$ given by (3.170) and (3.171) are affine in z_x. Hence, $\Psi_1 < 0$ and $\Psi_2 < 0$ for all $-C \leq z_x \leq C$ if these inequalities hold in the vertices $z_x = \pm C$, i.e., if LMIs (3.167) and (3.168) are feasible.

Step 3: Now we derive sufficient LMI-based conditions to guarantee that $\dot{V}(t) - 2\alpha_0 V(t) \leq 0$ for $[t_k, t_k + \delta)$.

Differentiating $V(t)$ along (3.165) subject to (3.147), we have

$$\begin{aligned}
&\dot{V}(t) - 2\alpha_0 V(t) = 2P_1 \int_0^L z(x,t) z_t(x,t) dx \\
&-2\alpha_0 P_1 \int_0^L z^2(x,t) dx - 2\alpha_0 P_2 \nu \int_0^L z_{xx}^2(x,t) dx \\
&+2P_2 \nu \int_0^L z_{xx}(x,t) z_{xxt}(x,t) dx + R\tfrac{4h^2}{\pi^2} \sum_{j=1}^{N} \int_{\Omega_j} [\rho_{jt}(t)]^2 dx \\
&-Re^{-2\alpha h} \sum_{j=1}^{N} \int_{\Omega_j} [\rho_j(t)]^2 dx - 2(\alpha + \alpha_0) V_R(t).
\end{aligned}$$

We further apply the descriptor method, where the left-hand side of the following equation

$$\begin{aligned}
2 \int_0^L &[P_4 z(x,t) + P_2 z_t(x,t)][-z_t(x,t) - z_{xx}(x,t) \\
&-z(x,t) z_x(x,t) - \nu z_{xxxx}(x,t)] dx = 0
\end{aligned}$$

with some $P_4 > 0$ is added to $\dot{V}$.

Using (3.175), we obtain

$$\dot{V}(t) - 2\alpha_0 V(t) \leq \int_0^L \eta_0^T \Psi_0 \eta_0 dx, \ t \in [t_k, t_k + \delta),$$

where $\eta_0 = \mathrm{col}\{z(x,t), z_t(x,t), z_{xx}(x,t)\}$.

Step 4: From Step 1-Step 3, we obtain if $\|z_0\|_V < \sqrt{\frac{P_2\nu}{L}}C$, then the feasibility of LMIs (3.167)–(3.169) implies that any strong solution of (3.165), (3.147) initialized with z_0 admits a priori estimate

$$\begin{aligned} &V(t) \le e^{2\alpha_0(t-t_k)}V(t_k), \ \forall t \in [t_k, t_k+\delta), \\ &V(t) \le e^{-2\alpha(t-t_k-\delta)}V(t_k+\delta), \ \forall t \in [t_k+\delta, t_{k+1}). \end{aligned} \tag{3.186}$$

Since $\alpha_1 < \alpha$ and $t_{k+1} - t_k \ge h_0$, (3.173) implies

$$(\alpha_1 - \alpha)(t_{k+1} - t_k) \le (\alpha_1 - \alpha)h_0 \le -(\alpha_0 + \alpha)\delta,$$

which together with (3.186) leads to

$$V(t_{k+1}) \le e^{2\alpha_0\delta}e^{-2\alpha(t_{k+1}-t_k-\delta)}V(t_k) \le e^{-2\alpha_1(t_{k+1}-t_k)}V(t_k). \tag{3.187}$$

From (3.186) it follows

$$\begin{aligned} &V(t) \le e^{2\alpha_0\delta}V(t_k), \ t \in [t_k, t_k+\delta); \\ &V(t) \le V(t_k+\delta) \le e^{2\alpha_0\delta}V(t_k), \ t \in [t_k+\delta, t_{k+1}). \end{aligned}$$

Therefore, for $t \in [t_k, t_{k+1})$

$$\begin{aligned} &V(t) \le e^{2\alpha_0\delta}V(t_k) \le e^{2\alpha_0\delta-2\alpha_1(t_k-t_{k-1})}V(t_{k-1}) \le e^{2\alpha_0\delta-2\alpha_1(t-t_{k-1}-h)}V(t_{k-1}) \\ &\le e^{2\alpha_0\delta-2\alpha_1(t-t_{k-2}-h)}V(t_{k-2}) \le \cdots \le e^{2\alpha_0\delta-2\alpha_1(t-h)}V(0). \end{aligned}$$

Hence,

$$V(t) \le e^{-2\alpha_1(t-h)+2\alpha_0\delta}V(0), \ \forall t \ge 0.$$

The latter bound guarantees the existence of these strong solutions for all $t \in [0, t_1]$. Then using step method [11], we conclude that the strong solution exists for all $t \ge 0$.

We will prove next that (3.184) holds. On one hand, for $t = 0$, inequality (3.184) holds by hypothesis in Theorem 3.5. On the other hand, let (3.184) be false for some $t > 0$ and let t^* be the smallest instant such that $V(t^*) \ge \frac{P_2\nu}{L}C^2$. Since V is continuous in time, we have $V(t^*) = \frac{P_2\nu}{L}C^2$ and $V(t) < \frac{P_2\nu}{L}C^2$ for $t \in [0, t^*)$. Since $z_x(0, t) = 0$, the Sobolev inequality (Lemma 1.7) implies: $\max_{x\in[0,L]} |z_x(x, t)|^2 \le L\|z_{xx}(\cdot, t)\|^2_{L^2} \le \frac{L}{P_2\nu}V(t) \le \frac{L}{P_2\nu}V(0) = \frac{L}{P_2\nu}\|z_0\|^2_V < C^2$ for $t \in [0, t^*)$. Thus, the feasibility of LMIs (3.167)–(3.169) guarantee that (3.185) is true for all $t \in [0, t^*)$ Hence, by continuity, $V(t) \le V(0) < \frac{P_2\nu}{L}C^2$ for all $t \in [0, t^*]$, which contradicts the definition of t^*. Therefore, (3.184) holds.

□

Remark 3.8 The LMIs in Theorem 3.5 are always feasible for appropriate decision variables and small enough α, α_0, h, δ, C, ε, P_2 and large enough N such that P_2K and $\frac{K}{N-1}$ are small. Indeed, consider Ψ_1, Ψ_2 and Ψ_0 given by (3.170), (3.171) and (3.172) respectively. We will show that strict inequalities (3.167)–(3.169) hold with

$\alpha = h = C = 0$. Then LMIs (3.167)–(3.169) hold with small enough $\alpha > 0, h > 0$ and $C > 0$. Set $R = \lambda_2 > 2K$, $P_1 = P_3 = P_4 = 1$, $\lambda_1 = K$ such $\psi_{12} = P_1 - P_3 = 0$ and $P_3 K - \lambda_1 = 0$. By applying Schur complement, for small enough h and large enough N, we obtain

$$\Psi_1|_{z_x=0} < 0 \iff \begin{bmatrix} \psi_{11} + \frac{2(\frac{K}{N-1})^2}{\lambda_2+2\frac{K}{N-1}} & \psi_{13} \\ * & \psi_{33} + \frac{P_2}{2} \end{bmatrix} < 0,$$

$$\Psi_2 < 0 \iff \lambda_2 - 2K > P_2 K^2,$$
$$\begin{bmatrix} \tilde{\psi}_{11} + \frac{P_2 K^2}{2 - \frac{2P_2K^2}{\lambda_2 - 2K}} & \psi_{13} + \frac{P_2 K}{2 - \frac{2P_2K^2}{\lambda_2 - 2K}} \\ * & \psi_{33} + \frac{P_2}{2 - \frac{2P_2K^2}{\lambda_2 - 2K}} \end{bmatrix} < 0,$$

$$\Psi_0 < 0 \iff \begin{bmatrix} -2\alpha_0 & 0 & -1 \\ * & -2P_2 & -P_2 \\ * & * & -2\alpha_0 P_2 \nu - 2\nu \end{bmatrix} < 0,$$

where

$$\psi_{11} = -\lambda_3 \frac{\pi^4}{L^4} - \frac{K}{N-1}, \quad \psi_{13} = -1 - \frac{\lambda_2(\frac{L}{N} + \varepsilon)^2}{2\pi^2},$$
$$\psi_{33} = -2\nu + \lambda_3, \quad \tilde{\psi}_{11} = -\lambda_3 \frac{\pi^4}{L^4} - K.$$

Then the latter LMIs are feasible with appropriate $\nu - \sqrt{\nu^2 - \frac{L^4}{\pi^4}} < \lambda_3 < \nu + \sqrt{\nu^2 - \frac{L^4}{\pi^4}}$, small enough α_0, ε, P_2 and large enough N such that $P_2 K^2$, $\frac{K}{N-1}$ are small.

Remark 3.9 Technically our Lyapunov-based approach employs the following novel tools compared to [3]:

(a) The inequality (3.158) (that results from the switching law) and (3.178) that allow to derive feasible LMIs.
(b) For the moving time we have derived a new LMI that guarantees a bound on V in step 3 of the Theorem 3.5 proof and the bound on the dwelling time δ in step 4.
(c) For point-like measurements that were not considered in [3], we use a mean value theorem (above (3.161)) and further Wirtinger's inequality in (3.176), which is different from the direct application of the Poincaré inequality in [3] for the averaged measurements. Moreover, we employ the Wirtinger-based Lyapunov functional for sampled-data control with ρ_j which includes c_j, which leads to simpler LMIs comparatively to Lyapunov functional of [3] with $c_j = 1$.

Under the averaged state measurements (3.153) we have the following result.

Theorem 3.6 *Consider the closed-loop system* (3.155) *subject to* (3.147) *and the switching law* (3.157) *with* $c_j = 1$. *Given positive scalars* h, α, K *and tuning parameter* $C > 0$, $\alpha_0 > 0$ *such that* $\alpha h_0 > (\alpha_0 + \alpha)\delta$, *let there exist scalars* $R > 0$, $P_n > 0$, $\lambda_n \geq 0$ $(n = 1, 2, 3)$ *that satisfy the LMIs:*

$$\Theta_1|_{z_x=\pm C} < 0, \tag{3.188}$$

$$\Theta_2|_{z_x=\pm C} < 0, \tag{3.189}$$

$$\Psi_0|_{z_x=\pm C} < 0, \tag{3.190}$$

where

$$\Theta_1 = \Psi_1 + \Pi, \tag{3.191}$$

$$\Theta_2 = \Psi_2 + \Pi, \tag{3.192}$$

where Ψ_1, Ψ_2, Ψ_0 *are given by* (3.170), (3.171), (3.172) *respectively, and*

$$\Pi = \left[\begin{array}{ccc|c} 0 & 0 & \frac{\lambda_2}{2\pi^2}\left(\frac{2L\varepsilon}{N} + \varepsilon^2\right) & \\ 0 & R\frac{4h^2}{\pi^2}\left(1 - \frac{L}{N\varepsilon}\right) & 0 & O_{3\times 2} \\ 0 & 0 & 0 & \\ \hline & * & & 0_{2\times 2} \end{array}\right] \tag{3.193}$$

Then for any initial function $z_0 \in H_0^2(0, L)$ *subject to* $\|z_0\|_V < \sqrt{\frac{P_2 \nu}{L}} C$, *the closed-loop system* (3.155) *subject to* (3.147) *is exponentially stable with a decay rate* $\alpha_1 > 0$ *which is subject to* (3.173).

Proof For the case of switched controller under the averaged measurements, by arguments of Theorem 3.5, the well-posedness of (3.155) subject to (3.147) can be established via the step method.

Denote

$$\tilde{f}_j(x, t) \triangleq z(x, t) - \frac{\int_{\Omega_j} z(x, t)dx}{|\Omega_j|}, \quad \tilde{\rho}_j(t) \triangleq \frac{\int_{\Omega_j} \int_{t_k}^{t} z_s(x, s)dsdx}{|\Omega_j|},$$

where $|\Omega_j| = \dfrac{L}{N}$.

Then the switching controller (3.154) via the switching law (3.157) with $c_j = 1$ can be rewritten as

$$u_{\sigma_k}(t) = -K[z(x, t) - \tilde{f}_{\sigma_k}(x, t) - \tilde{\rho}_{\sigma_k}(t)]. \tag{3.194}$$

We choose the Lyapunov function V with $\tilde{\rho}_j$ instead of ρ_j. Differentiating V along the solution of the closed-loop system (3.155) subject to (3.147), we get (3.174) with $\tilde{\rho}_j$ instead of ρ_j. The substitution $f_j \to \tilde{f}_j$ and $\rho_j \to \tilde{\rho}_j$ in Theorem leads to the following changes:

$$\int_{\Omega_j} [\tilde{\rho}_{jt}(t)]^2 dx = \frac{1}{|\Omega_j|}\left(\int_{\Omega_j} z_t(x,t)dx\right)^2 \leq \int_{\Omega_j} z_t^2(x,t)dx,$$

$$\begin{aligned} &-\frac{\lambda_1}{N-1}\sum_{j\neq\sigma_k}^{N}\int_{\Omega_j}[z(x,t)-\tilde{f}_j(x,t)-\tilde{\rho}_j(t)]^2dx \\ &+\lambda_1\int_{\Omega_{\sigma_k}}[z(x,t)-\tilde{f}_{\sigma_k}(x,t)-\tilde{\rho}_{\sigma_k}(t)]^2dx \geq 0, \end{aligned} \tag{3.195}$$

$$\frac{\lambda_2 L^2}{N^2\pi^2}\|z_x(\cdot,t)\|^2_{L^2(0,L)} - \lambda_2\sum_{j=1}^{N}\|\tilde{f}_j(\cdot,t)\|^2_{L^2(\Omega_j)} \geq 0 \tag{3.196}$$

for any $\lambda_1 \geq 0$, $\lambda_2 \geq 0$.

Set $\tilde{\eta}_1 = \mathrm{col}\{z(x,t), z_t(x,t), z_{xx}(x,t), \tilde{f}_j(x,t), \tilde{\rho}_j(x,t)\}, \tilde{\eta}_2 = \mathrm{col}\{z(x,t), z_t(x,t), z_{xx}(x,t), \tilde{f}_{\sigma_k}(x,t), \tilde{\rho}_{\sigma_k}(x,t)\}$, $\eta_0 = \mathrm{col}\{z(x,t), z_t(x,t), z_{xx}(x,t)\}$. Applying descriptor method and adding the left-hand sides of (3.195)–(3.196) and (3.180) to $\dot{V}$, we finally obtain

$$\begin{aligned} \dot{V}(t) + 2\alpha V(t) &\leq \sum_{j\neq\sigma_k}^{N}\int_{\Omega_j}\tilde{\eta}_1^T\Theta_1\tilde{\eta}_1dx + \int_{\Omega_{\sigma_k}}\tilde{\eta}_2^T\Theta_2\tilde{\eta}_2dx, \\ &\qquad\qquad t \in [t_k+\delta, t_{k+1}), \\ \dot{V}(t) - 2\alpha_0 V(t) &\leq \int_0^L \eta_0^T\Psi_0\eta_0dx, \qquad t \in [t_k, t_k+\delta), \end{aligned}$$

where Θ_l $(l = 1, 2)$ and Ψ_0 are given by (3.191), (3.192) and (3.172), respectively.

Hence, $\dot{V}(t) + 2\alpha V(t) \leq 0$, $\dot{V}_0(t) - 2\alpha V_0(t) \leq 0$, if $\Theta_l < 0 (l = 1, 2)$ and $\Psi_0 < 0$ hold for all $-C \leq z_x \leq C$. Matrices $\Theta_l (l = 1, 2)$ and Ψ_0 given by (3.191), (3.192) and (3.172) are affine in z_x. Hence, $\Theta_l < 0 (l = 1, 2)$ and $\Psi_0 < 0$ for all $-C \leq z_x \leq C$ if these inequalities hold in the vertices $z_x = \pm C$, i.e. if LMIs (3.188)–(3.190) are feasible.

Remark 3.10 For the negligible moving time with $\delta \to 0$, the LMI conditions of Theorem 3.5 are reduced to (3.167) and (3.168), and the LMI conditions of Theorem 2 are reduced to (3.188) and (3.189).

3.3.3 Extensions

The proposed method is efficient for various classes of PDEs. In this section we will discuss its extension to KSE under the periodic boundary conditions and to the reaction-diffusion equation.

Consider (3.155) under the periodic boundary conditions:

$$\frac{\partial^m z}{\partial x^m}(0,t) = \frac{\partial^m z}{\partial x^m}(L,t),\ t > 0,\ m = 0, 1, 2, 3 \tag{3.197}$$

and the switching law (3.157). The well-posedness under the periodic boundary conditions can be established similar to the case of Dirichlet boundary conditions. Denote

$$H^2_{per}(0, L) \triangleq \{g \in H^2(0, L) | g(0) = g(L), g'(0) = g'(L)\}.$$

We employ the extension of Sobolev's inequality [3, 32].

Lemma 3.2 *Let $z \in H^1(0, L)$ be a scalar function. Then*

$$\max_{x\in[0,L]} |z(x,t)|^2 \leq (1 + \frac{1}{L})\|z(\cdot,t)\|^2_{L^2(0,L)} + \|z_x(\cdot,t)\|^2_{L^2(0,L)}.$$

Note that Wirtinger's inequality cannot be applied for the case of the periodic boundary conditions (3.197). Therefore, from Lemma 3.2 we directly obtain the following result:

- If the LMI conditions of Theorem 3.5 /Theorem 3.6 with $\lambda_3 = 0$ hold, then for the initial function $z_0 \in H^2_{per}(0, L)$ satisfying $\|z_0\|_V < \sqrt{\frac{L}{(L+1)M}}C$, there exists a unique strong solution of the corresponding closed-loop system and the closed-loop system is stable, where $M \triangleq \max\left\{\frac{1}{2P_1}, \frac{3}{2P_2\nu}\right\}$.

Remark 3.11 In the companion conference paper [30], the presented switching control sampled-data control was presented for the semilinear heat equation

$$z_t(x,t) = \frac{\partial}{\partial x}[a(x)z_x(x,t)] + \varphi(z(x,t))z(x,t) + b_{\sigma_k}(x)u_{\sigma_k}(t),$$

under the averaged measurements and the Dirichlet boundary conditions $z(0, t) = z(1, t) = 0,\ t > 0$. The functions a and φ are of class C^1 and may be unknown. Global sampled-data stabilization by switching was achieved under assumption that the inequalities $a(x) \geq a_0 > 0,\ \varphi_m \leq \varphi \leq \varphi_M$ hold with known bounds a_0, φ_m and φ_M.

3.3.4 Numerical Example

In this section, we present a numerical example which verifies the effectiveness of the proposed method for KSE. For the heat equation see example in [33].

3.3.4.1 Case of the Dirichlet Boundary Conditions

Consider KSE (3.146) under the Dirichlet boundary conditions (3.147) with $L = 2\pi$ and instability parameter $\nu = 0.8 < 1$. The initial function is chosen as follows:

$$z(x, 0) = z_0(x) = 0.015(1 - \cos x)\sin x.$$

Figure 3.9 demonstrates the profile of the open-loop system initialized by $z_0(x)$. It is shown that the unforced system is unstable.

We choose $K = 25, \alpha_0 = 0.32, \alpha = 0.07,\ C = 1$ and verify LMIs of Theorems 3.5 and 3.6 under point-like and averaged measurements. The results are given in Table 3.2, which shows that as N increases, the maximum value of h increases. Since the averaged measurements contain more information than the point-like measurements, the corresponding switched closed-loop system under the averaged measurements preserves the stability for larger h. Note that e.g. for $N = 15$ and $\varepsilon = \pi/30$ the closed-loop system preserves the exponential stability within a given domain of initial conditions $\|z_0\|_V < 0.743$ for $0.0059 \le t_{k+1} - t_k \le h \le 0.006$. Note also that for the choice of K we used trials and errors to minimize N for $h = 0$ and $C = 1$. Then, we increased h and the corresponding value of N.

We further provide numerical simulations of the solutions to the corresponding closed-loop systems with $N = 15,\ K = 25,\ \alpha = 0.07$:

(a) A finite difference method was utilized for numerical simulations of the solution to the closed-loop system (3.165) subject to (3.147) under the output-feedback

$$u_{\sigma_k}(t) = \begin{cases} 0, & t \in [t_k, t_k + \delta), \\ -25 \int_{\Omega_{\sigma_k}} c_{\sigma_k}(x) z(x, t_k) dx, & t \in [t_k + \delta, t_{k+1}) \end{cases}$$

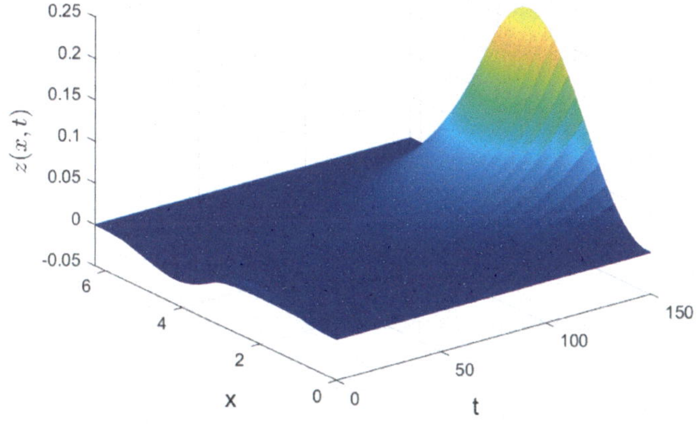

Fig. 3.9 State of unforced system: Dirichlet boundary conditions

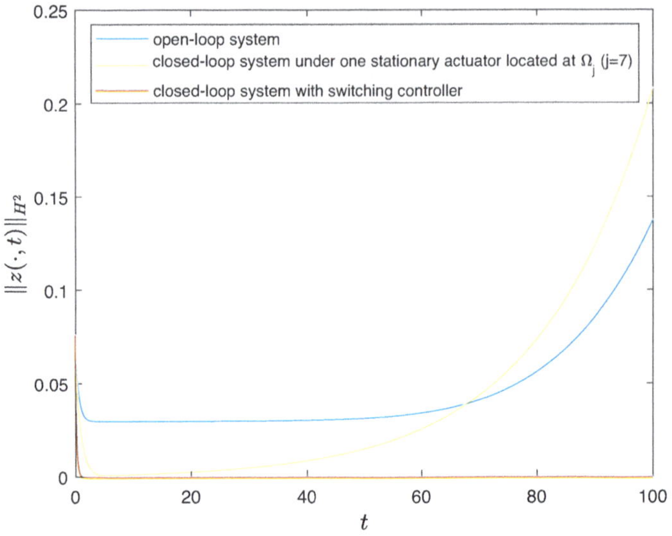

Fig. 3.10 H^2-norm $\|z(\cdot,t)\|_{H^2}$ of the open-loop system, closed-loop system under one stationary actuator located at Ω_7 and switched closed-loop system: Dirichlet boundary conditions

via the switching law (3.154), (3.157) with $t_{k+1} - t_k = 0.006, \varepsilon = \pi/30, |\Omega_{\sigma_k}| = \frac{2\pi}{15}$, and the moving time for sensors and actuators $\delta = 0.001$. The steps in time and space were set as $dx = \pi/30$ and $dt = 10^{-5}$, respectively. Figure 3.10 shows the time evolution of $\|z(\cdot,t)\|_{H^2}$ of the closed-loop and open-loop systems. Note that the closed-loop system with switching controller is stable, whereas the closed-loop system under only one stationary actuator is unstable.

(b) Next, a finite difference method is utilized to compute the numerical solution of the closed-loop system (3.155) subject to (3.147) under the output-feedback

$$u_{\sigma_k}(t) = \begin{cases} 0, & t \in [t_k, t_k + \delta), \\ -\frac{375}{2\pi} \int_{\Omega_{\sigma_k}} z(x, t_k)dx, & t \in [t_k + \delta, t_{k+1}) \end{cases}$$

via the switching law (3.154), with $t_{k+1} - t_k = 0.030$, $|\Omega_{\sigma_k}| = \frac{2\pi}{15}$, and the moving time for sensors and actuators $\delta = 0.005$. Set the steps in time and space as $dx = \pi/30$ and $dt = 10^{-5}$, respectively. Figure 3.11 shows the profile of the closed-loop system. The locations of sensor/actuator under the switching control law are given in Fig. 3.12.

3.3.4.2 Case of the Periodic Boundary Conditions

We proceed further with the case of periodic boundary conditions (3.197) under the same initial conditions. For the switching controller (3.154) via (3.157), by using the

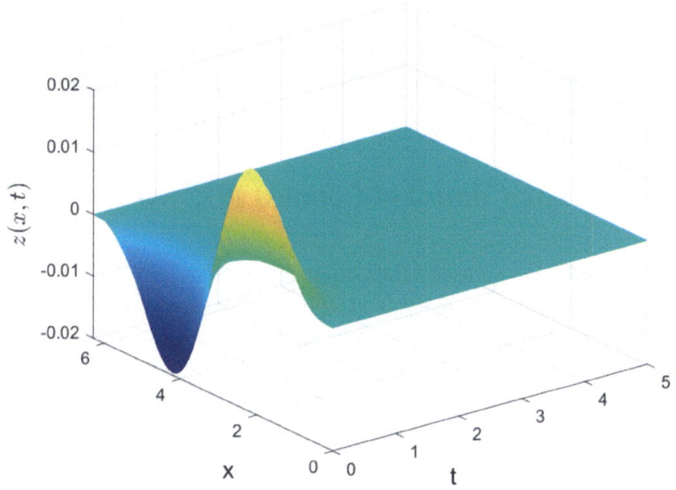

Fig. 3.11 State response of closed-loop system under the averaged measurements and Dirichlet boundary conditions

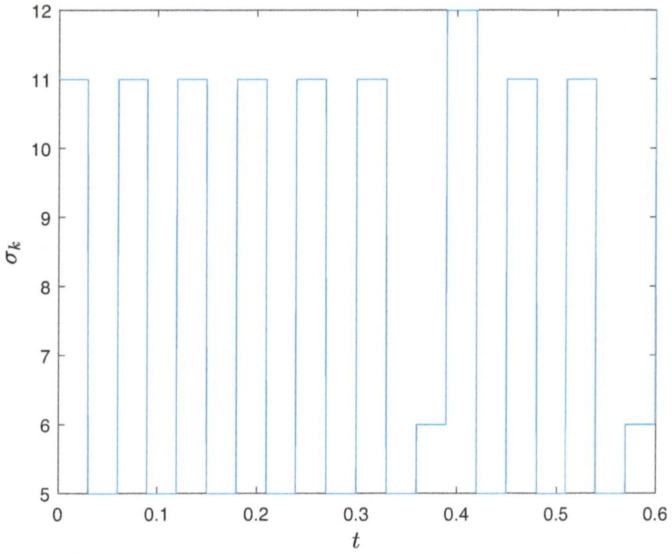

Fig. 3.12 Sensor/actuator locations for $N = 15$, $t_{k+1} - t_k = 0.030$

Yalmip, we verify LMI conditions of Theorem 3.5 with $\lambda_3 = 0$, $N = 15$, $K = 25$, $C = 1, \alpha = 0.07, \alpha_0 = 0.32, \delta = 0.001$, and $\varepsilon = \pi/30$. We find that the closed-loop system (3.165), (3.197) preserves exponential stability within a given domain of initial conditions $\|z_0\|_V < 0.773$ for $0.0059 \leq t_{k+1} - t_k \leq h \leq 0.006$. In the numerical simulation, a finite difference method is utilized to compute the numerical solution of the closed-loop system (3.165), (3.197) under the output-feedback (3.154) via the switching law (3.157) with $|\Omega_{\sigma_k}| = \frac{2\pi}{15}$ and $t_{k+1} - t_k = 0.006$. Numerical simulations of the solutions confirm the theoretical results that follow from LMIs.

3.4 Sampled-Data Control of Coupled PDE Systems

This section is devoted to sampled-data control design of a PDE-PDE cascade system. These PDEs are governed by heat equations with different reaction coefficients. In order to stabilize the cascaded heat-heat system, we start with the design of continuous-time feedback control law. Then sampled-data control is further proposed for practical reasons. Sufficient conditions are derived for guaranteeing the exponential stability and well-posedness of the corresponding closed-loop system via Lyapunov method and modal decomposition method. Numerical examples illustrate the efficiency of the proposed method. For stabilization of a more general higher-order underactuated system of heat equations see [34].

3.4.1 Problem Formulation

Consider the unstable cascaded heat-heat system

$$\begin{cases} u_t(x,t) = u_{xx}(x,t) + au(x,t), x \in (0,1), t > 0, \\ z_t(x,t) = z_{xx}(x,t) + bz(x,t), x \in (0,1), t > 0, \\ u_x(0,t) = sz(0,t), \\ u(1,t) = 0, \\ z_x(1,t) = 0, \\ z_x(0,t) = U(t), \\ u(x,0) = u_0(x), \\ z(x,0) = z_0(x), \end{cases} \tag{3.198}$$

where a, b are positive reaction coefficients, $s \in \mathbb{R}$ is a nonzero parameter, $U(t)$ is the boundary control input, $(u(x,t), z(x,t))^T$ is the state of the coupled heat-heat equations. For sufficiently large a and b, it is straightforward to demonstrate that the open-loop system (3.198) (without control) is unstable (see Fig. 3.13 below).

Our goal is to design a boundary sampled-data control law and find a sufficient condition that guarantees the exponential stability of the corresponding closed-loop system, on the basis of the modal decomposition approach and Lyapunov method.

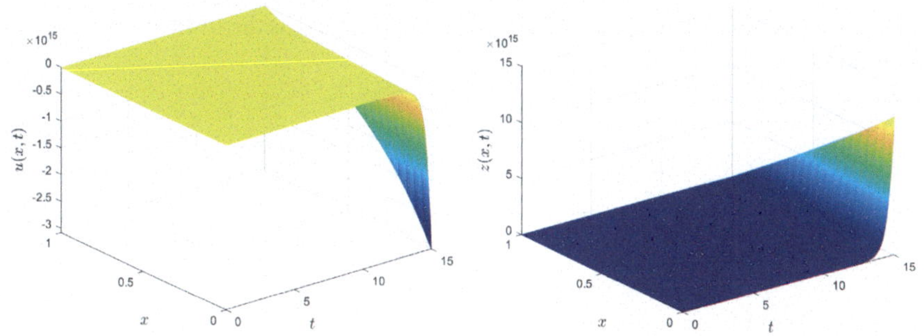

(a) The state $u(x,t)$ (b) The state $z(x,t)$

Fig. 3.13 The states of open-loop system with $a = \frac{1}{2}$, $b = 6$

3.4.2 Continuous-Time Feedback Control

In this section, we will concentrate on the continuous-time feedback control design of the cascaded heat-heat system (3.198) via modal decomposition approach. The system (3.198) can be divided into two subsystems that are cascaded through boundary interconnection as follows:

$$\begin{cases} u_t(x,t) = u_{xx}(x,t) + au(x,t), x \in (0,1), t > 0, \\ u_x(0,t) = sz(0,t), \\ u(1,t) = 0, \\ u(x,0) = u_0(x), \end{cases} \tag{3.199}$$

$$\begin{cases} z_t(x,t) = z_{xx}(x,t) + bz(x,t), x \in (0,1), t > 0, \\ z_x(1,t) = 0, \\ z_x(0,t) = U(t), \\ z(x,0) = z_0(x). \end{cases} \tag{3.200}$$

Let $\mathcal{H} = L^2(0,1)$ be the Hilbert space with the inner product $\langle Z_1, Z_2 \rangle = \int_0^1 Z_1 Z_2 dx$, $\forall Z_i \in \mathcal{H}$, $i = 1, 2$. The subsystem operators $\mathcal{A}_1 : D(\mathcal{A}_1) \subset \mathcal{H} \to \mathcal{H}$ and $\mathcal{A}_2 : D(\mathcal{A}_2) \subset \mathcal{H} \to \mathcal{H}$ are defined by

$$\begin{cases} \mathcal{A}_1 f = f'' + af, \ \forall f \in D(\mathcal{A}_1) \\ D(\mathcal{A}_1) = \left\{ f \in H^2(0,1) \,\middle|\, f'(0) = f(1) = 0 \right\}, \end{cases}$$

$$\begin{cases} \mathcal{A}_2 g = g'' + bg, \ \forall g \in D(\mathcal{A}_2) \\ D(\mathcal{A}_2) = \left\{ g \in H^2(0,1) \,\middle|\, g'(0) = g'(1) = 0 \right\}. \end{cases}$$

Then the subsystems (3.199) and (3.200) can be expressed as the following evolution equations

$$\frac{d}{dt}u(\cdot,t) = \mathcal{A}_1 u(\cdot,t) + s\mathcal{B}z(0,t), \tag{3.201}$$

$$\frac{d}{dt}z(\cdot,t) = \mathcal{A}_2 z(\cdot,t) + \mathcal{B}U(t), \tag{3.202}$$

where $\mathcal{B} = -\delta(x)$.

By solving eigenvalue problems, we obtain the following eigenvalues and related eigenfunctions, respectively:

$$\left\{\Phi_{1,n_1}\right\} = \left\{\sqrt{2}\cos\left(\tfrac{(2n_1+1)\pi x}{2}\right)\Big|n_1 \in \mathbb{N}\right\}, \tag{3.203}$$

$$\left\{\lambda_{1,n_1}\right\} = \left\{a - \tfrac{(2n_1+1)^2\pi^2}{4}\Big|n_1 \in \mathbb{N}\right\}, \tag{3.204}$$

$$\left\{\Phi_{2,n_2}\right\} = \left\{\sqrt{2}\cos(n_2\pi x)\Big|n_2 \in \mathbb{N}\right\}, \tag{3.205}$$

$$\left\{\lambda_{2,n_2}\right\} = \left\{b - n_2^2\pi^2\Big|n_2 \in \mathbb{N}\right\}. \tag{3.206}$$

Moreover, $\{\Phi_{1,n_1}\}$ and $\{\Phi_{2,n_2}\}$ form two bases for $\mathcal{H}$.

In what follows, the solutions of (3.201) and (3.202) can be presented in a new form via the bases:

$$u(x,t)=\sum_{n_1=0}^{\infty} X_{1,n_1}(t)\Phi_{1,n_1}(x),\ z(x,t)=\sum_{n_2=0}^{\infty} X_{2,n_2}(t)\Phi_{2,n_2}(x),$$

where $X_{1,n_1}(t) \triangleq \langle u, \Phi_{1,n_1}\rangle,\ X_{2,n_2}(t) \triangleq \langle z, \Phi_{2,n_2}\rangle$.

Differentiating under the integral sign, integrating by parts and using the boundary conditions, we obtain the projection of (3.198) on $\{\Phi_{1,n_1}\}_{n_1\geq 0}$, $\left\{\Phi_{2,n_2}\right\}_{n_2\geq 0}$:

$$\begin{cases} \dot{X}_{1,n_1}(t) = \lambda_{1,n_1}X_{1,n_1}(t) - 2s\sum\limits_{n_2=0}^{\infty} X_{2,n_2}(t),\ n_1 \in \mathbb{N}, \\ \dot{X}_{2,n_2}(t) = \lambda_{2,n_2}X_{2,n_2}(t) - \sqrt{2}U(t),\ n_2 \in \mathbb{N}, \\ X_{1,n_1}(0) = \sqrt{2}\int_0^1 u_0(x)\cos\left(\frac{(2n_1+1)\pi x}{2}\right)dx,\ n_1 \in \mathbb{N}, \\ X_{2,n_2}(0) = \sqrt{2}\int_0^1 z_0(x)\cos(n_2\pi x)dx,\ n_2 \in \mathbb{N}. \end{cases} \tag{3.207}$$

Let $\delta_1, \delta_2 > 0$ be desired decay rates. Since $\lim\limits_{n_1\to\infty} \lambda_{1,n_1} = -\infty$, $\lim\limits_{n_2\to\infty} \lambda_{2,n_2} = -\infty$, there exist $N_1, N_2 \in \mathbb{N}$ such that $\lambda_{1,n_1} \leq -\delta_1$, $\forall n_1 > N_1$, $\lambda_{2,n_2} \leq -\delta_2$, $\forall n_2 > N_2$, where

$$N_1 = \begin{cases} 0, & 0 < a < \frac{\pi^2}{4}, \\ \left\lfloor \frac{2\sqrt{a}-\pi}{2\pi} \right\rfloor, & a \geq \frac{\pi^2}{4}, \end{cases} \tag{3.208}$$

$$N_2 = \lfloor \tfrac{\sqrt{b}}{\pi} \rfloor, \ b > 0. \tag{3.209}$$

The Eq. (3.207) can be equivalently represented by

$$\begin{pmatrix} \dot{X}_1(t) \\ \dot{\overline{X}}_1(t) \\ \dot{X}_2(t) \\ \dot{\overline{X}}_2(t) \end{pmatrix} = \begin{pmatrix} A_1 & \mathbf{0} & -2sC_1 & -2sC_2 \\ \mathbf{0} & \overline{A}_1 & -2sC_3 & -2sC_4 \\ \mathbf{0} & \mathbf{0} & A_2 & \mathbf{0} \\ \mathbf{0} & \mathbf{0} & \mathbf{0} & \overline{A}_2 \end{pmatrix} \begin{pmatrix} X_1(t) \\ \overline{X}_1(t) \\ X_2(t) \\ \overline{X}_2(t) \end{pmatrix} - \begin{pmatrix} \mathbf{0} \\ \mathbf{0} \\ D \\ \overline{D} \end{pmatrix} U(t), \tag{3.210}$$

where the column vectors $\overline{X}_1(t)$, $\overline{X}_2(t)$, $\overline{D}$ are infinite vectors, and $\overline{A}_1$, $\overline{A}_2$, C_2, C_3, C_4 are infinite matrix.

Moreover,

$$\begin{aligned}
X_1(t) &\triangleq \left(X_{1,0}(t), X_{1,1}(t), \ldots, X_{1,N_1}(t)\right)^T \in R^{(N_1+1)\times 1}, \\
\overline{X}_1(t) &\triangleq \left(X_{1,N_1+1}(t), \ldots, X_{1,n_1}(t), \ldots\right)^T, n_1 = N_1 + 1, \ldots \\
X_2(t) &\triangleq \left(X_{2,0}(t), X_{2,1}(t), \ldots, X_{2,N_2}(t)\right)^T \in R^{(N_2+1)\times 1}, \\
\overline{X}_2(t) &\triangleq \left(X_{2,N_2+1}(t), \ldots, X_{2,n_2}(t), \ldots\right)^T, n_2 = N_2 + 1, \ldots \\
A_1 &\triangleq \text{diag}\left\{\lambda_{1,0}, \lambda_{1,1} \ldots, \lambda_{1,N_1}\right\} \in R^{(N_1+1)\times(N_1+1)}, \\
\overline{A}_1 &\triangleq \text{diag}\left\{\lambda_{1,N_1+1}, \ldots, \lambda_{1,n_1}, \ldots\right\}, n_1 = N_1 + 1, \ldots \\
A_2 &\triangleq \text{diag}\left\{\lambda_{2,0}, \lambda_{2,1} \ldots, \lambda_{2,N_2}\right\} \in R^{(N_2+1)\times(N_2+1)}, \\
\overline{A}_2 &\triangleq \text{diag}\left\{\lambda_{2,N_2+1}, \ldots, \lambda_{2,n_2}, \ldots\right\}, n_2 = N_2 + 1, \ldots \\
C_1 &\triangleq \begin{pmatrix} 1 & 1 & \cdots & 1 \\ \vdots & \vdots & & \vdots \\ 1 & 1 & \cdots & 1 \end{pmatrix} \in R^{(N_1+1)\times(N_2+1)}, \\
C_2 &\triangleq \left(1_{n_1 n_2}\right) = \begin{pmatrix} 1 & 1 & \cdots \\ \vdots & \vdots & \\ 1 & 1 & \cdots \end{pmatrix}, n_1 = 0, \ldots, N_1; \ n_2 = N_2 + 1, \ldots
\end{aligned}$$

$$C_3 \triangleq \left(1_{n_1 n_2}\right) = \begin{pmatrix} 1 & \cdots & 1 \\ 1 & \cdots & 1 \\ \vdots & & \vdots \end{pmatrix}, n_1 = N_1 + 1, \ldots;\ n_2 = 0, \ldots, N_2.$$

$$C_4 \triangleq \left(1_{n_1 n_2}\right) = \begin{pmatrix} 1 & 1 & \cdots \\ 1 & 1 & \cdots \\ \vdots & \vdots & \ddots \end{pmatrix}, n_1 = N_1 + 1, \ldots; n_2 = N_2 + 1, \ldots$$

$$D \triangleq \left(\sqrt{2}, \sqrt{2}, \ldots, \sqrt{2}\right)^T \in R^{(N_2+1)\times 1},$$

$$\overline{D} \triangleq \left(\sqrt{2}_{n_2 1}\right) = \left(\sqrt{2}, \sqrt{2}, \ldots\right)^T, n_2 = N_2 + 1, \ldots$$

From (3.210), it follows that $\dot{X}_2(t) = A_2 X_2(t) - DU(t)$, where A_2 is unstable. Since the pair $(A_2, -D)$ is stabilizable, there exists $K \in \mathbb{R}^{1\times(N_2+1)}$ such that $A_2 + DK$ is Hurwitz. Now, we design the following boundary control law

$$U(t) = -KX_2(t) + W(t), \tag{3.211}$$

where $W(t)$ will be determined later.

In order to find $W(t)$, a new transformation is introduced to decouple X_1 and X_2 (c.f. (3.210)):

$$\begin{pmatrix} Y_1(t) \\ \overline{Y}_1(t) \\ Y_2(t) \\ \overline{Y}_2(t) \end{pmatrix} = \begin{pmatrix} I & \mathbf{0} & T & \mathbf{0} \\ \mathbf{0} & I & \mathbf{0} & \mathbf{0} \\ \mathbf{0} & \mathbf{0} & I & \mathbf{0} \\ \mathbf{0} & \mathbf{0} & \mathbf{0} & I \end{pmatrix} \begin{pmatrix} X_1(t) \\ \overline{X}_1(t) \\ X_2(t) \\ \overline{X}_2(t) \end{pmatrix}, \tag{3.212}$$

where $\overline{Y}_1(t)$, $\overline{Y}_2(t)$ are infinite column vectors with explicit forms as follows

$$\begin{aligned} Y_1(t) &= \left(Y_{1,0}(t), Y_{1,1}(t), \ldots, Y_{1,N_1}(t)\right)^T \in \mathbb{R}^{(N_1+1)\times 1}, \\ \overline{Y}_1(t) &= \left(Y_{1,N_1+1}(t), \ldots, Y_{1,n_1}(t), \ldots\right)^T, \ n_1 = N_1 + 1, \ldots \\ Y_2(t) &= \left(Y_{2,0}(t), Y_{2,1}(t), \ldots, Y_{2,N_2}(t)\right)^T \in \mathbb{R}^{(N_2+1)\times 1}, \\ \overline{Y}_2(t) &= \left(Y_{2,N_2+1}(t), \ldots, Y_{2,n_2}(t), \ldots\right)^T, \ n_2 = N_2 + 1, \ldots \end{aligned}$$

and $T \in \mathbb{R}^{(N_1+1)\times(N_2+1)}$ is the solution of the Sylvester equation:

$$-A_1 T - 2sC_1 + T(A_2 + DK) = 0. \tag{3.213}$$

Note that $A_2 + DK$ is Hurwitz and A_1 is unstable. It means that $A_2 + DK$ and A_1 have no common eigenvalues. From Theorem 8.2.1 of [35], the Sylvester equation (3.213) has a unique solution T in $\mathbb{R}^{(N_1+1)\times(N_2+1)}$.

Substituting (3.211) into (3.210), and applying the transformation (3.212), we obtain that (3.210) is equivalent to

$$\begin{pmatrix} \dot{Y}_1(t) \\ \dot{\overline{Y}}_1(t) \\ \dot{Y}_2(t) \\ \dot{\overline{Y}}_2(t) \end{pmatrix} = \begin{pmatrix} A_1 & \mathbf{0} & \mathbf{0} & -2sC_2 \\ \mathbf{0} & \overline{A}_1 & -2sC_3 & -2sC_4 \\ \mathbf{0} & \mathbf{0} & A_2+DK & \mathbf{0} \\ \mathbf{0} & \mathbf{0} & \overline{D}K & \overline{A}_2 \end{pmatrix} \begin{pmatrix} Y_1(t) \\ \overline{Y}_1(t) \\ Y_2(t) \\ \overline{Y}_2(t) \end{pmatrix} - \begin{pmatrix} TD \\ \mathbf{0} \\ D \\ \overline{D} \end{pmatrix} W(t),$$

which implies $\dot{Y}_1(t) = A_1 Y_1(t) - 2sC_2\overline{Y}_2(t) - TDW(t)$.

Lemma 3.3 *Consider the system*

$$\begin{cases} \dot{Y}_1(t) = A_1 Y_1(t) - 2sC_2\overline{Y}_2(t) - TDW(t), \\ Y_1(0) = Y_1^* \in \mathbb{R}^{(N_1+1)\times 1}, \end{cases} \tag{3.214}$$

where A_1, C_2, D are given by (3.210) and Y_1^* is initial value of $Y_1(t)$. Let $T = (a_{ij}) \in \mathbb{R}^{(N_1+1)\times(N_2+1)}$ be the solution of the Sylvester equation (3.213). If $b - a$ is not the solution of the equation

$$\sum_{j=0}^{N_2} \frac{1}{-4x+[4j^2-(2i+1)^2]\pi^2} = 0, \ i = 0, 1, \ldots, N_1, \tag{3.215}$$

then

$$\sum_{j=0}^{N_2} a_{ij} \neq 0, \quad i = 0, 1, \ldots, N_1. \tag{3.216}$$

Proof Let (3.216) be false. Denote the matrix $B \triangleq (b_{ij})$, $i = 0, 1, \ldots, N_1$; $j = 0, 1, \ldots, N_2$. And $b_{ij} \triangleq a_{ij}(\lambda_{1,i} - \lambda_{2,j}) - \sqrt{2}k_j \sum_{j=0}^{N_2} a_{ij}$. From the Sylvester equation (3.213), it should be noticed that

$$a_{ij}(\lambda_{1,i} - \lambda_{2,j}) - \sqrt{2}k_j \sum_{j=0}^{N_2} a_{ij} = -2s.$$

For arbitrary $(i+1)$-th row,

$$\begin{cases} a_{i0}(\lambda_{1,i} - \lambda_{2,0}) - \sqrt{2}k_0 \sum_{j=0}^{N_2} a_{ij} = -2s, \\ a_{i1}(\lambda_{1,i} - \lambda_{2,1}) - \sqrt{2}k_1 \sum_{j=0}^{N_2} a_{ij} = -2s, \\ \vdots \\ a_{iN_2}(\lambda_{1,i} - \lambda_{2,N_2}) - \sqrt{2}k_{N_2} \sum_{j=0}^{N_2} a_{ij} = -2s, \end{cases}$$

which yields

$$\begin{cases} a_{i0}(\lambda_{1,i} - \lambda_{2,0}) = -2\,s, \\ a_{i1}(\lambda_{1,i} - \lambda_{2,1}) = -2\,s, \\ \quad\vdots \\ a_{iN_2}(\lambda_{1,i} - \lambda_{2,N_2}) = -2\,s. \end{cases}$$

Now the discussion is divided into two cases.
Case 1: If there exists j such that $\lambda_{1,i} - \lambda_{2,j} = 0$, then $0 = -2s$, which leads to a contradiction with $s \neq 0$.
Case 2: If $\lambda_{1,i} - \lambda_{2,j} \neq 0$ holds for any j, then

$$\begin{cases} a_{i0} = -2\,s\frac{1}{\lambda_{i,0}-\lambda_{2,0}}, \\ a_{i1} = -2\,s\frac{1}{\lambda_{i,0}-\lambda_{2,1}}, \\ \quad\vdots \\ a_{iN_2} = -2\,s\frac{1}{\lambda_{i,0}-\lambda_{2,N_2}}, \end{cases}$$

which yields $\sum\limits_{j=0}^{N_2} a_{ij} = -2s \sum\limits_{j=0}^{N_2} \frac{1}{\lambda_{1,i}-\lambda_{2,j}}$. From $s \neq 0$ and $\sum\limits_{j=0}^{N_2} a_{ij} = 0$, it follows that $\sum\limits_{j=0}^{N_2} \frac{1}{\lambda_{1,i}-\lambda_{2,j}} = 0$. Recalling the eigenvalues given by (3.204) and (3.206), we have $\sum\limits_{j=0}^{N_2} \frac{1}{4a-4b+[4j^2-(2i+1)^2]\pi^2} = 0$, which contradicts to the assumption of $b - a$. Hence, for both cases, $\sum\limits_{j=0}^{N_2} a_{ij} \neq 0$. □

Lemma 3.4 *Consider the system (3.214). Let $T = (a_{ij}) \in \mathbb{R}^{(N_1+1)\times(N_2+1)}$ be the solution of the Sylvester equation (3.213). Assume that $b - a$ is not the solution of the Eq. (3.215). Then there exists $K_1 \in \mathbb{R}^{1\times(N_1+1)}$ such that $A_1 + TDK_1$ is Hurwitz.*

Proof There exists $K_1 \in \mathbb{R}^{1\times(N_1+1)}$ such that $A_1 + TDK_1$ is Hurwitz $\Longleftrightarrow$ The system (3.214) is stabilizable $\Longleftrightarrow$ The system (3.214) is controllable $\Longleftrightarrow$

$$\operatorname{rank}\Big(TD,\, A_1TD,\, A_1^2TD, \ldots, A_1^{N_1}TD\Big) = N_1 + 1, \tag{3.217}$$

where

$$\begin{aligned}&\left(TD, A_1TD, A_1^2TD, \ldots, A_1^{N_1}TD\right)\\&=\begin{pmatrix}\sqrt{2}\sum\limits_{j=0}^{N_2}a_{0j} & \sqrt{2}\lambda_{1,0}\sum\limits_{j=0}^{N_2}a_{0j} & \cdots & \sqrt{2}\lambda_{1,0}^{N_1}\sum\limits_{j=0}^{N_2}a_{0j}\\ \sqrt{2}\sum\limits_{j=0}^{N_2}a_{1j} & \sqrt{2}\lambda_{1,1}\sum\limits_{j=0}^{N_2}a_{1j} & \cdots & \sqrt{2}\lambda_{1,1}^{N_1}\sum\limits_{j=0}^{N_2}a_{1j}\\ \vdots & \vdots & & \vdots\\ \sqrt{2}\sum\limits_{j=0}^{N_2}a_{N_1j} & \sqrt{2}\lambda_{1,N_1}\sum\limits_{j=0}^{N_2}a_{N_1j} & \cdots & \sqrt{2}\lambda_{1,N_1}^{N_1}\sum\limits_{j=0}^{N_2}a_{N_1j}\end{pmatrix}.\end{aligned}$$

Then (3.217) is equivalent to

$$\begin{aligned}&\begin{vmatrix}\sqrt{2}\sum\limits_{j=0}^{N_2}a_{0j} & \sqrt{2}\lambda_{1,0}\sum\limits_{j=0}^{N_2}a_{0j} & \cdots & \sqrt{2}\lambda_{1,0}^{N_1}\sum\limits_{j=0}^{N_2}a_{0j}\\ \sqrt{2}\sum\limits_{j=0}^{N_2}a_{1j} & \sqrt{2}\lambda_{1,1}\sum\limits_{j=0}^{N_2}a_{1j} & \cdots & \sqrt{2}\lambda_{1,1}^{N_1}\sum\limits_{j=0}^{N_2}a_{1j}\\ \vdots & \vdots & & \vdots\\ \sqrt{2}\sum\limits_{j=0}^{N_2}a_{N_1j} & \sqrt{2}\lambda_{1,N_1}\sum\limits_{j=0}^{N_2}a_{N_1j} & \cdots & \sqrt{2}\lambda_{1,N_1}^{N_1}\sum\limits_{j=0}^{N_2}a_{N_1j}\end{vmatrix}\\&=(\sqrt{2})^{N_1+1}\sum_{j=0}^{N_2}a_{0j}\cdots\sum_{j=0}^{N_2}a_{N_1j}\begin{vmatrix}1 & \lambda_{1,0} & \cdots & \lambda_{1,0}^{N_1}\\ 1 & \lambda_{1,1} & \cdots & \lambda_{1,1}^{N_1}\\ \vdots & \vdots & & \vdots\\ 1 & \lambda_{1,N_1} & \cdots & \lambda_{1,N_1}^{N_1}\end{vmatrix}\\&=(\sqrt{2})^{N_1+1}\sum_{j=0}^{N_2}a_{0j}\cdots\sum_{j=0}^{N_2}a_{N_1j}\prod_{0\le l<i\le N_1}(\lambda_{1,i}-\lambda_{1,l})\neq 0.\end{aligned}\tag{3.218}$$

Since $\lambda_{1,0}, \ldots, \lambda_{1,N_1}$ are different, we conclude that (3.218) holds if and only if $\sum\limits_{j=0}^{N_2} a_{ij} \neq 0$ $(i = 0, 1, \ldots, N_1)$ is satisfied. By Lemma 3.3, we find that if $b - a$ is not the solution of the Eq. (3.215), then $\sum\limits_{j=0}^{N_2} a_{ij} \neq 0 \quad (i = 0, 1, \ldots, N_1)$. Thus, there exists $K_1 \in \mathbb{R}^{1\times(N_1+1)}$ such that $A_1 + TDK_1$ is Hurwitz. □

Based on the above results Lemmas 3.3 and 3.4, we construct the following controller

$$U(t) = -KX_2(t) - K_1X_1(t) - K_1TX_2(t), \tag{3.219}$$

where K, K_1 such that $A_2 + DK$, $A_1 + TDK_1$ are Hurwitz.

By selecting the controller (3.219), we arrive at the closed-loop system

$$\begin{pmatrix} \dot{X}_1(t) \\ \dot{\overline{X}}_1(t) \\ \dot{X}_2(t) \\ \dot{\overline{X}}_2(t) \end{pmatrix} = \begin{pmatrix} A_1 & \mathbf{0} & -2sC_1 & -2sC_2 \\ \mathbf{0} & \overline{A}_1 & -2sC_3 & -2sC_4 \\ DK_1 & \mathbf{0} & A_2 + DK + DK_1 T & \mathbf{0} \\ \overline{D}K_1 & \mathbf{0} & \overline{D}K + \overline{D}K_1 T & \overline{A}_2 \end{pmatrix} \begin{pmatrix} X_1(t) \\ \overline{X}_1(t) \\ X_2(t) \\ \overline{X}_2(t) \end{pmatrix}. \tag{3.220}$$

Note that the transformation (3.212) transforms the closed-loop system (3.220) into

$$\begin{pmatrix} \dot{Y}_1(t) \\ \dot{\overline{Y}}_1(t) \\ \dot{Y}_2(t) \\ \dot{\overline{Y}}_2(t) \end{pmatrix} = \begin{pmatrix} A_1 + TDK_1 & \mathbf{0} & \mathbf{0} & -2sC_2 \\ \mathbf{0} & \overline{A}_1 & -2sC_3 & -2sC_4 \\ DK_1 & \mathbf{0} & A_2 + DK & \mathbf{0} \\ \overline{D}K_1 & \mathbf{0} & \overline{D}K & \overline{A}_2 \end{pmatrix} \begin{pmatrix} Y_1(t) \\ \overline{Y}_1(t) \\ Y_2(t) \\ \overline{Y}_2(t) \end{pmatrix}. \tag{3.221}$$

Therefore, the Eq. (3.221) can be represented as follows

$$\begin{cases} \dot{Y}(t) = FY(t) + S \sum\limits_{n_2=N_2+1}^{\infty} Y_{2,n_2}(t), \\ \dot{Y}_{1,n_1}(t) = \lambda_{1,n_1} Y_{1,n_1}(t) - 2s \sum\limits_{n_2=N_2+1}^{\infty} Y_{2,n_2}(t) + \tilde{K}_1 Y(t), n_1 > N_1, \\ \dot{Y}_{2,n_2}(t) = \lambda_{2,n_2} Y_{2,n_2}(t) + \tilde{K}_2 Y(t), n_2 > N_2, \end{cases} \tag{3.222}$$

where

$$F \triangleq \begin{pmatrix} A_1 + TDK_1 & \mathbf{0} \\ DK_1 & A_2 + DK \end{pmatrix},$$

$$\tilde{K}_1 \triangleq \Big(\underbrace{0, 0, \ldots, 0}_{N_1+1}, \underbrace{-2s, -2s, \ldots, -2s}_{N_2+1}\Big),$$

$$\tilde{K}_2 \triangleq \left(\sqrt{2}K_1 \; \sqrt{2}K\right), Y(t) \triangleq \Big(Y_1(t), Y_2(t)\Big)^T,$$

$$S \triangleq \Big(\underbrace{-2s, -2s, \ldots, -2s}_{N_1+1}, \underbrace{0, 0, \ldots, 0}_{N_2+1}\Big)^T.$$

Now we derive stability conditions for the system (3.222) via Lyapunov approach. Consider the Lyapunov function candidate

$$V_1(t) = Y^T(t)PY(t) + \sum_{n_1=N_1+1}^{\infty} Y_{1,n_1}^2(t) + \sum_{n_2=N_2+1}^{\infty} Y_{2,n_2}^2(t), \tag{3.223}$$

where the matrix $0 < P \in \mathbb{R}^{(N_1+N_2+2)\times(N_1+N_2+2)}$.

Differentiating $V_1(t)$ along the solution to the closed-loop system (3.222), we have

$$\begin{aligned}
&\dot{V}_1(t) + 2\delta V_1(t) = Y^T(t)\Big(PF + F^T P + 2\delta P\Big)Y(t)\\
&+2\sum_{n_1=N_1+1}^{\infty}\Big(\lambda_{1,n_1} + \delta\Big)Y_{1,n_1}^2(t) + 2\sum_{n_2=N_2+1}^{\infty}\Big(\lambda_{2,n_2} + \delta\Big)Y_{2,n_2}^2(t)\\
&-4s\sum_{n_1=N_1+1}^{\infty} Y_{1,n_1}(t)\sum_{n_2=N_2+1}^{\infty} Y_{2,n_2}(t) + 2\sum_{n_1=N_1+1}^{\infty} Y_{1,n_1}(t)\tilde{K}_1 Y(t)\\
&+2\sum_{n_2=N_2+1}^{\infty} Y_{2,n_2}(t)\tilde{K}_2 Y(t) + 2Y^T PS\sum_{n_2=N_2+1}^{\infty} Y_{2,n_2}(t).
\end{aligned} \tag{3.224}$$

Next we need to evaluate the terms $\sum\limits_{n_1=N_1+1}^{\infty} Y_{1,n_1}(t)\tilde{K}_1 Y(t)$, $\sum\limits_{n_2=N_2+1}^{\infty} Y_{2,n_2}(t)\tilde{K}_2 Y(t)$ from (3.224). The Cauchy-Schwarz's and Young's inequalities yield

$$\sum_{n_1=N_1+1}^{\infty} 2Y_{1,n_1}(t)\tilde{K}_1 Y(t) \le \tfrac{1}{\alpha_1}\sum_{n_1=N_1+1}^{\infty} Y_{1,n_1}^2(t) + \alpha_1\Big\|\tilde{K}_1 Y(t)\Big\|^2, \tag{3.225}$$

$$\sum_{n_2=N_2+1}^{\infty} 2Y_{2,n_2}(t)\tilde{K}_2 Y(t) \le \tfrac{1}{\alpha_2}\sum_{n_2=N_2+1}^{\infty} Y_{2,n_2}^2(t) + \alpha_2\Big\|\tilde{K}_2 Y(t)\Big\|^2, \tag{3.226}$$

for any $\alpha_1 > 0, \alpha_2 > 0$.

Denote $\xi(t) = \sum\limits_{n_1=N_1+1}^{\infty} Y_{1,n_1}(t), \eta(t) = \sum\limits_{n_2=N_2+1}^{\infty} Y_{2,n_2}(t), \zeta(t) = \mathrm{col}\Big\{Y(t), \xi(t), \eta(t)\Big\}$.
By using the Cauchy-Schwarz inequality, we obtain

$$\begin{aligned}
\xi^2(t) &= \Big[\sum_{n_1=N_1+1}^{\infty} \frac{1}{(-\lambda_{1,n_1})^{\frac{1}{2}}}(-\lambda_{1,n_1})^{\frac{1}{2}} Y_{1,n_1}(t)\Big]^2\\
&\le M_1\sum_{n_1=N_1+1}^{\infty}\Big|\lambda_{1,n_1}\Big| Y_{1,n_1}^2(t),
\end{aligned} \tag{3.227}$$

$$\begin{aligned}
\eta^2(t) &= \Big[\sum_{n_2=N_2+1}^{\infty} \frac{1}{(-\lambda_{2,n_2})^{\frac{1}{2}}}(-\lambda_{2,n_2})^{\frac{1}{2}} Y_{2,n_2}(t)\Big]^2\\
&\le M_2\sum_{n_2=N_2+1}^{\infty}\Big|\lambda_{2,n_2}\Big| Y_{2,n_2}^2(t),
\end{aligned} \tag{3.228}$$

for some constants $M_1 > 0$ and $M_2 > 0$.

Taking into account all the above estimates (3.224)–(3.228), we have

$$\dot{V}_1(t) + 2\delta V_1(t) + \beta_1\Big[M_1 \sum_{n_1=N_1+1}^{\infty} |\lambda_{1,n_1}| Y_{1,n_1}^2(t) - \xi^2(t)\Big]$$
$$+\beta_2\Big[M_2 \sum_{n_2=N_2+1}^{\infty} |\lambda_{2,n_2}| Y_{2,n_2}^2(t) - \eta^2(t)\Big] \leq \zeta^T(t)\Psi\zeta(t) \tag{3.229}$$
$$+2 \sum_{n_1=N_1+1}^{\infty} W_{1,n_1} Y_{1,n_1}^2(t) + 2 \sum_{n_2=N_2+1}^{\infty} W_{2,n_2} Y_{2,n_2}^2(t).$$

If $W_{1,n_1} = \lambda_{1,n_1} + \delta + \frac{1}{2\alpha_1} + \frac{M_1\beta_1|\lambda_{1,n_1}|}{2} < 0, n_1 > N_1$, $W_{2,n_2} = \lambda_{2,n_2} + \delta + \frac{1}{2\alpha_2} + \frac{M_2\beta_2|\lambda_{2,n_2}|}{2} < 0, n_2 > N_2$,

$$\Psi = \begin{pmatrix} \Psi_0 & 0 & PS \\ 0 & -\beta_1 & -2s \\ S^T P & -2s & -\beta_2 \end{pmatrix} < 0, \tag{3.230}$$

then $\dot{V}_1(t) + 2\delta V_1(t) \leq 0$ holds, where $\Psi_0 = PF + F^T P + 2\delta P + \alpha_1 \tilde{K}_1^T \tilde{K}_1 + \alpha_2 \tilde{K}_2^T \tilde{K}_2$.

The monotonicity of $\left\{\lambda_{1,n_1}\right\}_{n_1=0}^{\infty}$ and Schur Complement Theorem imply that $W_{1,n_1} < 0$ for all $n_1 > N_1$ iff

$$\begin{pmatrix} \lambda_{1,N_1+1} + \delta + \frac{M_1\beta_1|\lambda_{1,N_1+1}|}{2} & \frac{1}{\sqrt{2}} \\ \frac{1}{\sqrt{2}} & -\alpha_1 \end{pmatrix} < 0. \tag{3.231}$$

Similarly, the monotonicity of $\left\{\lambda_{2,n_2}\right\}_{n_2=0}^{\infty}$ and Schur Complement Theorem imply that $W_{2,n_2} < 0$ for all $n_2 > N_2$ iff

$$\begin{pmatrix} \lambda_{2,N_2+1} + \delta + \frac{M_2\beta_2|\lambda_{2,N_2+1}|}{2} & \frac{1}{\sqrt{2}} \\ \frac{1}{\sqrt{2}} & -\alpha_2 \end{pmatrix} < 0. \tag{3.232}$$

Summarizing the above, we arrive at the following main result.

Theorem 3.7 *Assume that $b-a$ is not a solution of the Eq. (3.215). Let $T \in \mathbb{R}^{(N_1+1)\times(N_2+1)}$ be the solution of the Sylvester equations (3.213). Let $A_1 + TDK_1$ and $A_2 + TK$ be Hurwitz. Let M_1, M_2 be positive constants satisfying $M_1 > \sum_{n_1=N_1+1}^{\infty} \frac{1}{-a+n_1^2\pi^2}$, and $M_2 > \sum_{n_2=N_2+1}^{\infty} \frac{1}{-b+n_2^2\pi^2}$. Given the decay rate $\delta > 0$, if there exist a positive definite matrix $P \in \mathbb{R}^{(N_1+N_2+2)\times(N_1+N_2+2)}$, and positive scalars α_1, α_2, β_1, β_2 such that (3.230)–(3.232) hold, then for any initial functions $u_0 \in \mathcal{H}$, $z_0 \in \mathcal{H}$, the closed-loop system (3.220) is exponentially stable in the sense that*

$$\|u(\cdot,t)\|_{\mathcal{H}}^2 + \|z(\cdot,t)\|_{\mathcal{H}}^2 \leq M_0 e^{-2\delta t}\Big(\|u_0\|_{\mathcal{H}}^2 + \|z_0\|_{\mathcal{H}}^2\Big), t \geq 0, \tag{3.233}$$

for some constant $M_0 > 0$.

Proof From (3.229), we obtain that if (3.230)–(3.232) hold, then

$$V_1(t) \le e^{-2\delta t} V_1(0), \quad \forall t \ge 0. \tag{3.234}$$

Note that the transformation (3.212) is invertible. Hence,

$$m\|X_1\|^2 \le \|Y_1\|^2 \le M\|X_1\|^2. \tag{3.235}$$

for some constants $m > 0$, $M > 0$.

From (3.223) and (3.235), we obtain

$$\begin{aligned} &\min\left\{m, 1\right\} \min\left\{\lambda_{min}(P), 1\right\}\left(\|u(\cdot,t)\|^2_{\mathcal{H}} + \|z(\cdot,t)\|^2_{\mathcal{H}}\right) \le V_1(t) \\ &\le \max\left\{M, 1\right\} \max\left\{\lambda_{max}(P), 1\right\}\left(\|u(\cdot,t)\|^2_{\mathcal{H}} + \|z(\cdot,t)\|^2_{\mathcal{H}}\right), \end{aligned} \tag{3.236}$$

Therefore, (3.234) and (3.236) imply (3.233). □

Remark 3.12 Note that $\sum\limits_{n_2=0}^{\infty} X_{2,n_2}(t) = \sum\limits_{n_2=0}^{N_2} X_{2,n_2}(t) + \sum\limits_{n_2=N_2+1}^{\infty} X_{2,n_2}(t)$. Since $\sum\limits_{n_2=0}^{N_2} X_{2,n_2}(t) < \infty$, here we just need to prove $\sum\limits_{n_2=N_2+1}^{\infty} X_{2,n_2}(t) < \infty$. From $N_2 = \left\lfloor \frac{\sqrt{b}}{\pi} \right\rfloor$ (c.f. (3.209)), it follows that for $n_2 \ge N_2 + 1$, $b - n_2^2\pi^2 < 0$ holds. By solving the second equation in (3.207), for $n_2 = N_2 + 1, N_2 + 2, \ldots$, we have $X_{2,n_2}(t) = X_{2,n_2}(0)e^{(b-n_2^2\pi^2)t} - \sqrt{2}\int_0^t e^{(b-n_2^2\pi^2)(t-s)}U(s)ds$. A direct computation via the Cauchy-Schwarz's and Hölder inequalities leads to

$$\left| \sum_{n_2=N_2+1}^{\infty} X_{2,n_2}(t) \right| \le \sum_{n_2=N_2+1}^{\infty} |X_{2,n_2}(0)| + \sqrt{2}\|U\|_\infty \sum_{n_2=N_2+1}^{\infty} \frac{1}{-b + n_2^2\pi^2}.$$

Therefore, $\sum\limits_{n_2=N_2+1}^{\infty} X_{2,n_2}(t) < \infty$.

Remark 3.13 If $b - a$ is the solution of the Eq. (3.215), then from the proof of Lemma 3.4 it follows that (A_1, TD) maybe not controllable, let alone the stabilization.

3.4.3 Sampled-Data Control

This section is devoted to suggesting the sampled-data controller for stabilizing the cascaded PDE-PDE system. Meanwhile, well-posedness proof and stability analysis will be established.

Let $0 = t_0 < t_1 < \cdots < t_k < \cdots$, $\lim_{k\to\infty} t_k = \infty$ be sampling time instants. The sampling intervals in time may be variable but bounded, $0 \le t_{k+1} - t_k \le h$, $\forall k \ge 0$, where h is the corresponding upper bound.

Selecting the sampled-data controller

$$U(t) = -K_1X_1(t_k) - (K + K_1T)X_2(t_k),\ t_k \le t < t_{k+1}, \tag{3.237}$$

we arrive at the following closed-loop system

$$\begin{aligned}\begin{pmatrix}\dot{X}_1(t)\\ \dot{\overline{X}}_1(t)\\ \dot{X}_2(t)\\ \dot{\overline{X}}_2(t)\end{pmatrix} &= \begin{pmatrix}A_1 & \mathbf{0} & -2sC_1 & -2sC_2\\ \mathbf{0} & \overline{A}_1 & -2sC_3 & -2sC_4\\ \mathbf{0} & \mathbf{0} & A_2 & \mathbf{0}\\ \mathbf{0} & \mathbf{0} & \mathbf{0} & \overline{A}_2\end{pmatrix}\begin{pmatrix}X_1(t)\\ \overline{X}_1(t)\\ X_2(t)\\ \overline{X}_2(t)\end{pmatrix}\\ &\quad + \begin{pmatrix}\mathbf{0} & \mathbf{0} & \mathbf{0} & \mathbf{0}\\ \mathbf{0} & \mathbf{0} & \mathbf{0} & \mathbf{0}\\ DK_1 & \mathbf{0} & DK + DK_1T & \mathbf{0}\\ \overline{D}K_1 & \mathbf{0} & \overline{D}K + \overline{D}K_1T & \mathbf{0}\end{pmatrix}\begin{pmatrix}X_1(t_k)\\ \overline{X}_1(t_k)\\ X_2(t_k)\\ \overline{X}_2(t_k)\end{pmatrix},\ t \in [t_k, t_{k+1}).\end{aligned} \tag{3.238}$$

In what follows, we will establish the well-posedness and the stability analysis of the closed-loop system (3.238). Via the transformation (3.212), the system (3.238) is converted into the following form:

$$\begin{aligned}\begin{pmatrix}\dot{Y}_1(t)\\ \dot{\overline{Y}}_1(t)\\ \dot{Y}_2(t)\\ \dot{\overline{Y}}_2(t)\end{pmatrix} &= \begin{pmatrix}A_1 & \mathbf{0} & -2sC_1 + TA_2 & -2sC_2\\ \mathbf{0} & \overline{A}_1 & -2sC_3 & -2sC_4\\ \mathbf{0} & \mathbf{0} & A_2 & \mathbf{0}\\ \mathbf{0} & \mathbf{0} & \mathbf{0} & \overline{A}_2\end{pmatrix}\begin{pmatrix}Y_1(t)\\ \overline{Y}_1(t)\\ Y_2(t)\\ \overline{Y}_2(t)\end{pmatrix}\\ &\quad + \begin{pmatrix}TDK_1 & \mathbf{0} & TDK + TDK_1T & \mathbf{0}\\ \mathbf{0} & \mathbf{0} & \mathbf{0} & \mathbf{0}\\ DK_1 & \mathbf{0} & DK + DK_1T & \mathbf{0}\\ \overline{D}K_1 & \mathbf{0} & \overline{D}K + \overline{D}K_1T & \mathbf{0}\end{pmatrix}\begin{pmatrix}Y_1(t_k)\\ \overline{Y}_1(t_k)\\ Y_2(t_k)\\ \overline{Y}_2(t_k)\end{pmatrix},\ t \in [t_k, t_{k+1}).\end{aligned} \tag{3.239}$$

Without loss of generality, we first consider the system (3.239) on the first time interval $[0, t_1)$. It follows that

$$\begin{cases} \dot{Y}_{1,n_1}(t) = \lambda_{1,n_1} Y_{1,n_1}(t) - 2s \sum\limits_{n_2=0}^{\infty} Y_{2,n_2}(t) + \sum\limits_{n_1=0}^{N_1} \sum\limits_{n_2=0}^{N_2} a_{n_1 n_2} \lambda_{2,n_2} \times Y_{2,n_2}(t) + \tilde{F}_1, \\ \qquad\qquad\qquad\qquad\qquad\qquad\qquad\qquad\qquad\qquad 0 \le n_1 \le N_1, \\ \dot{Y}_{1,n_1}(t) = \lambda_{1,n_1} Y_{1,n_1}(t) - 2s \sum\limits_{n_2=0}^{\infty} Y_{2,n_2}(t), \qquad n_1 > N_1, \\ \dot{Y}_{2,n_2}(t) = \lambda_{2,n_2} Y_{2,n_2}(t) + \tilde{F}_2, \qquad n_2 \in \mathbb{N}, \\ Y_{1,n_1}(0) = X_{1,n_1}(0) + \sum\limits_{n_2=0}^{N_2} a_{n_1 n_2} X_{2,n_2}(0), \qquad 0 \le n_1 \le N_1, \\ Y_{1,n_1}(0) = X_{1,n_1}(0), \qquad n_1 > N_1, \\ Y_{2,n_2}(0) = X_{2,n_2}(0), \qquad n_2 \in \mathbb{N}, \end{cases}$$

where

$$\tilde{F}_1 \triangleq \sqrt{2} \sum_{n_2=0}^{N_2} a_{n_1 n_2} \Big(K_1 Y_1(0) + (K + K_1 T) Y_2(0) \Big),$$

$$\tilde{F}_2 \triangleq \sqrt{2} K_1 Y_1(0) + \sqrt{2}(K + K_1 T) Y_2(0).$$

A direct computation yields

$$\begin{cases} Y_{1,n_1}(t) = Y_{1,n_1}(0) e^{(a - \frac{(2n_1+1)^2\pi^2}{4})t} + \int_0^t e^{(a - \frac{(2n_1+1)^2\pi^2}{4})(t-s)} \tilde{F}_3(s) ds, \ 0 \le n_1 \le N_1, \\ Y_{1,n_1}(t) = Y_{1,n_1}(0) e^{(a - \frac{(2n_1+1)^2\pi^2}{4})t} + \int_0^t e^{(a - \frac{(2n_1+1)^2\pi^2}{4})(t-s)} \tilde{F}_4(s) ds, \ n_1 > N_1, \\ Y_{2,n_2}(t) = Y_{2,n_2}(0) e^{(b - n_2^2\pi^2)t} + \tilde{F}_2 \int_0^t e^{(b - n_2^2\pi^2)(t-s)} ds, \quad n_2 \in \mathbb{N}, \end{cases}$$

where

$$\begin{aligned} \tilde{F}_3(t) \triangleq & -2s \sum_{n_2=0}^{\infty} \Big(Y_{2,n_2}(0) e^{(b - n_2^2\pi^2)t} + \tilde{F}_2 \int_0^t e^{(b - n_2^2\pi^2)(t-s)} ds \Big) \\ & + \sum_{n_1=0}^{N_1} \sum_{n_2=0}^{N_2} a_{n_1 n_2} (b - n_2^2\pi^2) \Big(Y_{2,n_2}(0) e^{(b - n_2^2\pi^2)t} \\ & + \tilde{F}_2 \int_0^t e^{(b - n_2^2\pi^2)(t-s)} ds \Big) + \tilde{F}_1, \\ \tilde{F}_4(t) \triangleq & -2s \sum_{n_2=0}^{\infty} \Big(Y_{2,n_2}(0) e^{(b - n_2^2\pi^2)t} + \tilde{F}_2 \int_0^t e^{(b - n_2^2\pi^2)(t-s)} ds \Big). \end{aligned}$$

Remark 3.14 From $\sum\limits_{n_2=0}^{\infty} Y_{2,n_2}(t) = \sum\limits_{n_2=0}^{N_2} Y_{2,n_2}(t) + \sum\limits_{n_2=N_2+1}^{\infty} Y_{2,n_2}(t)$, it follows that $\sum\limits_{n_2=0}^{N_2} Y_{2,n_2}(t) < \infty$, so we just need to prove $\sum\limits_{n_2=N_2+1}^{\infty} Y_{2,n_2}(t) < \infty$. A direct computation yields

$$
\left|\sum_{n_2=N_2+1}^{\infty} Y_{2,n_2}(t)\right| \le \frac{1}{2} \sum_{n_2=N_2+1}^{\infty} Y_{2,n_2}^2(0)
+\frac{1}{2} \sum_{n_2=N_2+1}^{\infty} e^{2(b-n_2^2\pi^2)t} + \tilde{F}_2^2 \int_0^t \sum_{n_2=N_2+1}^{\infty} e^{2(b-n_2^2\pi^2)(t-s)}ds.
$$

Note that $Y_{2,n_2}(0) \in l_{N_1+1}^2(\mathbb{N})$. Thus, $\sum\limits_{n_2=N_2+1}^{\infty} Y_{2,n_2}(t) < \infty$.

Note that (3.239) can be rewritten as follows

$$
\begin{cases}
\dot{Y}(t) = F_1 Y(t) + F_2 Y(t_k) + S \sum\limits_{n_2=N_2+1}^{\infty} Y_{2,n_2}(t), \\
\dot{Y}_{1,n_1}(t) = \lambda_{1,n_1} Y_{1,n_1}(t) - 2s \sum\limits_{n_2=N_2+1}^{\infty} Y_{2,n_2}(t) + F_3 Y(t), n_1 > N_1, \\
\dot{Y}_{2,n_2}(t) = \lambda_{2,n_2} Y_{2,n_2}(t) + F_4 Y(t_k), \quad n_2 > N_2,
\end{cases}
\tag{3.240}
$$

where

$$
F_1 \triangleq \begin{pmatrix} A_1 & -2sC_1 + TA_2 \\ \mathbf{0} & A_2 \end{pmatrix},
$$

$$
F_2 \triangleq \begin{pmatrix} TDK_1 & TDK + TDK_1T \\ DK_1 & DK + DK_1T \end{pmatrix},
$$

$$
F_3 \triangleq \Big(\underbrace{0, 0, \ldots, 0}_{N_1+1}, \underbrace{-2s, -2s, \ldots, -2s}_{N_2+1} \Big),
$$

$$
F_4 \triangleq \left(\sqrt{2}K_1 \ \sqrt{2}(K + K_1T)\right),
$$

and the notations of $Y(t)$, S are utilized in (3.222).

Now we derive stability conditions for the closed-loop system (3.240) via Lyapunov approach. Consider the Lyapunov functional candidate

$$
V(t) = V_1(t) + V_2(t), \ t \in [t_k, t_{k+1}), \tag{3.241}
$$

with

$$
V_2(t) = (t_{k+1} - t) \int_{t_k}^{t} e^{2\delta(s-t)} \dot{Y}^T(s) R \dot{Y}(s) ds,
$$

where the matrix $0 < R \in \mathbb{R}^{(N_1+N_2+2)\times(N_1+N_2+2)}$, and $V_1(t)$ is given by (3.223).

Differentiating $V(t)$ along the solution to the closed-loop system (3.240), we have

$$
\begin{aligned}
&\dot{V}(t) + 2\delta V(t) = Y^T(t)\Big(PF_1 + F_1^T P + 2\delta P\Big)Y(t)\\
&+2Y^T PS \sum_{n_2=N_2+1}^{\infty} Y_{2,n_2}(t) + 2\sum_{n_1=N_1+1}^{\infty}\Big(\lambda_{1,n_1}+\delta\Big)Y_{1,n_1}^2(t)\\
&+2\sum_{n_2=N_2+1}^{\infty}\Big(\lambda_{2,n_2}+\delta\Big)Y_{2,n_2}^2(t) - 4s\sum_{n_1=N_1+1}^{\infty} Y_{1,n_1}(t)\sum_{n_2=N_2+1}^{\infty} Y_{2,n_2}(t)\\
&+2Y^T(t)PF_2Y(t_k) + 2\sum_{n_1=N_1+1}^{\infty} Y_{1,n_1}(t)F_3Y(t)\\
&+2\sum_{n_2=N_2+1}^{\infty} Y_{2,n_2}(t)F_4Y(t_k) + (t_{k+1}-t)\dot{Y}^T(t)R\dot{Y}(t)\\
&-\int_{t_k}^{t} e^{2\delta(s-t)}\dot{Y}^T(s)R\dot{Y}(s)ds, \qquad \forall t\in[t_k, t_{k+1}).
\end{aligned}
$$

We now need to evaluate the terms $Y^T(t)PF_2Y(t_k)$, $\sum_{n_1=N_1+1}^{\infty} Y_{1,n_1}(t)F_3Y(t)$, $\sum_{n_2=N_2+1}^{\infty} Y_{2,n_2}(t)F_4Y(t_k)$, $\int_{t_k}^{t} e^{2\delta(s-t)}\dot{Y}^T(s)R\dot{Y}(s)ds$. Denote $v_1 \triangleq \frac{1}{t-t_k}\int_{t_k}^{t}\dot{Y}(s)ds$. It follows that $\lim_{t\to t_k^+} v_1 = \dot{Y}(t)$. It is straightforward to obtain that

$$
2Y^T(t)PF_2Y(t_k) = -2(t-t_k)Y^T(t)PF_2v_1(t) + 2Y^T(t)PF_2Y(t). \tag{3.242}
$$

The Cauchy-Schwarz's and Young's inequalities yield

$$
\sum_{n_1=N_1+1}^{\infty} 2Y_{1,n_1}(t)F_3Y(t) \le \frac{1}{\tilde{\alpha}_1}\sum_{n_1=N_1+1}^{\infty} Y_{1,n_1}^2(t) + \tilde{\alpha}_1\Big\|F_3Y(t)\Big\|^2, \tag{3.243}
$$

$$
\begin{aligned}
&\sum_{n_2=N_2+1}^{\infty} 2Y_{2,n_2}(t)F_4Y(t_k) \le \frac{1}{\tilde{\alpha}_2}\sum_{n_2=N_2+1}^{\infty} Y_{2,n_2}^2(t) + Y^2(t)\\
&+\tilde{\alpha}_2\Big\|F_4\Big\|^2(t-t_k)^2v_1^2(t),
\end{aligned} \tag{3.244}
$$

for any $\tilde{\alpha}_1 > 0, \tilde{\alpha}_2 > 0$.

By utilizing Jensen's inequality, we have

$$
\int_{t_k}^{t}\dot{Y}^T(s)R\dot{Y}(s)ds \ge (t-t_k)v_1^T R v_1, \tag{3.245}
$$

From descriptor method, it follows that

$$
\begin{aligned}
&0 = 2\Big[Y^T(t)P_2^T + \dot{Y}^T(t)P_3^T\Big]\Big[(F_1+F_2)Y(t) - (t-t_k)\\
&\times F_2v_1 + S\sum_{n_2=N_2+1}^{\infty} Y_{2,n_2}(t) - \dot{Y}(t)\Big],
\end{aligned} \tag{3.246}
$$

with some $(N_1 + N_2 + 2) \times (N_1 + N_2 + 2)$-matrices P_2, P_3.

Denote $\tilde{\zeta}_0(t) = \text{col}\Big\{Y(t), \dot{Y}(t), \xi(t), \eta(t)\Big\}$, $\tilde{\zeta}(t) = \text{col}\Big\{Y(t), \dot{Y}(t), v_1(t), \xi(t), \eta(t)\Big\}$, where $\xi(t)$, $\eta(t)$ are given by (3.227) and (3.228). Taking into account all the above estimates (3.227), (3.228), (3.242)–(3.246), we deduce that

$$\begin{aligned}
&\dot{V}(t) + 2\delta V(t) + \tilde{\beta}_1\Big[M_1 \sum_{n_1=N_1+1}^{\infty} |\lambda_{1,n_1}| Y_{1,n_1}^2(t) - \xi^2(t)\Big] \\
&+\tilde{\beta}_2\Big[M_2 \sum_{n_2=N_2+1}^{\infty} |\lambda_{2,n_2}| Y_{2,n_2}^2(t) - \eta^2(t)\Big] \leq \frac{t_{k+1}-t}{h}\tilde{\zeta}_0^T \Psi_1 \tilde{\zeta}_0 \\
&+\frac{t-t_k}{h}\tilde{\zeta}^T \Psi_2 \tilde{\zeta} + 2\sum_{n_1=N_1+1}^{\infty} \tilde{W}_{1,n_1} Y_{1,n_1}^2(t) + 2\sum_{n_2=N_2+1}^{\infty} \tilde{W}_{2,n_2} Y_{2,n_2}^2(t), \quad \forall t \in [t_k, t_{k+1}),
\end{aligned}$$

for some constants $\tilde{\beta}_1, \tilde{\beta}_2 > 0$.

If $\tilde{W}_{1,n_1} = \lambda_{1,n_1} + \delta + \frac{1}{2\tilde{\alpha}_1} + \frac{M_1\tilde{\beta}_1|\lambda_{1,n_1}|}{2} < 0, n_1 > N_1$, $\tilde{W}_{2,n_2} = \lambda_{2,n_2} + \delta + \frac{1}{2\tilde{\alpha}_2} + \frac{M_2\tilde{\beta}_2|\lambda_{2,n_2}|}{2} < 0, n_2 > N_2$,

$$\Psi_1 = \begin{pmatrix} \Psi_{11} & \Psi_{12} & \mathbf{0} & \Psi_{14} \\ * & hR - P_3^T - P_3 & \mathbf{0} & P_3^T S \\ * & * & -\beta_1 & -2s \\ * & * & * & -\beta_2 \end{pmatrix} < 0, \tag{3.247}$$

$$\Psi_2 = \begin{pmatrix} \Psi_{11} & \Psi_{12} & -h(P+P_2)^T F_2 & \mathbf{0} & \Psi_{14} \\ * & -P_3^T - P_3 & -hP_3^T F_2 & \mathbf{0} & P_3^T S \\ * & * & \Psi_{33} & \mathbf{0} & \mathbf{0} \\ * & * & * & -\beta_1 & -2s \\ * & * & * & * & -\beta_2 \end{pmatrix} < 0, \tag{3.248}$$

then

$$\dot{V}(t) + 2\delta V(t) \leq 0, \quad \forall t \in [t_k, t_{k+1}). \tag{3.249}$$

Here

$$\begin{aligned}
&\Psi_{11} = (P+P_2)^T(F_1+F_2) + (F_1+F_2)^T(P+P_2) + 2\delta P + \alpha_1 F_3^T F_3 + I, \\
&\Psi_{12} = (F_1+F_2)^T P_3^T - P_2^T, \qquad \Psi_{14} = PS + P_2^T S, \\
&\Psi_{33} = \alpha_2 F_4^T F_4 h^2 - e^{-2\delta h} hR.
\end{aligned}$$

The monotonicity of $\Big\{\lambda_{1,n_1}\Big\}_{n_1=0}^{\infty}$ and Schur Complement Theorem imply that $\tilde{W}_{1,n_1} < 0$ for all $n_1 > N_1$ iff

$$\begin{pmatrix} \lambda_{1,N_1+1} + \delta + \dfrac{M_1\tilde{\beta}_1|\lambda_{1,N_1+1}|}{2} & \dfrac{1}{\sqrt{2}} \\ \dfrac{1}{\sqrt{2}} & -\tilde{\alpha}_1 \end{pmatrix} < 0. \tag{3.250}$$

Similarly, the monotonicity of $\left\{\lambda_{2,n_2}\right\}_{n_2=0}^{\infty}$ and Schur Complement Theorem imply that $\tilde{W}_{2,n_2} < 0$ for all $n_2 > N_2$ iff

$$\begin{pmatrix} \lambda_{2,N_2+1} + \delta + \dfrac{M_2\tilde{\beta}_2|\lambda_{2,N_2+1}|}{2} & \dfrac{1}{\sqrt{2}} \\ \dfrac{1}{\sqrt{2}} & -\tilde{\alpha}_2 \end{pmatrix} < 0. \tag{3.251}$$

Summarizing the above, we arrive at the main result.

Theorem 3.8 *Assume that $b - a$ is not a solution of the Eq. (3.215). Let $T \in \mathbb{R}^{(N_1+1)\times(N_2+1)}$ be the solution of the Sylvester equation (3.213). Let $A_1 + TDK_1$ and $A_2 + TK$ be Hurwitz. Let M_1, M_2 be positive constants satisfying $M_1 > \sum_{n_1=N_1+1}^{\infty} \frac{1}{-a+n_1^2\pi^2}$, and $M_2 > \sum_{n_2=N_2+1}^{\infty} \frac{1}{-b+n_2^2\pi^2}$. Given the decay rate $\delta > 0$, if there exist positive definite matrices R, P, P_2, $P_3 \in \mathbb{R}^{(N_1+N_2+2)\times(N_1+N_2+2)}$, and positive scalars $\tilde{\alpha}_1$, $\tilde{\alpha}_2$, $\tilde{\beta}_1$, $\tilde{\beta}_2$ such that (3.247)–(3.251) hold, then*
(i) For any initial functions $u_0 \in \mathcal{H}$, $z_0 \in \mathcal{H}$, the closed-loop system (3.238) is exponentially stable in the sense that

$$\|u(\cdot,t)\|_{\mathcal{H}}^2 + \|z(\cdot,t)\|_{\mathcal{H}}^2 \le \tilde{M}_0 e^{-2\delta t}\left(\|u_0\|_{\mathcal{H}}^2 + \|z_0\|_{\mathcal{H}}^2\right), t \ge 0, \tag{3.252}$$

with some constant $\tilde{M}_0 > 0$.
(ii) The closed-loop system (3.238) admits a strong solution.

Proof (i) Assume that the solution of (3.238) exists for all $t \ge 0$. From (3.249), we obtain that if (3.247)–(3.251) hold, then

$$V(t) \le e^{-2\delta t}V(0), \quad \forall t \ge 0. \tag{3.253}$$

Indeed, since $\dot{V}(t) + 2\delta V(t) \le 0$ for $t \in [t_k, t_{k+1})$, we have $V(t) \le e^{-2\delta(t-t_k)}V(t_k)$, $t \in [t_k, t_{k+1})$. From (3.223), we obtain that $V_1(t)$ is left continuous, i.e. $V_1(t_k^-) = V_1(t_k)$. From the definition of $V_2(t)$(c.f.(3.241)), we have $V_2(t_k) = 0$ and $V_2(t_k^-) = \lim_{t\to t_k^-}(t_k - t)\int_{t_{k-1}}^{t} e^{2\delta(s-t)}\dot{Y}^T(s)R\dot{Y}(s)ds = 0$. It follows that $V_2(t_k^-) = V_2(t_k) = 0$. Hence, $V(t_k^-) = V(t_k)$, which means that $V(t)$ is continuous at sampling times t_k. Therefore, for $t \in [t_k, t_{k+1})$,

$$V(t) \le e^{-2\delta(t-t_k)}V(t_k) \le e^{-2\delta(t-t_{k-1})}V(t_{k-1}) \le \cdots \le e^{-2\delta t}V(0).$$

Hence, $V(t) \le e^{-2\delta t}V(0)$, $\forall t \ge 0$.

Invoking Jensen's inequality, using (3.235) and (3.241), we obtain

$$\begin{aligned}V(t) &\geq \lambda_{\min}(P)\sum_{n_1=0}^{N_1} Y_{1,n_1}^2 + \lambda_{\min}(P)\sum_{n_2=0}^{N_2} Y_{2,n_2}^2 + \sum_{n_1=N_1+1}^{\infty} Y_{1,n_1}^2 \\ &+ \sum_{n_2=N_2+1}^{\infty} Y_{2,n_2}^2 + (t_{k+1}-t)e^{-2\delta h}(t-t_k)v_1^T(s)Rv_1 \\ &\geq \min\left\{m, 1\right\}\min\left\{\lambda_{min}(P), 1\right\}\left(\|u(\cdot,t)\|_{\mathcal{H}}^2 + \|z(\cdot,t)\|_{\mathcal{H}}^2\right).\end{aligned} \tag{3.254}$$

Therefore, (3.253) and (3.254) imply (3.252).
(ii) From (3.249), it follows that if (3.247)–(3.251) hold, then

$$V(t) \leq e^{-2\delta(t-t_k)}V(t_k), \quad \forall t \in [t_k, t_{k+1}).$$

Then the strong solution of (3.239) initialized with $\left(Y_1(0), \overline{Y}_1(0), Y_2(0), \overline{Y}_2(0)\right)^T \in \mathbb{R}^{N_1+1} \times l_{N_1+1}^2(\mathbb{N}) \times \mathbb{R}^{N_2+1} \times l_{N_2+1}^2(\mathbb{N})$ admits a priori estimate. From Theorem 6.23.5 of [10], it follows that the strong solution exists on the entire interval $[0, t_1]$. By applying the same arguments step-by-step for $[t_k, t_{k+1})$ $(k = 1, 2, \ldots)$, we conclude that the strong solution exists for all $t \geq 0$. □

Remark 3.15 The proposed method may be extended to the case with in-domain coupling. For instance, the approach is applicable to the following model:

$$\begin{cases} u_t(x,t) = u_{xx}(x,t) + au(x,t) + cz(x,t), x \in (0,1), t > 0, \\ z_t(x,t) = z_{xx}(x,t) + bz(x,t), x \in (0,1), t > 0, \\ u_x(0,t) = 0, \ u(1,t) = 0, \\ z_x(1,t) = 0, \ z_x(0,t) = U(t), \\ u(x,0) = u_0(x), \ z(x,0) = z_0(x), \end{cases}$$

where $c > 0$ is some constant.

3.4.4 Numerical Example

Here we present a numerical example which verifies the effectiveness of the proposed method for the unstable couple heat-heat system with different reaction coefficients. The finite difference method is applied to compute the displacement of system to illustrate the effect of the proposed feedback control law. The steps of space and time are taken as 0.05 and 0.002, respectively. Consider the system (3.198) with parameters $a = \frac{1}{2}$, $b = 6$, $s = 1$. It should be noticed that the reaction coefficients a and b are selected such that the sufficient condition of Lemma 3.4 is satisfied. The initial values are chosen as $z_0(x) = \sin x$, $u_0(x) = \cos x$. Figure 3.13 demonstrates

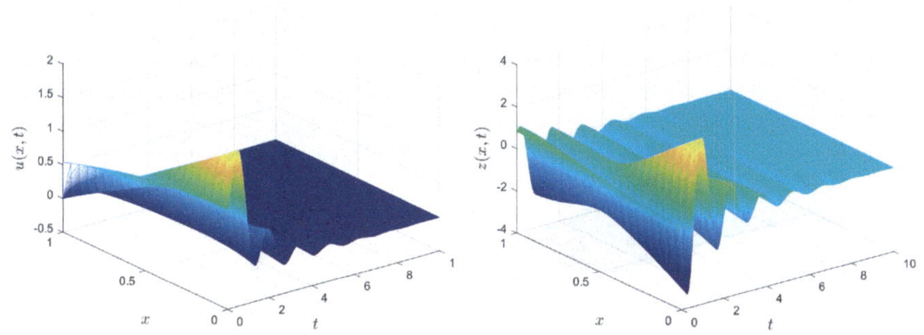

(a) The state $u(x,t)$ (b) The state $z(x,t)$

Fig. 3.14 The states of closed-loop system with $a = \frac{1}{2}$, $b = 6$ under continuous-time controller (3.219)

the profile of the open-loop system initialized by $u_0(x)$ and $z_0(x)$. It is shown that the unforced system is unstable. Utilizing Yalmip, we validate the LMI conditions of Theorem 3.7 for $\delta = 0$, $K_1 = -20$, $K = -10$. The feasible solutions are given as follows: $P_1 = 1.0468$, $P_2 = 0.0716$. Figure 3.14 shows that the closed-loop system under the continuous-time controller (3.219) is exponentially stable. Next, by using Yalmip we verify LMI conditions of Theorem 3.8 with same parameters δ, K_1 and K. The feasible solutions are obtained as follows:

$$R = \begin{pmatrix} 23.5925 & 17.1049 \\ 17.1049 & 44.0100 \end{pmatrix}, \quad P = \begin{pmatrix} 9.8346 & 13.7619 \\ 13.7619 & 22.2714 \end{pmatrix},$$

$$P_2 = \begin{pmatrix} 0.5318 & 0.4907 \\ 0.4907 & 0.5866 \end{pmatrix}, \quad P_3 = \begin{pmatrix} 1.3093 & 0.3509 \\ 0.3509 & 1.6072 \end{pmatrix}.$$

Figure 3.15 shows that the closed-loop system under the sample-data control law (3.237) preserves the exponential stability for $h \leq 0.04$. The number of sampling is 375. To make a comparison, Fig. 3.16 illustrates the energy of the closed-loop system under the continue-time controller (3.219) and sample-data control law (3.237) respectively. It is evident that the energy dissipates more rapidly for the case of continue-time controller than the case of sample-data controller. The simulations confirm our theoretical results.

3.5 Distributed Stabilization of KdVB with Input Delay

We consider distributed stabilization of 1D KdVB equation in the presence of constant input delay. The delay may be uncertain, but bounded by a known upper bound. On the basis of spatially distributed (either point or averaged) measurements,

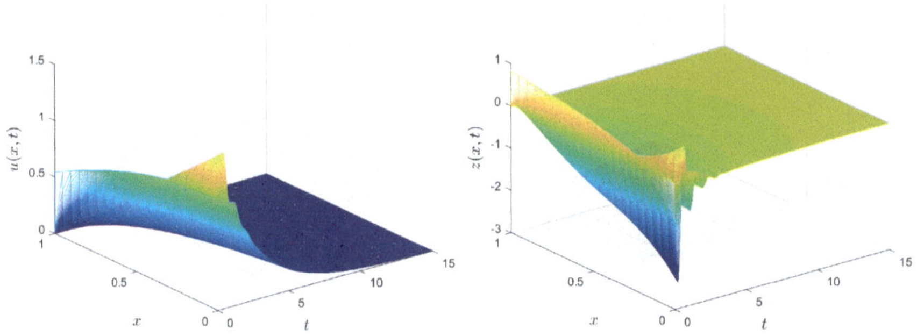

(a) The state $u(x,t)$ (b) The state $z(x,t)$

Fig. 3.15 The states of closed-loop system with $a = \frac{1}{2}, b = 6$ under sampled-data controller (3.237)

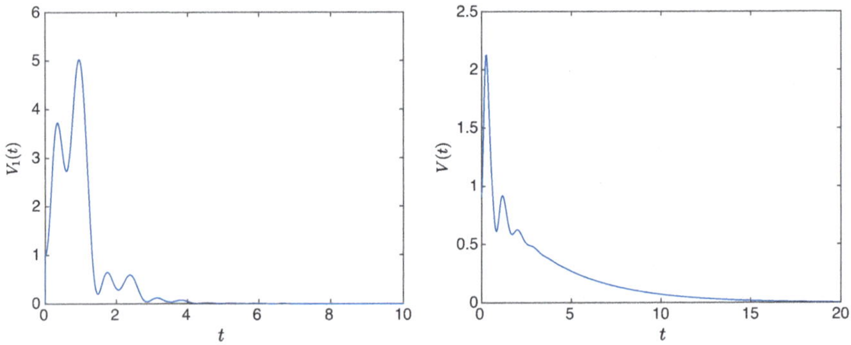

(a) Under continuous-time controller (3.219) (b) Under sampled-data controller (3.237)

Fig. 3.16 The energy of closed-loop system with $a = \frac{1}{2}, b = 6$

we design a regionally stabilizing controller applied through distributed in space shape functions. The existing Lyapunov-Krasovskii functionals for heat equation that depend on the state derivative are not applicable to KdVB equation because of the third order partial derivative. We suggest a new Lyapunov-Krasovskii functional that leads to regional stability conditions of the closed-loop system in terms of LMIs. By solving these LMIs, an upper bound on the delay that preserves regional stability can be found, together with an estimate on the set of initial conditions starting from which the state trajectories of the system are exponentially converging to zero. This estimate includes a priori Lyapunov-based bounds on the solutions of the open-loop system on the initial time interval of the length of delay. Numerical examples illustrate the efficiency of the method [32].

3.5.1 Problem Formulation

We consider the following KdVB equation involving both instability and dissipation under periodic boundary conditions:

$$\begin{cases} z_t(x,t) + z(x,t)z_x(x,t) - \beta z_{xx}(x,t) - \lambda z(x,t) + z_{xxx}(x,t) \\ = \sum\limits_{j=1}^{N} b_j(x)u_j(t-h),\ 0 < x < 1,\ t \geq 0, \\ z(0,t) = z(1,t),\ z_x(0,t) = z_x(1,t),\ z_{xx}(0,t) = z_{xx}(1,t), \\ z(x,0) = z_0(x), \end{cases} \tag{3.255}$$

where $x \in (0,1)$, β and λ are positive constants, $z(x,t)$ is the state of KdVB equation, $u_j(t) \in \mathbb{R}$, $(j = 1, 2, \ldots, N)$ are the control inputs, and $h > 0$ is a constant delay. The delay may be unknown, but bounded by a known bound $\bar{h} > 0$. The above model is inspired by [36] (see (2), (4b)). For $\lambda = 0$, the open-loop system (3.255) has constant solutions, whereas for $\lambda > 0$ the open-loop system may become unstable (see the examples in Sect. 3.5.5).

As in [1, 2, 7, 25], we assume that the points $0 = x_0 < x_1 < \cdots < x_N = 1$ divide [0, 1] into N intervals $\Omega_j = [x_{j-1}, x_j)$ and the width of each interval is upper bounded by some constant Δ:

$$0 < x_j - x_{j-1} = \Delta_j \leq \Delta.$$

The control inputs $u_j(t)$ enter (3.255) through the shape functions $b_j(x)$ such that

$$\begin{cases} b_j(x) = 0,\ x \notin \Omega_j, \\ b_j(x) = 1,\ otherwise, \end{cases} j = 1, \ldots, N. \tag{3.256}$$

Such shape functions correspond to actuation covering all the domain [0, 1].

We assume further that sensors provide either point

$$y_j(t) = z(\bar{x}_j, t),\quad \bar{x}_j = \frac{x_{j-1} + x_j}{2},\ t > 0, \tag{3.257}$$

or averaged

$$y_j(t) = \frac{\int_{x_{j-1}}^{x_j} z(\zeta, t)d\zeta}{\Delta_j},\ t > 0 \tag{3.258}$$

measurements of the state. In Sect. 3.5.4.2 we will also consider actuation and measurement concentrated on subdomains of [0, 1].

Our main objective is to design for (3.255) a regionally stabilizing distributed controller

$$u_j(t) = \begin{cases} -\mu y_j(t),\ j = 1, \ldots, N,\ t > 0, \\ 0, \qquad\qquad\qquad\quad t \leq 0, \end{cases} \tag{3.259}$$

where μ is a positive constant to be determined later, and $y_j(t)$ is given by (3.257) or (3.258).

Denote the characteristic function of the time interval $[0, h]$ by $\chi_{[0,h]}(t)$. Under the control law (3.259), the closed-loop system becomes

$$\begin{cases} z_t(x,t) + z(x,t)z_x(x,t) - \beta z_{xx}(x,t) - \lambda z(x,t) + z_{xxx}(x,t) \\ = -\mu \sum\limits_{j=1}^{N} b_j(x) > (1 - \chi_{[0,h]}(t))[z(x,t-h) - f_j(x,t-h)],\ 0 < x < 1,\ t \geq 0, \\ z(0,t) = z(1,t),\ z_x(0,t) = z_x(1,t),\ z_{xx}(0,t) = z_{xx}(1,t), \\ z(x,0) = z_0(x), \end{cases} \tag{3.260}$$

where for (3.257)

$$f_j(x,t-h) = \int_{\bar{x}_j}^{x} z_\zeta(\zeta, t-h)d\xi, \tag{3.261}$$

and for (3.258)

$$f_j(x,t-h) = \frac{\int_{x_{j-1}}^{x_j} [z(x,t-h) - z(\zeta,t-h)]d\zeta}{\Delta_j}. \tag{3.262}$$

3.5.2 Well-Posedness of (3.260) Subject to (3.261) or (3.262)

Define

$$\begin{cases} H^1_{per}(0,1) = \{g \in H^1(0,1) : g(0) = g(1)\}, \\ \|g\|^2_{H^1_{per}} = P_1 \int_0^1 g^2(x)dx + P \int_0^1 [g'(x)]^2 dx. \end{cases} \tag{3.263}$$

Here P_1 and P are positive constants that are related to the Lyapunov-Krasovskii functional (see (3.286) below) that will be used for stability analysis. It is obvious that $H^1_{per}(0,1)$ is a subspace of Sobolev space $H^1(0,1)$. Moreover, the norm $\|\cdot\|_{H^1_{per}(0,1)}$ is equivalent to $\|\cdot\|_{H^1(0,1)}$.

Definition 3.1 [21, 37] Given $T > 0$, a function $z \in C(0,T; H^1_{per}(0,1))$ such that $z_t \in L^\infty(0,T; L^2(0,1)) \cap L^2(0,T; H^1_{per}(0,1))$ is said to be a weak solution of the boundary value problem (3.260) subject to (3.261) or (3.262) on $[0,T]$ if for every ϕ from $H^1_{per}(0,1)$, the function $\int_0^1 z(\zeta,t)\phi(\zeta)d\zeta$ is absolutely continuous on $[0,T]$ and the relation

$$\begin{aligned} &\tfrac{d}{dt}\int_0^1 z(\zeta,t)\phi(\zeta)d\zeta + \int_0^1 z(\zeta,t)z_\zeta(\zeta,t)\phi(\zeta)d\zeta + \beta\int_0^1 z_\zeta(\zeta,t)\phi_\zeta(\zeta)d\zeta \\ &-\lambda\int_0^1 z(\zeta,t)\phi(\zeta)d\zeta - \int_0^1 z_{\zeta\zeta}(\zeta,t)\phi_\zeta(\zeta)d\zeta \\ &= -\mu\sum_{j=1}^{N}\int_{x_{j-1}}^{x_j}(1 - \chi_{[0,h]}(t))[z(\zeta,t-h) - f_j(\zeta,t-h)]\phi(\zeta)d\zeta \end{aligned} \tag{3.264}$$

holds for almost all $t \in [0,T]$.

Proposition 3.4 *Assume that the initial value $z_0 \in H^3(0,1) \cap H^1_{per}(0,1)$ satisfies the compatible conditions:*

$$z_0'(0) = z_0'(1), \;\; z_0''(0) = z_0''(1). \tag{3.265}$$

Then, for all $T > 0$, there exists a unique (weak) solution to the system (3.260) subject to (3.261) or (3.262) from the class

$$\begin{aligned} & z \in C(0,T; H^1_{per}(0,1)), \\ & z_t \in L^\infty(0,T; L^2(0,1)) \cap L^2(0,T; H^1_{per}(0,1)). \end{aligned}$$

The proof of Proposition 3.4 is based on the Galerkin approximation method. By arguments of Theorem 2.2 of [37] we establish the well-posedness of weak solution of (3.260) subject to (3.261) or (3.262).

Define $H^3_{per}(0,1) = \{g \in H^3(0,1) \cap H^1_{per}(0,1) : g'(0) = g'(1),\; g''(0) = g''(1)\}$ equipped with the norm of $H^3(0,1)$. Given $T > 0$. Suppose that $\{\phi_n\}_{n=1}^\infty$ is an orthonormal basis for $H^3_{per}(0,1)$. For any $N \in \mathbb{Z}^+$, define a finite-dimensional subspace of $H^1_{per}(0,1)$ by $V_N = \text{span}\{\phi_1, \phi_2, \ldots, \phi_N\}$. A Galerkin approximation solution to (3.260) is constructed as follows:

$$z^N(x,t) = \sum_{n=1}^N g_n^N(t)\phi_n(x), \tag{3.266}$$

which satisfies

$$\begin{cases} \langle z_t^N(\cdot,t), \phi\rangle + \langle z^N(\cdot,t) z_x^N(\cdot,t), \phi\rangle + \beta\langle z_x^N(\cdot,t), \phi_x\rangle \\ -\lambda\langle > z^N(\cdot,t), \phi\rangle - \langle z_{xx}^N(\cdot,t), \phi_x\rangle + \langle z_{xxx}^N, \phi\rangle \\ = -\mu \sum\limits_{j=1}^N \int_{\Omega_j} (1 - \chi_{[0,h]}(t))[z^N(x,t-h) - f_j^N(x,t-h)]\phi(x)dx \qquad \forall\, \phi \in V_N, \\ g_n^N(0) = \langle z_0, \phi_n\rangle, \; n = 1, \ldots, N. \end{cases} \tag{3.267}$$

Set $X(t) = (g_1^N(t), \ldots, g_n^N(t))^T$. From system (3.267), it follows that $X(t)$ satisfies a nonlinear ODE system:

$$\dot{X}(t) = AX(t) + F(X(t)),$$

where

$$\begin{aligned} A &= -\beta(\{\langle \phi_i', \phi_j'\rangle\}_{i,j=1}^N)^T + \lambda I_N + (\{\langle \phi_i'', \phi_j'\rangle\}_{i,j=1}^N)^T, \\ F(X(t)) &= (\langle X^T(\{\phi_i \phi_j'\}_{i,j=1}^N)X, \phi_k\rangle_{k=1}^N)^T \\ & -\mu \sum_{j=1}^N \int_{\Omega_j} (1 - \chi_{[0,h]}(t))[X(t-h) - (\langle f_j^N(t-h), \phi_k\rangle_{k=1}^N)^T]. \end{aligned}$$

Here $\langle \phi_i', \phi_j'\rangle\}_{i,j=1}^N$ and $\langle \phi_i'', \phi_j'\rangle\}_{i,j=1}^N$ are the matrices with the corresponding entrices. Note that the nonlinearity $F(X)$ is locally Lipschitz continuous in X. Hence, the existence of functions $g_1^N(t), \ldots, g_N^N(t)$ on some interval $[0, t_N)$ is ensured by the local Lipschitz condition. To extend these functions to any $t_N = T < \infty$ and to pass to the limit as $N \to \infty$, we need a-priori estimates. The proof is split into several lemmas.

Lemma 3.5 *For any $t \in [0, h]$, the following inequality holds:*

$$\max_{t\in[0,h]} \sup_N \left[\|z^N(\cdot,t)\|_{L^2}^2 + \int_0^t \|z_x^N(\cdot,s)\|_{L^2}^2 ds \right] < \infty. \tag{3.268}$$

Proof For $t \in [0, h]$, substituting $\phi = z^N(x,t)$ into (3.267), we have

$$\frac{1}{2}\frac{d}{dt}\|z^N(\cdot,t)\|_{L^2}^2 + \beta\|z_x^N(\cdot,t)\|_{L^2}^2 = \lambda\|z^N(\cdot,t)\|_{L^2}^2. \tag{3.269}$$

Note that $\langle z^N z_x^N, z^N\rangle$ and $\langle z_{xx}^N, z_x^N\rangle$ vanish after integration by parts. Then, by Gronwall inequality

$$\|z^N(\cdot,t)\|_{L^2}^2 + \int_0^t \|z_x^N(\cdot,s)\|_{L^2}^2 ds \le K\|z_0\|_{L^2}^2 \ \forall t \in [0,h], \tag{3.270}$$

where K is a constant independent on N. □

Lemma 3.6 *For any $t \in [0, h]$ and $z_0 \in H_{per}^3(0,1)$, the following inequality holds:*

$$\sup_N \|z_t^N(\cdot,0)\|_{L^2} < \infty. \tag{3.271}$$

Proof Set $t = 0$ in (3.267). Then

$$\begin{aligned}&\langle z_t^N(\cdot,0),\phi\rangle + \langle z^N(\cdot,0)z_x^N(\cdot,0),\phi\rangle - \beta\langle z_{xx}^N(\cdot,0),\phi\rangle \\ &-\lambda\langle > z^N(\cdot,0),\phi\rangle + \langle z_{xxx}^N(\cdot,0),\phi\rangle = 0.\end{aligned} \tag{3.272}$$

Substituting $\phi = z_t^N(x,0)$ into (3.272) we obtain

$$\begin{aligned}\|z_t^N(\cdot,0)\|_{L^2}^2 = &-\langle z^N(\cdot,0)z_x^N(\cdot,0) - \beta z_{xx}^N(\cdot,0) - \lambda z^N(\cdot,0) \\ &+z_{xxx}^N(\cdot,0), z_t^N(\cdot,0)\rangle.\end{aligned} \tag{3.273}$$

From (3.273) the Minkowski and Cauchy-Schwartz inequalities lead to

$$\begin{aligned}\|z_t^N(\cdot,0)\|_{L^2} &\le \|z^N(\cdot,0)z_x^N(\cdot,0)\|_{L^2} + \beta\|z_{xx}^N(\cdot,0)\|_{L^2} \\ &\quad +\lambda\|z^N(\cdot,0)\|_{L^2} + \|z_{xxx}^N(\cdot,0)\|_{L^2} \\ &\le K_1\|z^N(\cdot,0)\|_{H^3} + \|z^N(\cdot,0)\|_{L^\infty}\|z_x{}^N(\cdot,0)\|_{L^2}\end{aligned} \tag{3.274}$$

for some constant $K_1 > 0$.

$$\|z^N(\cdot,0)\|_{L^\infty}^2 \le \|z^N(\cdot,0)\|_{L^2}\|z_x{}^N(\cdot,0)\|_{L^2},$$

By using Lemma 1.8, we obtain

$$\|z^N(\cdot,0)\|_{L^\infty}^2 \le (1+\Gamma)\,\|z^N(\cdot,0)\|_{L^2}^2 + \frac{1}{\Gamma}\|z_x^N(\cdot,0)\|_{L^2}^2\ \forall \Gamma > 0. \tag{3.275}$$

Thus, (3.271) follows from (3.274) and (3.275). □

Lemma 3.7 *For any $t \in [0, h]$, the following inequality holds:*

$$\max_{t\in[0,h]} \sup_N \left[\|z_x^N(\cdot,t)\|_{L^2}^2 + \int_0^t \|z_{xx}^N(\cdot,s)\|_{L^2}^2 ds \right] < \infty. \tag{3.276}$$

Proof For $t \in [0, h]$, substituting $\phi = z_{xx}^N(x, t)$ into (3.267) and integrating by parts, we have

$$\begin{aligned} \frac{1}{2}\frac{d}{dt}\|z_x^N(\cdot,t)\|_{L^2}^2 = {} & \langle z^N(\cdot,t)z_x^N(\cdot,t), z_{xx}^N(\cdot,t)\rangle \\ & -\beta\|z_{xx}^N(\cdot,t)\|_{L^2}^2 + \lambda\|z_x^N(\cdot,t)\|_{L^2}^2. \end{aligned} \tag{3.277}$$

Note that

$$\langle z^N(\cdot,t)z_x^N(\cdot,t), z_{xx}^N(\cdot,t)\rangle \le \|z^N(\cdot,t)\|_{L^2}\|z_x^N(\cdot,t)\|_{L^\infty}\|z_{xx}^N(\cdot,t)\|_{L^2}.$$

From Lemma 3.5, it follows that

$$\langle z^N(\cdot,t)z_x^N(\cdot,t), z_{xx}^N(\cdot,t)\rangle \le K_2\|z_x^N(\cdot,t)\|_{L^\infty}\|z_{xx}^N(\cdot,t)\|_{L^2}$$

holds for some constant $K_2 > 0$.

From Lemma 1.8, it follows that

$$\|z_x^N(\cdot,t)\|_{L^\infty}^2 \le (1+\Gamma)\|z_x^N(\cdot,t)\|_{L^2}^2 + \frac{1}{\Gamma}\|z_{xx}^N(\cdot,t)\|_{L^2}^2.$$

The latter inequality, together with Young inequality, yield

$$\begin{aligned} & \|z_x^N(\cdot,t)\|_{L^\infty}\|z_{xx}^N(\cdot,t)\|_{L^2} \\ & \le \left(\sqrt{1+\Gamma}\|z_x^N(\cdot,t)\|_{L^2} + \frac{1}{\sqrt{\Gamma}}\|z_{xx}^N(\cdot,t)\|_{L^2}\right)\|z_{xx}^N(\cdot,t)\|_{L^2} \\ & \le \frac{1+\Gamma}{\epsilon}\|z_x^N(\cdot,t)\|_{L^2}^2 + (\epsilon + \frac{1}{\sqrt{\Gamma}})\|z_{xx}^N(\cdot,t)\|_{L^2}^2\ \forall\epsilon > 0. \end{aligned}$$

Choose $\epsilon > 0$ and $\Gamma > 0$ such that $K_2(\epsilon + \frac{1}{\sqrt{\Gamma}}) < \beta$. Then, using the estimate (3.277), and exploiting the Gronwall lemma, we obtain

$$\|z_x^N(\cdot,t)\|_{L^2}^2 + \int_0^t \|z_{xx}^N(\cdot,s)\|_{L^2}^2 ds \le K_3\|z_x^N(\cdot,0)\|_{L^2}^2,$$

where K_3 is a constant independent on N. □

Lemma 3.8 *For any $t \in [0, h]$, the following inequality holds:*

$$\max_{t\in[0,h]} \sup_N \left[\|z_t^N(\cdot,t)\|_{L^2}^2 + \int_0^t \|z_{xt}^N(\cdot,s)\|_{L^2}^2 ds \right] < \infty. \tag{3.278}$$

Proof For $t \in [0, h]$, differentiating the first equation of (3.267) with respect to t, we have

$$\begin{aligned} &\langle z_{tt}^N(\cdot,t), \phi\rangle + \langle z_t^N(\cdot,t) z_x^N(\cdot,t), \phi\rangle + \langle z^N(\cdot,t) z_{xt}^N(\cdot,t), \phi\rangle \\ &+\beta\langle > z_{xt}^N(\cdot,t), \phi_x\rangle - \lambda\langle z_t^N(\cdot,t), \phi\rangle - \langle z_{xxt}^N(\cdot,t), \phi_x\rangle = 0. \end{aligned} \tag{3.279}$$

Substituting $\phi = z_t^N(x,t)$ into (3.279) we obtain

$$\begin{aligned} \frac{1}{2}\frac{d}{dt}\|z_t^N(\cdot,t)\|_{L^2}^2 &= \langle z^N(\cdot,t) z_{xt}^N(\cdot,t), z_t^N(\cdot,t)\rangle \\ &\quad -\beta\|z_{xt}^N(\cdot,t)\|_{L^2}^2 + \lambda\|z_t^N(\cdot,t)\|_{L^2}^2. \end{aligned} \tag{3.280}$$

Using Lemma 1.8, we obtain

$$\|z^N(\cdot,t)\|_{L^\infty} \le \sqrt{1+\Gamma}\|z^N(\cdot,t)\|_{L^2} + \frac{1}{\sqrt{\Gamma}}\|z_x^N(\cdot,t)\|_{L^2} \ \forall \Gamma > 0.$$

Hence,

$$\begin{aligned} &\langle z^N(\cdot,t) z_{xt}^N(\cdot,t), z_t^N(\cdot,t)\rangle \\ &\le \|z^N(\cdot,t)\|_{L^\infty} > \|z_{xt}^N(\cdot,t)\|_{L^2}\|z_t^N(\cdot,t)\|_{L^2} \\ &\le \left(\sqrt{1+\Gamma}\|z^N(\cdot,t)\|_{L^2} + \frac{1}{\sqrt{\Gamma}}\|z_x^N(\cdot,t)\|_{L^2}\right)\|z_{xt}^N(\cdot,t)\|_{L^2}\|z_t^N(\cdot,t)\|_{L^2}. \end{aligned} \tag{3.281}$$

From Lemmas 3.5 and 3.7, together with the latter inequality and Young's inequality we obtain

$$\begin{aligned} &\langle z^N(\cdot,t) z_{xt}^N(\cdot,t), z_t^N(\cdot,t)\rangle \\ &\le K_3\|z_{xt}^N(\cdot,t)\|_{L^2}\|z_t^N(\cdot,t)\|_{L^2} \\ &\le K_3[\epsilon_1\|z_{xt}^N(\cdot,t)\|_{L^2}^2 + \frac{1}{\epsilon_1}\|z_t^N(\cdot,t)\|_{L^2}^2] \ \forall \epsilon_1 > 0 \end{aligned} \tag{3.282}$$

for some constant $K_3 > 0$.

Choose $\epsilon_1 > 0$ such that $K_3\epsilon_1 < \beta$. Then from (3.280), (3.281) and (3.282), we get

$$\frac{1}{2}\frac{d}{dt}\|z_t^N(\cdot,t)\|_{L^2}^2 \le -(\beta - K_3\epsilon)\|z_{xt}^N(\cdot,t)\|_{L^2} + (\lambda + \frac{K_3}{\epsilon})\|z_t^N(\cdot,t)\|_{L^2}^2$$

Application of the Gronwall inequality and Lemma 3.6 yield (3.278). □

Proof of Proposition 3.4: Continuation
Given $T \in [0, h]$. From Lemmas 3.5-3.8 and Corollary 4.19 of [21], by arguments of [37] (see p. 178) we can extract a subsequence N_k, which is still denoted by N, such that

$$\begin{cases} z^N \rightharpoonup z \text{ in } L^\infty(0, T; H^1_{per}(0, 1)) \text{ weakstar, }^1 \\ z_t^N \rightharpoonup z_t \text{ in } L^\infty(0, T; L^2(0, 1)) \text{ weakstar, } \\ z_t^N \rightharpoonup z_t \text{ in } L^2(0, T; H^1_{per}(0, 1)) \text{ weak.}^2 \end{cases} \tag{3.283}$$

[1, 2]Since for any $\psi \in C_c^\infty(0, T)$ and $\phi \in H^1_{per}(0, 1)$,

$$\begin{aligned} &\int_0^T \langle z_t^N(\cdot, t), \phi\rangle \psi(t)dt + \int_0^T \langle z^N(\cdot, t) z_x^N(\cdot, t), \phi\rangle \psi(t)dt \\ &+\beta \int_0^T \langle z_x^N(\cdot, t), \phi_x\rangle \psi(t)dt - \lambda \int_0^T \langle z^N(\cdot, t), \phi\rangle \psi(t)dt \\ &- \int_0^T \langle z_{xx}^N(\cdot, t), \phi_x\rangle \psi(t)dt = 0, \end{aligned} \tag{3.284}$$

by (3.283), passing to the limit as $N \to \infty$ in (3.284), we obtain

$$\langle z_t, \phi\rangle + \langle z z_x, \phi\rangle + \beta\langle z_x, \phi_x\rangle - \lambda\langle z, \phi\rangle - \langle z_{xx}, \phi_x\rangle = 0, \ t \in [0, T] \text{ a.e.}$$

Therefore, there exists a weak solution to the system (3.260) subject to (3.261) or (3.262) for all $t \in [0, h]$.

We apply the same arguments step-by-step for $[h, 2h]$, $[2h, 3h]$, ... Let $T > 0$. Following this procedure, we obtain that there exists a weak solution of system (3.260) subject to (3.261) or (3.262) for all $t \in [0, T]$.

3.5.3 Regional Stability of the Delayed KdVB Equation

In order to derive stability conditions for the closed-loop system (3.260), we first establish stability conditions for (3.260) with $f_j = 0$:

$$\begin{cases} z_t(x, t) + z(x, t) z_x(x, t) - \beta z_{xx}(x, t) - \lambda z(x, t) + z_{xxx}(x, t) \\ = -\mu(1 - \chi_{[0,h]}(t)) z(x, t - h), 0 < x < 1, \ t \geq 0, \\ z(0, t) = z(1, t), \ z_x(0, t) = z_x(1, t), \ z_{xx}(0, t) = z_{xx}(1, t), \\ z(x, 0) = z_0(x). \end{cases} \tag{3.285}$$

[1] Definition: Let X be a Banach space. A sequence $f_n \in X^*$ converges weakly-$*$ ("weakly star") to f, written $f_n \overset{*}{\rightharpoonup} f$, if $f_n(x) \to f(x)$ for every $x \in X$.

[2] Definition: Let X be a Banach space. A sequence $x_n \in X$ converges weakly to x (in X), written $x_n \rightharpoonup x$ in X, if $f_n(x) \to f(x)$ for every $f \in X^*$.

Remark 3.16 Simple and efficient delay-dependent stability conditions that lead to LMIs for ODEs and heat equations are usually derived via Lyapunov-Krasovskii functionals V that depend on the state derivative (on z_t for heat equation) [11, 18]. However, such functionals for (4.19) would lead to positive term hz_t^2 in $\dot{V}$. By substituting z_t from (4.19) we will arrive at positive term $h\|z_{xxx}\|^2_{L_2}$ that cannot be compensated by using negative terms like $-\|z\|^2_{L_2}$, $-\|z_x\|^2_{L_2}$ or $-\|z_{xx}\|^2_{L_2}$.

To avoid using Lyapunov-Krasovskii functionals that depend on z_t, we introduce a novel augmented Lyapunov-Krasovskii functional that depends on the state only:

$$V(t) = V_{aug} + V_P + V_R + V_Q + V_S, \tag{3.286}$$

where

$$\begin{aligned}
V_{aug} &= \int_0^1 \theta^T \begin{bmatrix} P_1 & P_2 \\ P_2 & P_3 \end{bmatrix} \theta dx, \quad \theta = \begin{bmatrix} z(x,t) \\ \int_{t-h}^t z(x,s)ds \end{bmatrix}, \\
V_P &= P \int_0^1 z_x^2(x,t)dx, \\
V_R &= R \int_0^1 \int_{t-h}^t e^{-2\delta(t-s)}(s+h-t)z^2(x,s)dsdx, \\
V_Q &= Q \int_0^1 \int_{t-h}^t e^{-2\delta(t-s)} z^2(x,s)dsdx, \\
V_S &= S \int_0^1 \int_{t-h}^t e^{-2\delta(t-s)}(s+h-t)z_x^2(x,s)dsdx.
\end{aligned}$$

Here $P > 0$, $R > 0$, $Q > 0$, and $S > 0$. Moreover, we assume that for some some constant $\gamma > 0$ the following LMI holds:

$$\begin{bmatrix} P_1 - \gamma & P_2 \\ * & P_3 + Q\frac{e^{-2\delta h}}{h} \end{bmatrix} > 0. \tag{3.287}$$

Due to Jensen's inequality [17]

$$\begin{aligned}
&Q \int_0^1 \int_{t-h}^t e^{-2\delta(t-s)} z^2(x,s)dsdx \\
&\geq Qe^{-2\delta h} \int_0^1 \int_{t-h}^t z^2(x,s)dsdx \\
&\geq Q\frac{e^{-2\delta h}}{h} \int_0^1 \left[\int_{t-h}^t z(x,s)ds\right]^2 dx
\end{aligned}$$

LMI (3.287) guarantees the positivity of the Lyapunov-Krasovskii functional:

$$V(t) \geq \gamma \int_0^1 z^2(x,t)dx + P \int_0^1 z_x^2(x,t)dx. \tag{3.288}$$

Similar augmentations of Lyapunov-Krasovskii functionals were considered for ODEs in [38], but such functionals there depended on state derivatives. Note that when differentiating P_2-term, we arrive at z_{xx} term (see (3.309) below). Therefore, we choose V with non-zero V_P that leads to H^1 exponential stability. In $\dot{V}_P$ we have $-\|z_{xx}\|_{L^2}^2$ term (see (3.310) below) that allows to compensate z_{xx} term in (3.309). However, the last term in (3.310) is non-quadratic, leading to regional exponential stability. In order to find domain of attraction we derive stability conditions in terms of matrix inequalities that are affine in z ($\Xi < 0$, $\Lambda < 0$ where Ξ and Λ are given by (3.294) and (3.295) respectively). We find a bound on initial conditions that guarantees the bound $\max_{x\in[0,1]} |z(x,t)| \le C_1$ for all $t \ge 0$ for all solutions starting from these initial conditions.

Lemma 3.9 *Consider the system (3.285). Given positive scalars $\mu > \lambda$, h, δ and positive tuning parameters $\delta_1 > \lambda$, $C > 0$ and $C_1 > 0$. Let there exist scalars $\gamma > 0$, $\Gamma > 0$, $P_1 > 0$, $P_2 < 0$, $P > 0$, $Q > 0$, $R > 0$, $S > 0$, $P_3 \in \mathbb{R}$, and $\nu \in \mathbb{R}$ such that the LMIs (3.287) and*

$$\Xi\Big|_{z=C_1} \le 0, \ \Xi\Big|_{z=-C_1} \le 0, \tag{3.289}$$

$$\Lambda\Big|_{z=C_1} \le 0, \ \Lambda\Big|_{z=-C_1} \le 0, \tag{3.290}$$

$$\Phi\Big|_{z=C_1} \le 0, \ \Phi\Big|_{z=-C_1} \le 0, \tag{3.291}$$

$$\gamma \ge 1 + \Gamma, \ \begin{bmatrix} -P & 1 \\ 1 & -\Gamma \end{bmatrix} < 0, \tag{3.292}$$

hold, where

$$\Xi = \begin{bmatrix} -2P_1(\delta_1 - \lambda) & 0 & \nu \\ * & -2P_1\beta - 2P(\delta_1 - \lambda) + 2\nu & Pz \\ * & * & -2P\beta \end{bmatrix}, \tag{3.293}$$

$$\Phi = \begin{bmatrix} \phi_{11} & -P_1\mu - P_2 & 0 & \nu & \phi_{15} & 0 \\ * & -Qe^{-2\delta h} & 0 & P\mu & -P_2\mu - P_3 & 0 \\ * & * & \phi_{33} & Pz & -P_2 z & -P_2\beta \\ * & * & * & -2P\beta & 0 & P_2 \\ * & * & * & * & \phi_{55} & 0 \\ * & * & * & * & * & \phi_{66} \end{bmatrix}, \tag{3.294}$$

$$\Lambda = \begin{bmatrix} \Lambda_{11} & -P_2 & 0 & \nu & \Lambda_{15} & 0 \\ * & -Qe^{-2\delta h} & 0 & 0 & -P_3 & 0 \\ * & * & \Lambda_{33} & Pz & -P_2 z & -P_2\beta \\ * & * & * & -2P\beta & 0 & P_2 \\ * & * & * & * & \Lambda_{55} & 0 \\ * & * & * & * & * & \Lambda_{66} \end{bmatrix}, \tag{3.295}$$

$$\begin{aligned}
\phi_{11} &= 2P_2 + 2P_1\lambda + Rh + Q + 2\delta P_1, \\
\Lambda_{11} &= \phi_{11} - 2\delta_1 P_1, \\
\phi_{15} &= \Lambda_{15} = P_3 + 2\delta P_2 + P_2\lambda, \\
\phi_{33} &= -2P_1\beta + 2P\lambda + 2\nu + Sh + 2\delta P, \\
\Lambda_{33} &= \phi_{33} - 2\delta_1 P, \\
\phi_{55} &= \Lambda_{55} = -Re^{-2\delta h}\frac{1}{h} + 2\delta P_3, \\
\phi_{66} &= \Lambda_{66} = -Se^{-2\delta h}\frac{1}{h}.
\end{aligned} \tag{3.296}$$

Denote

$$\begin{aligned}
M = \max &\left\{ \left(P_1 + 2P_2h + P_3h^2 + (\frac{Rh^2}{2} + Qh) \right) P_1^{-1}, \right. \\
&\left. (P + \frac{Sh^2}{2})P^{-1} \right\} + (e^{2\delta_1 h} - 1).
\end{aligned} \tag{3.297}$$

If

$$MC^2 < C_1^2, \tag{3.298}$$

then for any initial state $z_0 \in H^3(0,1) \cap H^1_{per}(0,1)$ *satisfying the compatible conditions (3.265) with* $\|z_0\|_{H^1_{per}} < C$*, system (3.285) possesses a unique (weak) solution in the sense that for any* $T > 0$*,* $z(x,t) \in C(0,T; H^1_{per}(0,1))$*. Moreover, the solution of (3.285) satisfies*

$$V(t) \leq Me^{-2\delta(t-h)} \left[P_1 \int_0^1 z_0^2(x)dx + P \int_0^1 [z_0'(x)]^2 dx \right] \tag{3.299}$$

for all $t \geq h$.

Proof We divide the proof into three parts.

Step 1: By arguments of [18, 22], we first derive a simple bound on $V(h)$ in terms of z_0 such that $V(h) < C_1^2$.

Since the solution to (3.285) does not depend on the values of $z(x,t)$ for $t < 0$, we redefine the initial condition to be a function:

$$z(x,t) = z_0(x), \ t \leq 0. \tag{3.300}$$

Due to (3.300), we have

$$V_{aug}(0) = [P_1 + 2P_2h + P_3h^2] \int_0^1 z_0^2(x)dx.$$

Then

$$V(0) \leq \left[P_1 + 2P_2h + P_3h^2 + \frac{Rh^2}{2} + Qh\right]\int_0^1 z_0^2(x)dx \\ +(P + \frac{Sh^2}{2})\int_0^1 [z_0'(x)]^2dx. \tag{3.301}$$

We consider

$$V_0(t) = P_1 \int_0^1 z^2(x,t)dx + P\int_0^1 z_x^2(x,t)dx. \tag{3.302}$$

Given $\delta > 0$. Assume that there exists $\delta_1 > 0$ such that along (3.285)

$$\dot{V}_0(t) - 2\delta_1 V_0(t) \leq 0, \ t \in [0,h], \tag{3.303}$$

$$\dot{V}(t) + 2\delta V(t) - 2\delta_1 V_0(t) \leq 0, \ t \in [0,h], \tag{3.304}$$

then

$$V_0(t) \leq e^{2\delta_1 t} V_0(0), \ t \in [0,h], \tag{3.305}$$

$$V(t) \leq e^{-2\delta t} V(0) + (e^{2\delta_1 t} - 1)V_0(0), \ t \in [0,h]. \tag{3.306}$$

Substituting (3.301) into the inequality (3.306), together with the condition (3.298), we have

$$\begin{aligned} V(h) \leq &\left[\left(P_1 + 2P_2h + P_3h^2 + \frac{Rh^2}{2} + Qh\right)\int_0^1 z_0^2(x)dx \right. \\ &\left. +(P + \frac{Sh^2}{2})\int_0^1 [z_0'(x)]^2dx\right] e^{-2\delta h} \\ &+(e^{2\delta_1 h} - 1)\left[P_1\int_0^1 z_0^2(x)dx + P\int_0^1 [z_0'(x)]^2dx\right] \\ &\leq M\left[P_1\int_0^1 z_0^2(x)dx + P\int_0^1 [z_0'(x)]^2dx\right] < C_1^2, \end{aligned} \tag{3.307}$$

if

$$\|z_0\|^2_{H^1_{per}} = P_1\int_0^1 z_0^2(x)dx + P\int_0^1 [z_0'(x)]^2dx < C^2. \tag{3.308}$$

Step 2: We show next the LMIs (3.289) and (3.290) guarantee that (3.303) and (3.304) hold.

Differentiating V_{aug} along (3.285), where (3.264) is used with $\phi = z$ and $t \in [0,h]$, we obtain

$$
\begin{aligned}
&\dot{V}_{aug} = 2P_1 \int_0^1 z(x,t)z_t(x,t)dx \\
&+2P_2 \int_0^1 z_t(x,t) \int_{t-h}^t z(x,s)dsdx \\
&+2P_2 \int_0^1 z(x,t)[z(x,t) - z(x,t-h)]dx \\
&+2P_3 \int_0^1 \int_{t-h}^t z(x,s)ds[z(x,t) - z(x,t-h)]dx \\
&= -2P_1\beta \int_0^1 z_x^2(x,t)dx + 2P_1\lambda \int_0^1 z^2(x,t)dx \\
&-2P_2 \int_0^1 z(x,t)z_x(x,t) \int_{t-h}^t z(x,s)dsdx \\
&-2P_2\beta \int_0^1 z_x(x,t) \int_{t-h}^t z_x(x,s)dsdx \\
&+2P_2 \int_0^1 z_{xx}(x,t) \int_{t-h}^t z_x(x,s)dsdx \\
&+2P_2\lambda \int_0^1 z(x,t) \int_{t-h}^t z(x,s)dsdx \\
&+2P_2 \int_0^1 z(x,t)[z(x,t) - z(x,t-h)]dx \\
&+2P_3 \int_0^1 \int_{t-h}^t z(x,s)ds[z(x,t) - z(x,t-h)]dx.
\end{aligned}
\tag{3.309}
$$

We have

$$
\begin{aligned}
\dot{V}_P &= 2P \int_0^1 z_x(x,t)z_{xt}(x,t)dx = -2P \int_0^1 z_{xx}(x,t)z_t(x,t)dx \\
&= -2P\beta \int_0^1 z_{xx}^2(x,t)dx + 2P\lambda \int_0^1 z_x^2(x,t)dx \\
&+2P \int_0^1 z_{xx}(x,t)z_x(x,t)z(x,t)dx
\end{aligned}
\tag{3.310}
$$

and

$$
\dot{V}_Q + 2\delta V_Q = Q \int_0^1 z^2(x,t)dx - Q \int_0^1 e^{-2\delta h} z^2(x,t-h)dx. \tag{3.311}
$$

Further by applying Jensen's inequality we obtain

$$
\begin{aligned}
&\dot V_R + 2\delta V_R = Rh\int_0^1 z^2(x,t)dx\\
&-R\int_0^1\int_{t-h}^t e^{-2\delta(t-s)}z^2(x,s)dsdx\\
&\le Rh\int_0^1 z^2(x,t)dx - Re^{-2\delta h}\frac{1}{h}\int_0^1\left[\int_{t-h}^t z(x,s)ds\right]^2 dx,\\
&\dot V_S + 2\delta V_S = Sh\int_0^1 z_x^2(x,t)dx\\
&-S\int_0^1\int_{t-h}^t e^{-2\delta(t-s)}z_x^2(x,s)dsdx\\
&\le Sh\int_0^1 z_x^2(x,t)dx - Se^{-2\delta h}\frac{1}{h}\int_0^1\left[\int_{t-h}^t z_x(x,s)ds\right]^2 dx.
\end{aligned}
\tag{3.312}
$$

Additionally,

$$
2\nu\left[\int_0^1 z(x,t)z_{xx}(x,t)dx + \int_0^1 z_x^2(x,t)dx\right] = 0 \ \ \forall \nu\in\mathbb{R}. \tag{3.313}
$$

We add to $\dot V(t)+2\delta V(t)$ the left-hand side of (3.313). Then, by taking into account (3.309)–(3.312) we arrive at

$$
\dot V(t) + 2\delta V(t) - 2\delta_1 V_0(t) \le \int_0^1 \psi^\top(x,t)\Lambda\psi(x,t)dx,\, t\in[0,h], \tag{3.314}
$$

where

$$
\psi(x,t) = \mathrm{col}\{z(x,t), z(x,t-h), z_x(x,t), z_{xx}(x,t), \textstyle\int_{t-h}^t z(x,s)ds, \int_{t-h}^t z_x(x,s)ds\}, \tag{3.315}
$$

and Λ is given by (3.295).

Similarly, differentiating $V_0(t)$ along (3.285) and adding (3.313), we have

$$
\dot V_0(t) - 2\delta_1 V_0(t) = \int_0^1 \xi^\top(x,t)\Xi\xi(x,t)dx,\ \ t\in[0,h], \tag{3.316}
$$

where $\xi(x,t)=\mathrm{col}\{z(x,t), z_x(x,t), z_{xx}(x,t)\}$ and Ξ is given by (3.293).

As in [20], first we assume that

$$
z(x,t)\in(-C_1, C_1)\ \ \forall x\in[0,1],\ \ \forall t\in[0,h]. \tag{3.317}
$$

From (3.314) and (3.316), it follows that if $\Xi\le 0$ and $\Lambda\le 0$ for all $z\in(-C_1,C_1)$, then (3.303) and (3.304) hold. Matrices Ξ and Λ given by (3.293) and (3.295) are affine in z. Thus, $\Xi\le 0$ and $\Lambda\le 0$ for all $z\in(-C_1,C_1)$ if LMIs (3.289) and (3.290) are feasible. Therefore, (3.289) and (3.290) guarantee that (3.303) and (3.304) hold.

We prove next (3.317). If the LMIs (3.287) and (3.292) are feasible, then Lemma 1.8 and Schur complement theorem lead to

$$\begin{aligned}\max_{0\le x\le 1}|z(x,t)|^2 &\le (1+\Gamma)\int_0^1 z^2(x,t)dx+\frac{1}{\Gamma}\int_0^1[z_x(x,t)]^2dx\\ &\le \gamma\int_0^1 z^2(x,t)dx+P\int_0^1[z_x(x,t)]^2dx.\\ &\le P_1\int_0^1 z^2(x,t)dx+P\int_0^1[z_x(x,t)]^2dx\\ &= V_0(t)\qquad \forall t\in[0,h].\end{aligned} \tag{3.318}$$

Moreover, from (3.288), we have

$$\begin{aligned}\max_{0\le x\le 1}|z(x,t)|^2 &\le \gamma\int_0^1 z^2(x,t)dx+P\int_0^1[z_x(x,t)]^2dx\\ &\le V(t)\qquad \forall t\in[0,h].\end{aligned} \tag{3.319}$$

Therefore, it is sufficient to show that

$$V_0(t)<C_1^2,\qquad V(t)<C_1^2 \tag{3.320}$$

for all $t\in[0,h]$.

Indeed, for $t=0$, the inequalities (3.320) hold. From the definition of M, we obtain $M>e^{2\delta_1 h}>1$. Since $\|z_0\|_{H^1_{per}}<C$ and $MC^2<C_1^2$ are satisfied, we obtain $V_0(0)<C^2<C_1^2$. On the other hand, from (3.301) we have $V(0)\le MV_0(0)<MC^2<C_1^2$.

Let (3.320) be false for some $h^*\in(0,h]$. Then

$$V_0(h^*)\ge C_1^2>C^2>V_0(0),$$

or

$$V(h^*)\ge C_1^2>V(0).$$

Since V_0 and V are continuous, there must exist $t^*\in(0,h^*]$ such that

$$V_0(t)<C_1^2\ \forall t\in[0,t^*)\text{ and }V_0(t^*)=C_1^2, \tag{3.321}$$

or

$$V(t)<C_1^2\ \forall t\in[0,t^*)\text{ and }V(t^*)=C_1^2. \tag{3.322}$$

The first relation of (3.321), together with the feasibility of (3.289), guarantees that $\dot V_0(t)-2\delta_1 V_0(t)\le 0$ on $[0,t^*)$. Therefore, $V_0(t^*)\le e^{2\delta_1 t^*}V_0(0)\le e^{2\delta_1 h}C^2<MC^2<C_1^2$. This contradicts the second relation of (3.321). On the other hand, the first relation of (3.322), together with the feasibility of (3.290), guarantees that $\dot V(t)+2\delta V(t)-2\delta_1 V_0(t)\le 0$ on $[0,t^*)$. Hence, (3.306) holds and $V(t^*)\le$

$MC^2 < C_1^2$. This contradicts the second relation of (3.322). Thus, (3.320) and consequently, (3.303) and (3.304) are true on $[0, h]$.

Step 3: We continue to find sufficient conditions in terms of LMIs to guarantee $\dot{V}(t) + 2\delta V(t) \leq 0$ for all $t > h$.

Differentiating V along (3.285) and integrating by parts, we obtain (3.311) and (3.312). For $t > h$, (3.309)–(3.310) become

$$
\begin{aligned}
\dot{V}_{aug} = & -2P_1\beta \int_0^1 z_x^2(x,t)dx + 2P_1\lambda \int_0^1 z^2(x,t)dx \\
& -2P_1\mu \int_0^1 z(x,t)z(x,t-h)dx \\
& -2P_2 \int_0^1 z(x,t)z_x(x,t) \int_{t-h}^t z(x,s)dsdx \\
& -2P_2\beta \int_0^1 z_x(x,t) \int_{t-h}^t z_x(x,s)dsdx \\
& +2P_2 \int_0^1 z_{xx}(x,t) \int_{t-h}^t z_x(x,s)dsdx \\
& +2P_2\lambda \int_0^1 z(x,t) \int_{t-h}^t z(x,s)dsdx \\
& -2P_2\mu \int_0^1 z(x,t-h) \int_{t-h}^t z(x,s)dsdx \\
& +2P_2 \int_0^1 z(x,t)[z(x,t) - z(x,t-h)]dx \\
& +2P_3 \int_0^1 \int_{t-h}^t z(x,s)ds[z(x,t) - z(x,t-h)]dx,
\end{aligned} \tag{3.323}
$$

and

$$
\begin{aligned}
\dot{V}_P = & -2P\beta \int_0^1 z_{xx}^2(x,t)dx + 2P\lambda \int_0^1 z_x^2(x,t)dx \\
& +2P \int_0^1 z_{xx}(x,t)z_x(x,t)z(x,t)dx \\
& +2P\mu \int_0^1 z_{xx}(x,t)z(x,t-h)dx.
\end{aligned} \tag{3.324}
$$

By adding to $\dot{V}(t) + 2\delta V(t)$ the equality (3.313), and using (3.311)–(3.312), (3.323), (3.324), we obtain

$$
\dot{V}(t) + 2\delta V(t) \leq \int_0^1 \psi^\top(x,t)\Phi\psi(x,t)dx, \tag{3.325}
$$

where $\psi(x,t)$ is given by (3.315).

From Step 1-Step 2, we obtain if $\|z_0\|_{H_{per}^1} < C$, then the feasibility of LMIs (3.289), (3.290), (3.298) imply that $V(h) < C_1^2$. Furthermore, if LMIs (3.287), (3.292) hold, then by using (3.288), (3.318) we obtain

$$\max_{0\le x\le 1}|z(x,h)|^2 \le (1+\Gamma)\int_0^1 z^2(x,h)dx + \frac{1}{\Gamma}\int_0^1 z_x^2(x,h)dx$$
$$\le \gamma\int_0^1 z^2(x,h)dx + P\int_0^1 z_x^2(x,h)dx \le V(h) < C_1^2.$$

Matrix Φ given by (3.294) is affine in z. Hence, $\Phi \le 0$ for all $z \in (-C_1, C_1)$ if LMIs (3.291) are satisfied. Therefore, (3.291) guarantees $\dot V(t) + 2\delta V(t) \le 0$, which implies

$$V(t) \le e^{-2\delta(t-h)}V(h)\ \forall t \ge h. \tag{3.326}$$

Using (3.307) and (3.326), we obtain (3.299). □

Remark 3.17 Given any $\mu > \lambda$, the LMIs in Lemma 3.9 are always feasible for appropriate decision variables, large enough δ_1 and small enough h. Indeed, consider $\Xi = \{\Xi_{ij}\}$, $\Phi = \{\phi_{ij}\}$ and $\Lambda = \{\Lambda_{ij}\}$ given by (3.293), (3.294) and (3.295) respectively. Note that for any $P > 0$, LMI (3.292) holds with appropriate $\Gamma > 0$. We will show that strict inequalities (3.289)–(3.291) hold with $\delta = 0$ and $C_1 = 0$. Then LMIs (3.289)–(3.291) hold with small enough $\delta > 0$ and $C_1 > 0$. Choose $R = S = 1$, $\nu = 0$, $P_1 = 1$, $P_2 = -\mu$, $P_3 = \mu^2$ such that $\phi_{12} = -P_1\mu - P_2 = 0$ and $\phi_{25} = -P_2\mu - P_3 = 0$. Hence, $\Xi_{|z=0} < 0$ for large enough $\delta_1 > \lambda$. By applying Schur complement, we obtain

$$\Phi\Big|_{z=0} = \begin{bmatrix} \phi_{11} & 0 & 0 & 0 & \mu(\mu-\lambda) & 0 \\ * & -Q & 0 & P\mu & 0 & 0 \\ * & * & \phi_{33} & 0 & 0 & \mu\beta \\ * & * & * & -2P\beta & 0 & -\mu \\ * & * & * & * & -\frac{1}{h} & 0 \\ * & * & * & * & * & -\frac{1}{h} \end{bmatrix} \le 0 \Longleftrightarrow$$
$$\begin{bmatrix} \phi_{11} & 0 & 0 & \mu(\mu-\lambda) & 0 \\ * & \phi_{33} & 0 & 0 & \mu\beta \\ * & * & -2P\beta + (P\mu)^2Q^{-1} & 0 & -\mu \\ * & * & * & -\frac{1}{h} & 0 \\ * & * & * & * & -\frac{1}{h} \end{bmatrix} \le 0, \tag{3.327}$$

where $\phi_{11} = -2(\mu-\lambda) + h + Q$, $\phi_{33} = -2\beta + 2P\lambda + h$. Choose $Q = \mu - \lambda$. For small enough h, LMI (3.327) is feasible with appropriate (small enough) $P > 0$, whereas $\Lambda_{|z=0} < 0$ holds for large enough $\delta_1 > \lambda$.

3.5.4 *Regional Delayed Stabilization of KdVB Equation*

3.5.4.1 Controller Distributed on the Whole Domain

Our main objective is to provide sufficient conditions for regional stability of the closed-loop system (3.260) subject to (3.261) or (3.262) with b_j defined by (3.256). For this purpose we define a Lyapunov-Krasovskii functional of the form:

$$V_1(t) = V(t) + W \int_0^1 \int_{t-h}^t e^{-2\delta(t-s)} z_x^2(x,s) ds dx, \tag{3.328}$$

where $W > 0$ and $V(t)$ is given by (3.286).

Theorem 3.9 *Consider the system (3.260) subject to (3.261) or (3.262). Given positive scalars Δ, $\mu > \lambda$, h, δ, and positive tuning parameters $\delta_1 > \lambda$, $C > 0$ and $C_1 > 0$. Let there exist scalars $\gamma > 0$, $\Gamma > 0$, $P_1 > 0$, $P_2 < 0$, $P > 0$, $Q > 0$, $R > 0$, $S > 0$, $W > 0$, $\alpha > 0$, $P_3 \in \mathbb{R}$ and $\nu \in \mathbb{R}$ such that the LMIs (3.287), (3.289), (3.292) and*

$$-We^{-2\delta h} + \alpha \le 0, \tag{3.329}$$

$$\Sigma\Big|_{z=C_1} \le 0, \ \Sigma\Big|_{z=-C_1} \le 0, \tag{3.330}$$

$$\Theta\Big|_{z=C_1} \le 0, \ \Theta\Big|_{z=-C_1} \le 0, \tag{3.331}$$

hold, where Ξ is given by (3.293),

$$\Sigma = \Lambda + \text{diag}\{0, 0, W, 0, 0, 0\}, \tag{3.332}$$

and

$$\Theta = \left[\begin{array}{cccccc:c} & & & \Phi & & & * \\ \hdashline P_1\mu & 0 & 0 & 0 & -P\mu & P_2\mu & -\alpha\frac{\pi^2}{\Delta^2} \end{array}\right] + \text{diag}\{0, 0, W, 0, 0, 0, 0\}$$

with Φ defined by (3.294).

Denote

$$\begin{aligned} M_1 = \max &\left\{\left(P_1 + 2P_2h + P_3h^2 + (\frac{Rh^2}{2} + Qh)\right) P_1^{-1}, \right. \\ &\left.(P + \frac{Sh^2}{2} + Wh)P^{-1}\right\} + (e^{2\delta_1 h} - 1). \end{aligned} \tag{3.333}$$

If

$$M_1 C^2 < C_1^2, \tag{3.334}$$

then for any initial state $z_0 \in H^3(0,1) \cap H^1_{per}(0,1)$ satisfying the compatible conditions (3.265) and $\|z_0\|_{H^1_{per}} < C$, system (3.260) subject to (3.261) or (3.262) possesses a unique (weak) solution in the sense that for any $T > 0$, $z(x,t) \in C(0,T; H^1_{per}(0,1))$. Moreover, the solution of (3.260) subject to (3.261) or (3.262) satisfies

$$V_1(t) \leq M_1 e^{-2\delta(t-h)} \left[P_1 \int_0^1 z_0^2(x)dx + P \int_0^1 [z_0'(x)]^2 dx \right] \tag{3.335}$$

for all $t \geq h$.

Proof Consider the system (3.260) subject to (3.261) or (3.262). Since the solution to the system (3.260) subject to (3.261) or (3.262) does not depend on the values of $z(x,t)$ for $t < 0$, we redefine the initial condition to be constant:

$$z(x,t) = z_0(x), \ t \leq 0.$$

Next the substitution $V(t) \to V_1(t)$ in Lemma 3.9 leads to the following changes:

$$V(0) \to V_1(0), \ M \to M_1, \ V(h) \to V_1(h),$$

where

$$\begin{aligned} V_1(0) \leq & \left(P_1 + 2P_2 h + P_3 h^2 + (\frac{Rh^2}{2} + Qh) \right) \int_0^1 z_0^2(x)dx \\ & + (P + \frac{Sh^2}{2} + Wh) \textstyle\int_0^1 [z_0'(x)]^2 dx. \end{aligned} \tag{3.336}$$

Here V_1 and M_1 are given by (3.328) and (3.333) respectively.

By Lemma 3.9, the LMIs (3.287), (3.289) and (3.292) guarantee (3.303). We show next that the feasibility of the LMIs (3.329), (3.330) and (3.331) guarantee

$$\dot{V}_1(t) + 2\delta V_1(t) - 2\delta_1 V_0(t) \leq 0 \ \forall t \in [0,h] \tag{3.337}$$

and

$$\dot{V}_1(t) + 2\delta V_1(t) \leq 0 \ \forall t > h. \tag{3.338}$$

Then from (3.303), (3.337), it follows that if (3.334) holds, then for initial state satisfying $\|z_0\|_{H^1_{per}} < C$, the following inequality holds:

$$V_1(h) \leq M_1 \left[P_1 \int_0^1 z_0^2(x)dx + P \int_0^1 [z_0'(x)]^2 dx \right] < C_1^2. \tag{3.339}$$

Differentiating V_1 along (3.260), we find by employing (3.314)

$$
\begin{aligned}
&\dot{V}_1(t) + 2\delta V_1(t) - 2\delta_1 V_0(t) \\
&= \dot{V}(t) + 2\delta V(t) - 2\delta_1 V_0(t) + W \int_0^1 z_x^2(x,t)dx \\
&-We^{-2\delta h} \int_0^1 z_x^2(x,t-h)dx \\
&\leq \int_0^1 \psi^{\top}(x,t)\Sigma\psi(x,t)dx - We^{-2\delta h}\int_0^1 z_x^2(x,t-h)dx,
\end{aligned}
\tag{3.340}
$$

where $\psi(x,t)$ is defined by (3.315) and Σ is given by (3.332).

Due to the affinity of Σ in $z \in (-C_1, C_1)$, we conclude that (3.330) guarantees (3.337).

Note that (3.260) comparatively to (3.285) has an additional term $-\mu \sum_{j=1}^{N} b_j(x) f_j(x, t-h)$ for $t > h$. Then from (3.325) and the substitution $\mu z(x,t-h) \to \mu[z(x,t-h) - f_j(x,t-h)]$, we have

$$
\begin{aligned}
&\dot{V}_1(t) + 2\delta V_1(t) \leq \int_0^1 \psi^T(x,t)\Phi\psi(x,t)dx \\
&+2P_1\mu \sum_{j=1}^{N} \int_{x_{j-1}}^{x_j} z(x,t) f_j(x,t-h)dx \\
&+2P_2\mu \sum_{j=1}^{N} \int_{x_{j-1}}^{x_j} \int_{t-h}^{t} z(x,s)ds f_j(x,t-h)dx \\
&-2P\mu \sum_{j=1}^{N} \int_{x_{j-1}}^{x_j} z_{xx}(x,t) f_j(x,t-h)dx \\
&+W\int_0^1 z_x^2(x,t)dx - We^{-2\delta h}\int_0^1 z_x^2(x,t-h)dx \ \forall t > h,
\end{aligned}
\tag{3.341}
$$

where $\psi(x,t)$ is defined by (3.315) and Φ is given by (3.294).

For the case of the point measurements, $f_j(x,t-h)$ is given by (3.261). Wirtinger's inequality yields

$$
\begin{aligned}
&\int_{x_{j-1}}^{x_j} f_j^2(x,t-h)dx = \int_{x_{j-1}}^{x_j} \left[\int_{\bar{x}_j}^{x} z_\zeta(\zeta,t-h)d\xi\right]^2 dx \\
&= \int_{x_{j-1}}^{\bar{x}_j} \left[\int_{\bar{x}_j}^{x} z_\zeta(\zeta,t-h)d\xi\right]^2 dx \int_{\bar{x}_j}^{x_j} \left[\int_{\bar{x}_j}^{x} z_\zeta(\zeta,t-h)d\xi\right]^2 dx \\
&\leq \frac{\Delta^2}{\pi^2} \int_{x_{j-1}}^{x_j} z_x^2(x,t-h)dx \ \forall t > h.
\end{aligned}
$$

For the case of the averaged measurements, $f_j(x,t-h)$ is given by (3.262). Since $\int_{x_{j-1}}^{x_j} f_j(x,t-h)dx = 0$, and $\frac{d}{dx} f_j(x,t-h) = z_x(x,t-h)$, the Poincaré inequality (Lemma 1.1) leads to

$$\int_{x_{j-1}}^{x_j} f_j^2(x,t-h)dx \leq \frac{\Delta^2}{\pi^2}\int_{x_{j-1}}^{x_j} z_x^2(x,t-h)dx \;\forall t > h.$$

Hence, in both cases the following inequality

$$\alpha\sum_{j=1}^{N}\int_{x_{j-1}}^{x_j}\left[z_x^2(x,t-h) - \frac{\pi^2}{\Delta^2}f_j^2(x,t-h)\right]dx \geq 0 \tag{3.342}$$

holds for some constant $\alpha > 0$.

Set $\eta = \mathrm{col}\{z(x,t), z(x,t-h), z_x(x,t), z_{xx}(x,t), \int_{t-h}^{t} z(x,s)ds, \int_{t-h}^{t} z_x(x,s) ds, f_j(x,t-h)\}$. We add (3.313) to (3.341) and find

$$\begin{aligned}
&\dot{V}_1(t) + 2\delta V_1(t)\\
&\leq \dot{V}_1(t) + 2\delta V_1(t) + \alpha\sum_{j=1}^{N}\int_{x_{j-1}}^{x_j}\left[z_x^2(x,t-h)\right.\\
&\left.-\frac{\pi^2}{\Delta^2}f_j^2(x,t-h)\right]dx\\
&\leq \sum_{j=1}^{N}\int_{x_{j-1}}^{x_j}\eta^{\top}\Theta\eta dx - (We^{-2\delta h} - \alpha)\int_0^1 z_x^2(x,t-h)dx.
\end{aligned}$$

Since $V(h) < C_1^2$, we obtain that if $-We^{-2\delta h} + \alpha \leq 0$ and $\Theta \leq 0$ hold for all $z \in (-C_1, C_1)$, then (3.338) holds. The matrix Θ is affine with respect to z. Therefore, $\Theta \leq 0$ for all $z \in (-C_1, C_1)$ if the LMIs (3.331) are satisfied. Hence, the LMIs (3.329), (3.331) yield (3.338), which implies

$$V_1(t) \leq e^{-2\delta(t-h)}V_1(h) \qquad \forall t \geq h. \tag{3.343}$$

Inequalities (3.343) and (3.339) imply (3.335) that completes the proof. □

Remark 3.18 Note that Σ and Θ are convex in h in the sense that if the conditions of Theorem 3.9 are satisfied for some $\bar{h} > 0$, then they are also satisfied with the same decision variables for all $h \in [0, \bar{h}]$. Hence, the feasibility of the LMIs in Theorem 3.9 guarantees the regional stability of (3.260) subject to (3.261) or (3.262) for all constant delays $h \in [0, \bar{h}]$. Hence, the delay in (3.255) may be uncertain, but upper-bounded by the known bound $\bar{h}$ meaning that there is robustness of the stability with respect to small enough delays.

Remark 3.19 Consider (3.260) subject to (3.261) or (3.262) without delay (i.e. with $h = 0$). For the energy function $E(t) = \frac{1}{2}\int_0^1 z^2(x,t)dx$, by employing (3.342) with $h = 0$ and applying S-procedure with $\alpha > 0$, the following inequalities hold for all $t \geq 0$

$$
\begin{aligned}
&\dot{E}(t) + 2\delta E(t) \\
&\le \dot{E}(t) + 2\delta E(t) + \alpha \sum_{j=1}^{N} \int_{x_{j-1}}^{x_j} \left[z_x^2(x,t) - \frac{\pi^2}{\Delta^2} f_j^2(x,t) \right] dx \\
&\le -(\beta - \alpha) \int_0^1 z_x^2 dx + \sum_{j=1}^{N} \int_{x_{j-1}}^{x_j} [z \ f_j] \Upsilon \begin{bmatrix} z \\ f_j \end{bmatrix} dx \le 0
\end{aligned}
$$

if $\alpha < \beta$ and

$$
\Upsilon \triangleq \begin{bmatrix} -\mu + \lambda + \delta & \frac{\mu}{2} \\ * & -\alpha \frac{\pi^2}{\Delta^2} \end{bmatrix} \le 0. \tag{3.344}
$$

Therefore, given positive scalars Δ and $\mu > \lambda$, if there exist $\alpha > 0$ and $\delta > 0$ such that $\alpha < \beta$ and the LMI (3.344) holds, then the closed-loop system is *globally exponentially stable* in the L^2-sense: $E(t) \le e^{-2\delta t} E(0) \quad \forall t \ge 0$.

3.5.4.2 Controller Distributed on Subdomains

In this subsection, we are concerned with the case that the actuation does not cover the whole domain [0, 1] and the averaged measurements are measured over the parts of the subdomains. As in [39], let

$$
0 \le \tilde{x}_1 < \tilde{x}_2 \le \tilde{x}_3 < \tilde{x}_4 \le \cdots \le \tilde{x}_{2N-1} < \tilde{x}_{2N} \le 1,
$$

$$
[\tilde{x}_{2j-1}, \tilde{x}_{2j}] \subset [x_{j-1}, x_j], \ j = 1, 2, \ldots, N.
$$

Define the shape functions

$$
\begin{cases} b_j(x) = 1, \ x \in [\tilde{x}_{2j-1}, \tilde{x}_{2j}], \\ b_j(x) = 0, \ otherwise, \end{cases} \quad j = 1, \ldots, N. \tag{3.345}
$$

Now we consider the system (3.255) under the averaged measurements

$$
y_j(t) = \frac{\int_{\tilde{x}_{2j-1}}^{\tilde{x}_{2j}} z(\xi, t) d\xi}{\tilde{\Delta}_j}, \quad \tilde{\Delta}_j = \tilde{x}_{2j} - \tilde{x}_{2j-1} \le \tilde{\Delta}. \tag{3.346}
$$

By applying the first mean value theorem, we obtain that there exists a point $\bar{x}_t^j \in [\tilde{x}_{2j-1}, \tilde{x}_{2j}]$ such that

$$
y_j(t)\tilde{\Delta}_j = \int_{\tilde{x}_{2j-1}}^{\tilde{x}_{2j}} z(\xi, t) d\xi = z(\bar{x}_t^j, t)\tilde{\Delta}_j. \tag{3.347}
$$

See Fig. 3.17, where

$$l_j = \max\{\tilde{x}_{2j} - x_{j-1}, x_j - \tilde{x}_{2j-1}\}. \tag{3.348}$$

By selecting the controller (3.259), we arrive at the closed-loop system

$$\begin{cases} z_t + zz_x - \beta z_{xxx} - \lambda z + z_{xxx} \\ = -\mu \sum_{j=1}^{N} b_j(x)(1 - \chi_{[0,h]}(t)) \left[z(\bar{x}_t^j, t) - f_j(t) \right], \\ z(0,t) = z(1,t), \ z_x(0,t) = z_x(1,t), \\ z_{xx}(0,t) = z_{xx}(1,t), \end{cases} \tag{3.349}$$

where

$$f_j(t) \triangleq y_j(t) - y_j(t-h) = \frac{\int_{\tilde{x}_{2j-1}}^{\tilde{x}_{2j}} [z(\xi,t) - z(\xi,t-h)]d\xi}{\tilde{\Delta}_j}.$$

For the shape function (3.345), we arrive at the following result.

Theorem 3.10 *Consider the closed-loop system (3.349) under the shape function (3.345). Denote*

$$l \triangleq \max_j l_j, \quad \bar{\Delta} \triangleq \min_j \frac{\tilde{\Delta}_j}{\Delta_j}. \tag{3.350}$$

Given positive scalars $\mu > \lambda$, h, δ, l, $\bar{\Delta}$ *and positive tuning parameters* $\delta_1 > \lambda$, $C > 0$ *and* $C_1 > 0$. *Let there exist scalars* $\gamma > 0$, $\Gamma > 0$, $P_1 > 0$, $P_2 < 0$, $P > 0$, $Q > 0$, $R > 0$, $S > 0$, $\alpha_i > 0$ $(i = 0, 1, 2, 3)$, $P_3 \in \mathbb{R}$ *and* $\nu \in \mathbb{R}$ *such that the LMIs (3.287), (3.289), (3.290), (3.292) and*

$$\tilde{\Theta}|_{z=C_1} \leq 0, \ \tilde{\Theta}|_{z=-C_1} \leq 0 \tag{3.351}$$

hold, where

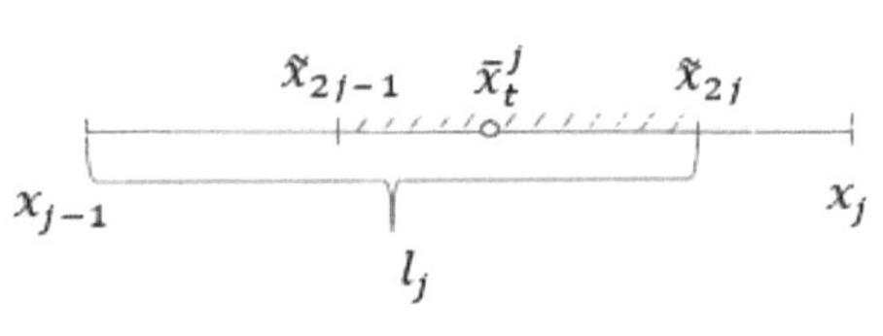

Fig. 3.17 Subdomain $[\tilde{x}_{2j-1}, \tilde{x}_{2j}]$, point $\bar{x}_t^j$ and l_j

$$\tilde{\Theta} = \begin{bmatrix} \tilde{\theta}_{11} & \tilde{\theta}_{12} & 0 & \nu & \tilde{\theta}_{15} & 0 & \tilde{\theta}_{17} & 0 & 0 & 0 \\ * & \tilde{\theta}_{22} & 0 & 0 & -P_3 & 0 & 0 & 0 & 0 & 0 \\ * & * & \tilde{\theta}_{33} & Pz & -P_2 z & -P_2\beta & 0 & 0 & 0 & 0 \\ * & * & * & \tilde{\theta}_{44} & 0 & P_2 & 0 & 0 & 0 & 0 \\ * & * & * & * & \tilde{\theta}_{55} & 0 & 0 & 0 & 0 & 0 \\ * & * & * & * & * & \tilde{\theta}_{66} & 0 & 0 & 0 & 0 \\ * & * & * & * & * & * & \tilde{\theta}_{77} & \tilde{\theta}_{78} & P\mu & -P_2\mu \\ * & * & * & * & * & * & * & \tilde{\theta}_{88} & -P\mu & P_2\mu \\ * & * & * & * & * & * & * & * & -\alpha_2 & 0 \\ * & * & * & * & * & * & * & * & * & -\alpha_1 \end{bmatrix},$$

$$\begin{aligned}
&\tilde{\theta}_{11} = 2P_2 + 2P_1\lambda + Rh + Q + 2\delta P_1 + \alpha_0 - \alpha_3 \frac{\pi^2}{4l^2}, \\
&\tilde{\theta}_{12} = -P_2 - \alpha_0,\ \tilde{\theta}_{15} = P_2\lambda + P_3 + 2\delta P_2,\ \tilde{\theta}_{17} = \alpha_3 \frac{\pi^2}{4l^2}, \\
&\tilde{\theta}_{22} = -Qe^{-2\delta h} + \alpha_0,\ \tilde{\theta}_{33} = -2P_1\beta + 2P\lambda + 2\nu + Sh + 2\delta P + \alpha_3, \\
&\tilde{\theta}_{44} = -2P\beta + \alpha_2,\ \tilde{\theta}_{55} = -Re^{-2\delta h}\frac{1}{h} + 2\delta P_3 + \alpha_1,\ \tilde{\theta}_{66} = -Se^{-2\delta h}\frac{1}{h}, \\
&\tilde{\theta}_{77} = -2P_1\mu\bar{\Delta} - \alpha_3 \frac{\pi^2}{4l^2},\ \tilde{\theta}_{78} = P_1\mu\bar{\Delta},\ \tilde{\theta}_{88} = -\alpha_0\bar{\Delta}.
\end{aligned}$$

If (3.298) with M given by (3.297) is satisfied, then for any initial state $z_0 \in H^3(0,1) \cap H^1_{per}(0,1)$ satisfying the compatible conditions (3.265) and $\|z_0\|_{H^1_{per}} < C$, the solution of the closed-loop system (3.349) satisfies (3.299) for all $t \geq h$.

Proof Consider V_0 given by (3.302) and V given by (3.286). For $t \leq h$, by Theorem 3.9, if LMIs (3.287), (3.289), (3.290) and (3.292) are feasible, then

$$\begin{aligned}
&\dot{V}_0(t) - 2\delta_1 V_0(t) \leq 0,\ t \in [0, h], \\
&\dot{V}(t) + 2\delta V(t) - 2\delta_1 V_0(t) \leq 0,\ t \in [0, h].
\end{aligned}$$

For $t > h$, differentiating V along (3.349) and integrating by parts, we have (3.311) and (3.312). Hence,

$$\begin{aligned}
\dot{V}_{aug}(t) = & -2P_1\beta \int_0^1 z_x^2(x,t)dx + 2P_1\lambda \int_0^1 z^2(x,t)dx \\
& -2P_1\mu \sum_{j=1}^N z^2(\bar{x}_t^j, t)\tilde{\Delta}_j + 2P_1\mu \sum_{j=1}^N f_j(t) z(\bar{x}_t^j, t)\tilde{\Delta}_j \\
& -2P_2 \int_0^1 z(x,t) z_x(x,t) \int_{t-h}^t z(x,s)dsdx \\
& -2P_2\beta \int_0^1 z_x(x,t) \int_{t-h}^t z_x(x,s)dsdx \\
& +2P_2 \int_0^1 z_{xx}(x,t) \int_{t-h}^t z_x(x,s)dsdx \\
& +2P_2\lambda \int_0^1 z(x,t) \int_{t-h}^t z(x,s)dsdx \\
& -2P_2\mu \sum_{j=1}^N [z(\bar{x}_t^j, t) - f_j(t)] \int_{x_{j-1}}^{x_j} b_j(x) \int_{t-h}^t z(x,s)dsdx \\
& +2P_2 \int_0^1 z(x,t)[z(x,t) - z(x,t-h)]dx \\
& +2P_3 \int_0^1 \int_{t-h}^t z(x,s)ds[z(x,t) - z(x,t-h)]dx,
\end{aligned} \tag{3.352}$$

and

$$\begin{aligned}
\dot{V}_P(t) = & -2P\beta \int_0^1 z_{xx}^2(x,t)dx + 2P\lambda \int_0^1 z_x^2(x,t)dx \\
& +2P \int_0^1 z_{xx}(x,t) z_x(x,t) z(x,t)dx \\
& +2P\mu \sum_{j=1}^N z(\bar{x}_t^j, t) \int_{x_{j-1}}^{x_j} b_j(x) z_{xx}(x,t)dx \\
& -2P\mu \sum_{j=1}^N f_j(t) \int_{x_{j-1}}^{x_j} b_j(x) z_{xx}(x,t)dx.
\end{aligned} \tag{3.353}$$

The Cauchy-Schwartz inequality leads to

$$\begin{aligned}
\int_{x_{j-1}}^{x_j} f_j^2(t)dx &= \Delta_j f_j^2(t) \\
&\leq \frac{\Delta_j}{\tilde{\Delta}_j} \int_{\tilde{x}_{2j-1}}^{\tilde{x}_{2j}} [z(\xi,t) - z(\xi,t-h)]^2 d\xi \\
&\leq \frac{\Delta_j}{\tilde{\Delta}_j} \int_{x_{j-1}}^{x_j} [z(\xi,t) - z(\xi,t-h)]^2 d\xi,
\end{aligned}$$

$$\int_{x_{j-1}}^{x_j} \left(b_j(x) \int_{t-h}^t z(x,s)ds \right)^2 dx \leq \int_{x_{j-1}}^{x_j} \left(\int_{t-h}^t z(x,s)ds \right)^2 dx,$$

and

$$\int_{x_{j-1}}^{x_j} \left(b_j(x)z_{xx}(x,t)\right)^2 dx \le \int_{x_{j-1}}^{x_j} z_{xx}^2(x,t)dx.$$

Hence, the following inequalities

$$\alpha_0 \sum_{j=1}^{N} \int_{x_{j-1}}^{x_j} \left[[z(x,t) - z(x,t-h)]^2 - \frac{\tilde{\Delta}_j}{\Delta_j} f_j^2(t) \right] dx \ge 0,$$

$$\alpha_1 \sum_{j=1}^{N} \int_{x_{j-1}}^{x_j} \left[\left(\int_{t-h}^{t} z(x,s)ds \right)^2 - \left(b_j(x) \int_{t-h}^{t} z(x,s)ds \right)^2 \right] dx \ge 0,$$

$$\alpha_2 \sum_{j=1}^{N} \int_{x_{j-1}}^{x_j} \left[z_{xx}^2(x,t) - \left(b_j(x)z_{xx}(x,t)\right)^2 \right] dx \ge 0$$

hold for some $\alpha_i > 0$ $(i = 0, 1, 2)$.

Wirtinger's inequality leads to

$$\begin{aligned}
&\int_{x_{j-1}}^{x_j} [z(x,t) - z(\bar{x}_t^j,t)]^2 dx \\
&= \int_{x_{j-1}}^{\bar{x}_t^j} [z(x,t) - z(\bar{x}_t^j,t)]^2 dx + \int_{\bar{x}_t^j}^{x_j} [z(x,t) - z(\bar{x}_t^j,t)]^2 dx \\
&\le \frac{4(\bar{x}_t^j - x_{j-1})^2}{\pi^2} \int_{x_{j-1}}^{\bar{x}_t^j} z_x^2(x,t)dx + \frac{4(x_j - \bar{x}_t^j)^2}{\pi^2} \int_{\bar{x}_t^j}^{x_j} z_x^2(x,t)dx.
\end{aligned}$$

Then from (3.348), we obtain

$$\int_{x_{j-1}}^{x_j} [z(x,t) - z(\bar{x}_t^j,t)]^2 dx \le \frac{4l^2}{\pi^2} \int_{x_{j-1}}^{x_j} z_x^2(x,t)dx. \tag{3.354}$$

Hence,

$$\alpha_3 \sum_{j=1}^{N} \int_{x_{j-1}}^{x_j} \left[z_x^2(x,t) - \frac{\pi^2}{4l^2} [z(x,t) - z(\bar{x}_t^j,t)]^2 \right] dx \ge 0$$

holds for some $\alpha_3 > 0$.

Therefore, for $t > h$, using (3.311), (3.312), (3.313), (3.352), (3.353) and applying the S-Procedure we obtain

$$
\begin{aligned}
&\dot{V} + 2\delta V \le \dot{V} + 2\delta V \\
&+\alpha_0 \sum_{j=1}^{N} \int_{x_{j-1}}^{x_j} \left[[z(x,t) - z(x,t-h)]^2 - \frac{\tilde{\Delta}_j}{\Delta_j} f_j^2(t) \right] dx \\
&+\alpha_1 \sum_{j=1}^{N} \int_{x_{j-1}}^{x_j} \left[\left(\int_{t-h}^{t} z(x,s)ds \right)^2 - \left(b_j(x) \int_{t-h}^{t} z(x,s)ds \right)^2 \right] dx \\
&+\alpha_2 \sum_{j=1}^{N} \int_{x_{j-1}}^{x_j} \left[z_{xx}^2(x,t) - \left(b_j(x) z_{xx}(x,t) \right)^2 \right] dx \\
&+\alpha_3 \sum_{j=1}^{N} \int_{x_{j-1}}^{x_j} \left[z_x^2(x,t) - \frac{\pi^2}{4l^2} [z(x,t) - z(\bar{x}_t^j, t)]^2 \right] dx \\
&\le \sum_{j=1}^{N} \int_{x_{j-1}}^{x_j} \tilde{\eta}^T \tilde{\Theta} \tilde{\eta} dx,
\end{aligned}
\tag{3.355}
$$

where

$$
\begin{aligned}
\tilde{\eta} = \operatorname{col}\{&z(x,t), z(x,t-h), z_x(x,t), z_{xx}(x,t), \\
&\textstyle\int_{t-h}^{t} z(x,s)ds, \int_{t-h}^{t} z_x(x,s)ds, z(\bar{x}_t^j, t), f_j(t), \\
&b_j z_{xx}(x,t), b_j \textstyle\int_{t-h}^{t} z(x,s)ds\}.
\end{aligned}
$$

Therefore, the LMIs (3.351) and the inequality (3.298) yield (3.299). □

Remark 3.20 Consider next the system (3.255) without delay (i.e. $h = 0$) under the shape functions (3.345), measurements (3.346) and the controller (3.259). In the non-delayed case, such a controller may globally exponentially stabilize the system in L^2 norm. Consider E given by Remark 3.19. Then from (3.352) we obtain

$$
\begin{aligned}
&\dot{E}(t) + 2\delta E(t) \\
&= -\beta \int_0^1 z_x^2(x,t)dx + (\lambda + \delta) \int_0^1 z^2(x,t)dx - \mu \sum_{j=1}^{N} z^2(\bar{x}_t^j, t) \tilde{\Delta}_j .
\end{aligned}
\tag{3.356}
$$

By using (3.354), the following inequality holds for all $t \ge 0$

$$
\begin{aligned}
&\dot{E}(t) + 2\delta E(t) \\
&\le -\beta \frac{\pi^2}{4l^2} \sum_{j=1}^{N} \int_{x_{j-1}}^{x_j} [z(x,t) - z(\bar{x}_t^j, t)]^2 dx \\
&+(\lambda + \delta) \int_0^1 z^2(x,t)dx - \mu \sum_{j=1}^{N} z^2(\bar{x}_t^j, t) \tilde{\Delta}_j \\
&\le \sum_{j=1}^{N} \int_{x_{j-1}}^{x_j} \left[z(x) \ z(\bar{x}_t^j) \right] \tilde{\Upsilon} \begin{bmatrix} z(x) \\ z(\bar{x}_t^j) \end{bmatrix} dx \le 0.
\end{aligned}
\tag{3.357}
$$

if

$$\tilde{\Upsilon} \triangleq \begin{bmatrix} -\beta\frac{\pi^2}{4l^2} + \lambda + \delta & \beta\frac{\pi^2}{4l^2} \\ * & -\beta\frac{\pi^2}{4l^2} - \mu\bar{\Delta} \end{bmatrix} \leq 0. \tag{3.358}$$

Therefore, given positive scalars $l < \sqrt{\frac{\beta\pi^2}{4\lambda}}$ and $\bar{\Delta}$ (c.f. (3.350)), if there exists $\delta > 0$ such that the LMI (3.358) hold, then the closed-loop system is globally exponentially stable in the L^2-sense: $E(t) \leq e^{-2\delta t}E(0), \ \forall t \geq 0$.

3.5.5 Simulation Examples

Example 3.1 Consider the system (3.255) under controller distributed on the whole domain with parameters $\beta = 0.5$ and $\lambda = 15$. Figure 3.18 demonstrates the time evolution of the L^2-norm $\|z(\cdot, t)\|^2_{L^2}$ for the open-loop system initialized by $z(x, 0) = 0.0025\sin(2\pi x)$, $0 \leq x \leq 1$. It is seen that the open-loop system is unstable.

For the control law (3.259) with the point measurements or averaged measurement, by using Yalmip we verify LMI conditions of Theorem 3.9 with $\mu = 20$, $\Delta = 0.1$, $\delta = 1$, $\delta_1 = 20$, $C = 0.044$, $C_1 = 0.05$. We find that the closed-loop system (3.260) subject to (3.261) or (3.262) preserves the exponential stability with a decay rate $\delta = 1$ for $h \leq 0.0019$ for any initial values satisfying $\|z_0\|_{H^1_{per}} < 0.044$. In order to

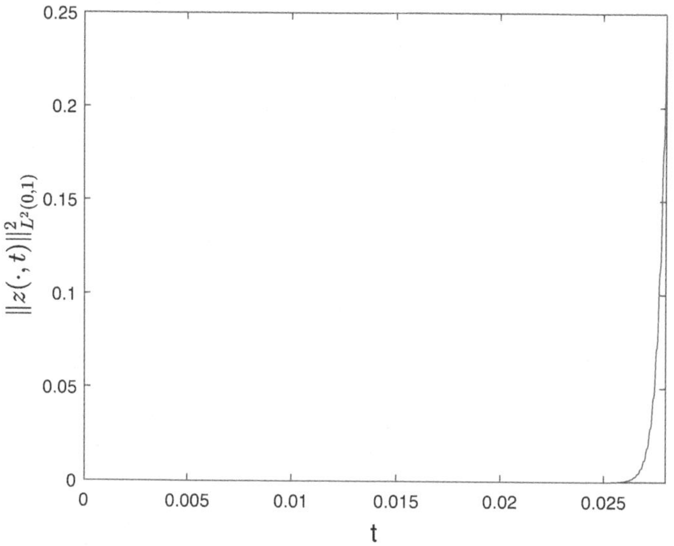

Fig. 3.18 Open-loop system (without control input) with $\lambda = 15$

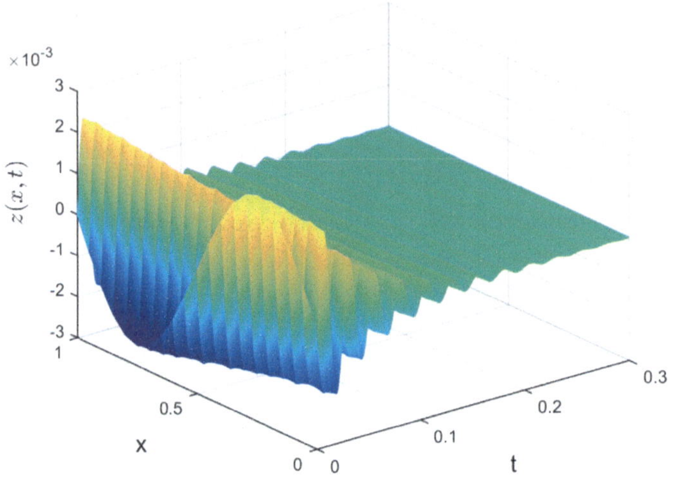

Fig. 3.19 State $z(x,t)$ with $h = 0.0019$ and $x_j - x_{j-1} = 0.1$ under the point measurements

obtain the estimate on the set of initial conditions (inside the domain of attraction) as large as possible, we minimize $P_1 + P$. The feasible solutions of the LMIs are given as follows: $P_1 = 387.3590$, $P = 5.5792$.

Next a finite difference method is applied to compute the state of the closed-loop system (3.260) subject to (3.261) to illustrate the effect of the proposed feedback control law (3.259) with the point measurements. We choose the same values of parameters and the initial condition $z_0(x) = 0.0025\sin(2\pi x)$, $0 \leq x \leq 1$. Hence,

$$\|z_0\|^2_{H^1_{per}} = 387.3590\|z_0\|^2 + 5.5792\|z_0'\|^2 < 0.044^2.$$

The steps of space and time are taken as 0.05 and 10^{-7}, respectively. Simulation of solutions under the controller

$$u_j(t) = \begin{cases} -20z(\bar{x}_j, t), & t > 0, \\ 0, & t \leq 0, \end{cases}$$

with $x_j - x_{j-1} = \Delta_j = \Delta = 0.1$, $\bar{x}_j = \dfrac{x_{j-1} + x_j}{2}$, $j = 1, \ldots, 10$, where the spatial domain is divided into ten sub-domains, shows that the closed-loop system is exponentially stable (see Fig. 3.19). Enlarging the value of h until 0.01, we find that the solution starting from the same initial condition is unbounded (see Fig. 3.20). The simulations of the solutions confirm the theoretical results. For the case of the averaged measurements, the simulation results are similar.

Example 3.2 Consider now the system (3.255) under the controller distributed on one subdomain. Here we choose parameters $\beta = 0.5$ and $\lambda = 0.2$. Figure 3.21 shows

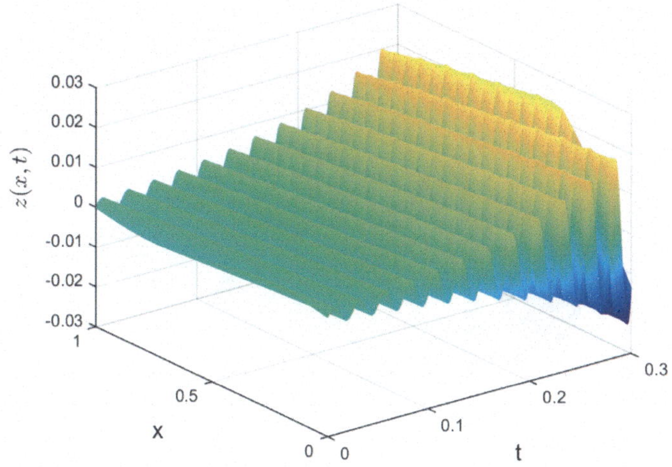

Fig. 3.20 State $z(x, t)$ with $h = 0.01$ and $x_j - x_{j-1} = 0.1$ under the point measurements

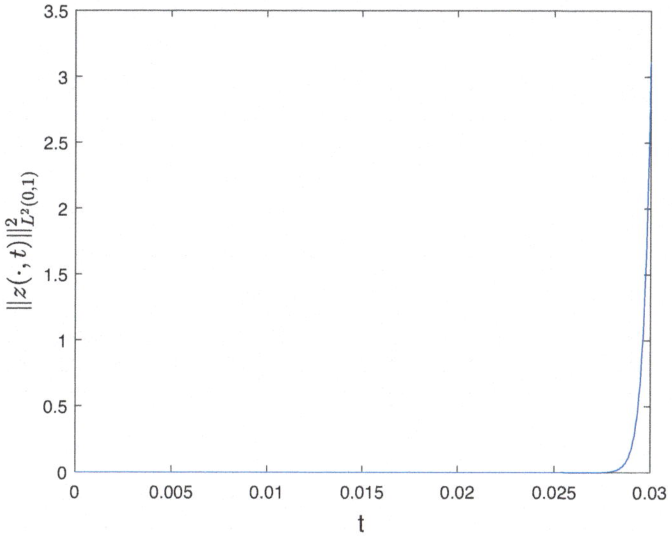

Fig. 3.21 Open-loop system (without control input) with $\lambda = 0.2$

that the corresponding open-loop system is unstable with the initial function $z_0(x) = 0.0025\sin(2\pi x)$, $0 \le x \le 1$.

For the control law (3.259) with the shape function (3.345) and the averaged measurement (3.346), by using Yalmip we verify LMI conditions of Theorem 3.10 with $\mu = 2$, $l = 0.8$, $\bar{\Delta} = 0.4$, $\delta = 0.14$, $\delta_1 = 20$, $C = 0.044$, $C_1 = 0.05$. We find that the closed-loop system (3.349) preserves the exponential stability with a decay rate $\delta = 0.14$ for $h \le 0.00638$ for any initial values satisfying $\|z_0\|_{H^1_{per}} < 0.044$. We

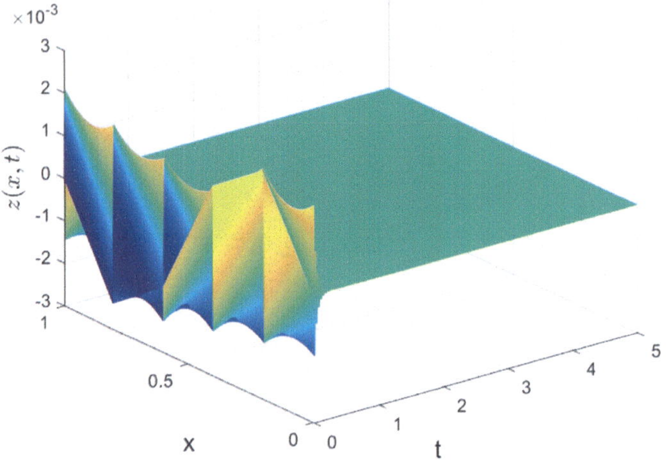

Fig. 3.22 State $z(x,t)$ with $[\tilde{x}_1, \tilde{x}_2] = [0.2, 0.6]$ under the averaged measurements

proceed further with the numerical simulations of the solutions of the closed-loop system (3.349) under the controller with $\mu = 2$, $\tilde{x}_1 = 0.2$, $\tilde{x}_2 = 0.6$. Let $x_0 = 0$, $x_1 = 1$. Hence, $[\tilde{x}_1, \tilde{x}_2] \subset [x_0, x_1]$, $l = 0.8$, $\bar{\Delta} = 0.4$. In this case, the control law is given by

$$u_1(t) = -2\int_{0.2}^{0.6} z(x,t)dx/0.4 = -5\int_{0.2}^{0.6} z(x,t)dx, \ t \geq 0. \tag{3.359}$$

The simulations show that the state of KdVB equation converges to zero (see Fig. 3.22). Simulations of the solutions confirm the theoretical results that follow from LMIs.

Next we consider the control law distributed on the whole domain:

$$u_1(t) = -2\int_0^1 z(x,t)dx, \ t \geq 0. \tag{3.360}$$

Using Theorem 3.9 with $\mu = 2$, $\Delta = 1$, $\delta_1 = 20$, $C = 0.044$, $C_1 = 0.05$, we find that the LMI conditions with the same $h = 0.00638$ leads to a larger decay rate $\delta = 0.8$.

3.6 Notes

The presented methods in this chapter give sufficient tools for sampled-data control of various PDE systems. By using Lyapunov method, stability conditions are derived for the corresponding closed-loop system in terms of LMIs. Furthermore, sampled-data control was extended in [40] for subpredictor-based stabilization of Takagi-Sugeno fuzzy PDEs with large uncertain delays. A recent advance for switching-based sampled-data control design was extended to high dimensional PDE systems [41].

References

1. Fridman, E., Bar Am, N.: Sampled-data distributed H_∞ control of transport reaction systems. SIAM J. Control. Optim. **51**, 1500–1527 (2013)
2. Fridman, E., Blighovsky, A.: Robust sampled-data control of a class of semilinear parabolic systems. Automatica **48**, 826–836 (2012)
3. Kang, W., Fridman, E.: Distributed sampled-data control of Kuramoto-Sivashinsky equation. Automatica **95**, 514–524 (2018)
4. Liu, W.J., Krstic, M.: Stability enhancement by boundary control in the Kuramoto-Sivashinsky equation. Nonlinear Anal. Ser. A Theory Methods **43**, 485–507 (2001)
5. Armaou, A., Christofides, P.D.: Wave suppression by nonlinear finite-dimensional control. Chem. Eng. Sci. **55**, 2627–2640 (2000)
6. Armaou, A., Christofides, P.D.: Feedback control of the Kuramoto-Sivashinsky Equation. Phys. D **137**, 49–61 (2000)
7. Lunasin, E., Titi, E.S.: Finite determining parameters feedback control for distributed nonlinear dissipative systems-a computational study. Evol. Equ. Control. Theory **6**, 535–557 (2017)
8. Tucsnak, M., Weiss, G.: Observation and Control for Operator Semigroups. Birkhäuser Advanced Texts: Basler Lehrbücher. Birkhäuser Verlag, Basel (2009)
9. Pazy, A.: Semigroups of Linear Operators and Applications to Partial Differential Equations. Springer, New York (1983)
10. Krasnosel'skii, M.A., Zabreyko, P.P., Pustylnik, E.I., Sobolevski, P.E.: Integral Operators in Spaces of Summable Functions. Springer, Dordrecht (2011)
11. Fridman, E.: Introduction to Time-Delay Systems: Analysis and Control. Birkhäuser, Basel (2014)
12. Bellman, R.E., Cooke, K.L.: Differential-Difference Equations. RAND Corporation (1963)
13. Henry, D.: Geometric Theory of Semilinear Parabolic Equations. Springer, New York (1981)
14. Fridman, E., Seuret, A., Richard, J.P.: Robust sampled-data stabilization of linear systems: an input delay approach. Automatica **40**, 1441–1446 (2004)
15. Mikheev, Yu., Sobolev, V., Fridman, E.: Asymptotic analysis of digital control systems. Autom. Remote. Control. **49**, 1175–1180 (1988)
16. Fridman, E.: New Lyapunov-Krasovskii functionals for stability of linear retarded and neutral type systems. Syst. & Control. Lett. **43**, 309–319 (2001)
17. Gu, K., Kharitonov, V.L., Chen, J.: Stability of Time-Delay Systems. Birkhäuser, Boston (2003)
18. Fridman, E., Orlov, Y.: Exponential stability of linear distributed parameter systems with time-varying delays. Automatica **45**(1), 194–201 (2009)
19. Am, Bar, N., Fridman, E.: Network-based distributed H_∞-filtering of parabolic systems. Automatica **50**, 3139–3146 (2014)
20. Selivanov, A., Fridman, E.: Sampled-data relay control of semilinear diffusion PDEs. In: Proceedings of the IEEE Conference on Decision and Control, pp. 4821–4826 (2016)
21. Robinson, J.C.: Infinite-Dimensional Dynamical Systems: An Introduction to Dissipative Parabolic PDEs and the Theory of Global Attractors. Cambridge University Press (2001)
22. Liu, K., Fridman, E.: Delay-dependent methods and the first delay interval. Syst. Control Lett. **64**, 57–63 (2014)
23. Kang, W., Fridman, E.: Sampled-data control of 2-D Kuramoto-Sivashinsky equation. IEEE Trans. Autom. Control. **67**(3), 1314–1326 (2022)

24. Tomlin, R., Gomes, S.N.: Point-actuated feedback control of multidimensional interfaces. IMA J. Appl. Math. (2019)
25. Azouani, A., Titi, E.S.: Feedback control of nonlinear dissipative systems by finite determining parameters–a reaction-diffusion paradigm. Evol. Equ. Control. Theory **3**, 579–594 (2014)
26. Selivanov, A., Fridman, E.: Delayed H_∞ control of 2-D diffusion systems under delayed pointlike measurements. Automatica **109**, 108541 (2019)
27. Liu, Z., Zheng, S.: Semigroups Associated with Dissipative Systems. Research Notes in Mathematics, vol. 398. Chapman & Hall/CRC, Boca Raton (1999)
28. Yakubovich, V.A.: S-procedure in nonlinear control theory. Vestink Leningr. Univ. **1**, 62–77 (1971)
29. Fridman, E.: A refined input delay approach to sampled-data control. Automatica **46**, 421–427 (2010)
30. Hetel, L., Fridman, E.: Robust sampled-data control of switched affine systems. IEEE Trans. Autom. Control. **58**, 2922–2928 (2013)
31. Selivanov, A., Fridman, E.: Observer-based input-to-state stabilization of networked control systems with large uncertain delays. Automatica **74**, 63–70 (2013)
32. Kang, W., Fridman, E.: Distributed stabilization of Korteweg-de Vries-Burgers equation in the presence of input delay. Automatica **100**, 260–273 (2019)
33. Kang, W., Fridman, E., Liu, C.X.: Stabilization for a semilinear heat equation by switching control. In: IEEE 60th Conference on Decision and Control (CDC), pp. 448–453 (2021)
34. Kitsos, C., Fridman, E.: Stabilization of underactuated linear coupled reaction-diffusion PDEs via distributed or boundary actuation. IEEE Trans. Autom. Control. **69**(7), 4183–4198 (2024)
35. Datta, B.: Numerical Methods for Linear Control Systems. Elsevier (2004)
36. Feng, B.F., Kawahara, T.: Stationary travelling-wave solutions of an unstable KdV-Burgers equation. Phys. D Nonlinear Phenom. **137**, 228–236 (2000)
37. Larkin, N.A.: Korteweg-de Vries and Kuramoto-Sivashinsky equations in bounded domains. J. Math. Anal. Appl. **297**, 169–185 (2004)
38. Seuret, A., Gouaisbaut, F.: Wirtinger-based integral inequality: application to time-delay systems. Automatica **49**, 2860–2866 (2013)
39. Wang, J.W., Wu, H.N.: Lyapunov-based design of locally collocated controllers for semi-linear parabolic PDE systems. J. Frankl. Inst. **351**, 429–441 (2014)
40. Kang, W., Zhang, J., Lam, H.K.: Subpredictor-based fuzzy control design for a reaction-diffusion equation. IEEE Trans. Fuzzy Syst. **30**(11), 4728–4740 (2022)
41. Kang, W., Fridman, E., Zhang, J., Wang, J.M.: Event-triggered control for stabilization of semi-linear N-D reaction-convection-diffusion PDE by switching. Int. J. Robust Nonlinear Control. (2024)

Chapter 4
Event-Triggered Control of PDE Systems

Chapter 3 presented time-triggered sampled-data control of PDEs via the time-delay approach to the corresponding Lyapunov functionals. In this chapter we consider event-triggered control for 1D nonlinear KdV equation, where the time-delay approach does not work (it seems to be impossible to construct an appropriate Lyapunov functionals because of the third spatial derivative in KdV equation). In this chapter, we introduce an innovative event-triggering mechanism to the sampled control law, aiming at optimizing control input updates by restricting them to meaningful instants. This approach not only alleviates computational burdens but also bolsters system efficiency. This approach also addresses the challenges of maintaining system stability and avoiding Zeno behavior. We establish sufficient conditions based on LMIs to guarantee regional exponential stability of the closed-loop system.

4.1 Preliminaries and Problem Formulation

4.1.1 State Feedback Control of a Nonlinear KdV Equation

Before proceeding to our problem's setting, let us explain the essential idea of the Lyapunov-based state feedback control for KdV equation. Consider the initial and boundary value problem

$$\begin{cases} \partial_t z + z\partial_x z + \partial_x z + \nu\partial_{xxx} z - \lambda z = f(x,t), & \forall x \in (0,L),\ t \geq 0, \\ z(0,t) = z(L,t) = 0,\ \partial_x z(L,t) = 0, & \forall\, t \geq 0, \\ z(x,0) = z^0(x), & \forall x \in (0,L), \end{cases} \tag{4.1}$$

with the initial state $z^0 \in L^2(0,L)$ and the source input $f \in L^1(0,T;L^2(0,L))$, where $\nu > 0$, $\lambda \geq 0$, and $z = z(x,t)$ is the state of the nonlinear KdV equation. For $\lambda > 0$, the open-loop system may be unstable (see the example below). Note that

W. Kang and E. Fridman, *Delay-Robust Control of Distributed Parameter Systems*, Advances in Delays and Dynamics 13, https://doi.org/10.1007/978-981-96-6344-6_4

destabilizing $\lambda > 0$ was considered in [1]. Also, $\lambda > 0$ may stand for the desired decay rate achieved after stabilization of (4.1) with $\lambda = 0$ (see Remark 4.3 below). By selecting the control law

$$f(x,t) = -Kz(x,t), \quad K > \lambda, \tag{4.2}$$

one obtains a closed-loop system that is globally exponentially stable, as it will be shown later. In this article, we would like to address the question of the robustness of this stability with respect to the presence of both an event triggering in time and a localized averaging in space of the feedback control law. Noticing that we cannot really apply infinite dimensional feedback control law, we will consider here a finite dimensional approximation of (4.2) that still stabilizes the system (see Sect. 4.3 below).

More precisely, we will consider that the control law will be implemented in such a way that for all $x \in (0, L)$, for all $t \in [t_k, t_{k+1})$,

$$f(x,t) = -K \sum_{j=1}^{N} \bar{z}_j(t_k) \mathbb{1}_{\Omega_j}(x), \quad K > \lambda, \tag{4.3}$$

where the sampling times t_k are following an appropriate event trigger law to be given later, while $\{\mathbb{1}_{\Omega_j}\}_j$ are the characteristic functions of the intervals $\{\Omega_j\}_j$ covering $(0, L)$, and $\bar{z}_j(t) = \frac{1}{|\Omega_j|}\int_{\Omega_j} z(x,t)dx$. We will also consider the case that the event-triggered controller does not cover the whole domain $[0, L]$, which is distributed on some parts of subdomains (see (4.35) in Sect. 4.5 below).

4.1.2 Well-Posedness and Exponential Stabilization

The proof of existence and regularity of solutions for the KdV equation has been investigated in many references, in particular in the field of controllability studies and even if several results rely on the smallness of the initial and source data (e.g. [2]), one can find in [3] the proof of the following general result :

Lemma 4.1 *For any $T > 0$, $L > 0$, if $z^0 \in L^2(0, L)$ and $f \in L^1(0, T; L^2(0, L))$, then the Cauchy problem (4.1) is well posed in the space $C([0, T]; L^2(0, L)) \cap L^2(0, T; H^1(0, L))$, meaning that there exists a unique solution z to the system (4.1) that satisfies, for a constant $c = c(T, L) > 0$,*

$$\|z\|_{L^\infty(0,T;L^2(0,L))} + \|z\|_{L^2(0,T;H^1(0,L))} \leq c\|z^0\|_{L^2(0,L)} + c\|f\|_{L^1(0,T;L^2(0,L))}.$$

The proof of this lemma is detailed in [3] and relies on a fixed point argument for the small time well-posedness (as also referenced and described in [2]) of the problem, that allows to handle the non-linearity $z\partial_x z$, and on clever a priori estimates of the local solution to extend arbitrarily the time frame and get a global existence and regularity result.

The proof of the well-posedness of the closed-loop system (4.1)–(4.2) stems from the same arguments and is not detailed here. Besides, it is easy to prove its exponential stability, stated here:

Lemma 4.2 *Let $L > 0$, $T > 0$, $K > \lambda$ and $z^0 \in L^2(0, L)$. The closed-loop KdV system (4.1)–(4.2) is exponentially stable in the sense that*

$$\|z(\cdot, t)\|^2_{L^2(0,L)} \leq e^{-2(K-\lambda)t} \|z^0\|^2_{L^2(0,L)}, \quad \forall t \geq 0.$$

Indeed : define the energy (that will act as a Lyapunov functional) of the solution of a KdV equation by

$$E(t) = \|z(\cdot, t)\|^2_{L^2(0,L)}, \qquad \forall t \geq 0. \tag{4.4}$$

Taking the time derivative of $E(t)$ along (4.1)–(4.2), we have, for any $t \geq 0$,

$$\begin{aligned}\dot{E}(t) &\leq -2(K - \lambda) \int_0^L |z(x, t)|^2 dx - \nu |\partial_x z(0, t)|^2 \\ &\leq -2(K - \lambda) E(t)\end{aligned}$$

implying $E(t) \leq e^{-2(K-\lambda)t} E(0), \ \forall t \geq 0$.

Furthermore, [2] gathers several internal stabilization results for nonlinear KdV equations, and specifically, the stabilization through a localized distributed internal damping $f(x, t) = -a(x)z(x, t)$ with $a \in L^\infty(0, L)$ such that $a(\cdot) \geq a_0 > \lambda$ in some subdomain ω of $(0, L)$, is actually also true, see e.g. [4, 5]. However, in our study, we focus on the case described by (4.3).

As already mentioned before, in this paper we will use a Lyapunov approach to deal with an event-triggered control of the KdV equation under averaged measurements. The next section is devoted to the description of our technical setting.

4.2 Problem Formulation and Main Result

We consider the following closed-loop KdV system:

$$\begin{cases} \partial_t z + z\partial_x z + \partial_x z + \nu \partial_{xxx} z - \lambda z = -K \sum_{j=1}^{N} \bar{z}_j(t_k) \mathbb{1}_{\Omega_j}(x), & \\ \qquad\qquad\qquad\qquad \text{in}(0, L) \times [t_k, t_{k+1}),\ k \in \mathbb{N}, & \\ z(0, t) = z(L, t) = 0,\ \partial_x z(L, t) = 0, & \forall\, t \geq 0, \\ z(x, 0) = z^0(x), & \forall x \in (0, L). \end{cases} \tag{4.5}$$

where the chosen control law for (4.1) is (4.3),

$$\bar{z}_j(t_k) = \frac{1}{|\Omega_j|}\int_{\Omega_j} z(x, t_k)dx. \tag{4.6}$$

This closed-loop system is defined under the following assumptions:

- Space averaging: As in [6–9], we assume that the points $0 = x_0 < x_1 < \cdots < x_N = L$ divide the interval $[0, L]$ into N intervals $\Omega_j = [x_{j-1}, x_j)$ covering it all. The width of each sub-interval is supposed to be upper bounded by some constant: $0 < x_j - x_{j-1} = |\Omega_j| \le \Delta$ and as expected, the characteristic functions $\mathbb{1}_{\Omega_j}(x)$ are such that
$$\begin{cases} \mathbb{1}_{\Omega_j}(x) = 0, \ x \notin \Omega_j, \\ \mathbb{1}_{\Omega_j}(x) = 1, \ \text{otherwise}, \end{cases} \quad j = 1, \cdots, N. \tag{4.7}$$
- Time sampling: The update instants satisfy $0 = t_0 < t_1 < \cdots < t_k < t_{k+1}$, $\lim_{k\to\infty} t_k = \infty$. We define the event trigger mechanism by the law
$$t_{k+1} = \inf\Big\{t \ge t_k \text{ such that } \|z(\cdot, t) - z(\cdot, t_k)\|^2_{L^2(0,L)} \ge \gamma E(t) + \gamma_0 E(0)e^{-2\theta t}\Big\} \tag{4.8}$$
where the energy E is defined by (4.4) as the $L^2(0, L)$−norm of the state, and γ, γ_0 and θ are positive constants to be determined.
It should be noticed that due to the term "$\gamma_0 E(0)e^{-2\theta t}$", here no dwell time is needed to be defined.
- Though the feedback is of finite dimension, both t_k and Δ depend on the initial data. The larger initial data is, the smaller t_k and Δ need to be.

Our main objective is to design a regionally stabilizing event-trigger controller

$$u_j(t) = -K\bar{z}_j(t_k)\mathbb{1}_{[t_k,t_{k+1})}(t)$$

that has a control gain $K > \lambda$ to be determined later. In other words, we aim at deriving sufficient conditions for regional exponential stability of the closed-loop system (4.5) and to find a bound on the domain of attraction.

Theorem 4.1 *Let $L > 0$, $T > 0$. Given a desired decay rate $\delta > 0$, a control gain $K > \lambda + \delta$, a length bound $\Delta > 0$, and positive tuning parameters λ_0, R, $\theta > \delta$, $\gamma_0 > 0$, assume that there exist positive scalars μ, λ_1, λ_2, γ, and Γ that solve the following optimization problem:*

$$\min \Gamma \quad \text{subject to}$$

$$-3\mu\nu + \lambda_1 + \lambda_2 + \frac{2}{3}\mu RL\sqrt{L} < 0, \tag{4.9}$$

$$\Phi = \begin{bmatrix} \phi_{11} & K(1+\mu L) & K(1+\mu L) \\ * & -\lambda_2 \frac{\pi^2}{\Delta^2} & 0 \\ * & * & -\lambda_0 \end{bmatrix} \prec 0, \tag{4.10}$$

$$(1+\mu L)\left(1+\frac{\lambda_0\gamma_0}{2(\theta-\delta)}\right) < R^2\Gamma, \tag{4.11}$$

where

$$\phi_{11} = -2K + 2\lambda + \mu + \lambda_0\gamma - \lambda_1\frac{\pi^2}{L^2} + 2\delta. \tag{4.12}$$

Then for any initial function $z^0 \in L^2(0, L)$ *satisfying* $\|z^0\|_{L^2(0,L)} < \frac{1}{\sqrt{\Gamma}}$, *the closed-loop system (4.5) under the event-triggering mechanism (4.8) is exponentially stable:*

$$E(t) \le \left(1+\frac{\lambda_0\gamma_0}{2(\theta-\delta)}\right)(1+\mu L)E(0)e^{-2\delta t} \tag{4.13}$$

for all $t \ge 0$. *Moreover, if the above LMIs hold with* $\delta = 0$, *then the closed-loop system is exponentially stable with a small enough decay rate.*

Remark 4.1 One could wish here that we do not make the assumption $K \ge \lambda + \delta$ on the gain we need to apply to stabilize our system, but we shall recall that the decay rate of the exponential stability of the system with continuous feedback law $-Kz$ is exactly $\delta = K - \lambda$ (Lemma 4.2) so that it's not reasonable to expect better when applying an approximated feedback law as we do.

Remark 4.2 If γ and γ_0 are small enough, then the event-triggering mechanism (4.8) gets more sensitive to the output change and transmits the signals more often, what makes the control more similar to the stabilizing continuous-time controller.

Remark 4.3 Consider (4.5) with $\lambda = 0$

$$\begin{cases} \partial_t z + z\partial_x z + \partial_x z + \nu\partial_{xxx} z = -K\sum_{j=1}^{N} \bar{z}_j(t_k)\mathbb{1}_{\Omega_j}(x), & \text{in}(0, L)\times[t_k, t_{k+1}),\ k \in \mathbb{N}, \\ z(0,t) = z(L,t) = 0,\ \partial_x z(L,t) = 0, & \forall\, t \ge 0, \\ z(x,0) = z^0(x), & \forall x \in (0, L), \end{cases} \tag{4.14}$$

where $\bar{z}_j(t_k)$ is given by (4.6).

Let $\tilde{z} = e^{\lambda t}z$. It is easy to see that $\tilde{z}$ is governed by

$$\begin{cases} \partial_t \tilde{z} + e^{-\lambda t}\tilde{z}\partial_x \tilde{z} + \partial_x \tilde{z} + \nu\partial_{xxx}\tilde{z} - \lambda\tilde{z} = -K\sum_{j=1}^{N} \hat{z}_j(t_k)\mathbb{1}_{\Omega_j}(x), \\ \qquad \text{in}(0, L)\times[t_k, t_{k+1}),\ k \in \mathbb{N}, \\ \tilde{z}(0,t) = \tilde{z}(L,t) = 0,\ \partial_x \tilde{z}(L,t) = 0, \quad \forall\, t \ge 0, \\ \tilde{z}(x,0) = z^0(x), \quad \forall x \in (0, L), \end{cases} \tag{4.15}$$

where $\hat{z}_j(t_k) = \frac{1}{|\Omega_j|}\int_{\Omega_j} \bar{z}(x, t_k)dx$.
From the proof of Theorem 4.1, it follows that LMIs of this Theorem guarantee stability of (4.15) since the nonlinear term "$e^{-\lambda t}\bar{z}\bar{z}_x$" with the multiplier $e^{-\lambda t} \le 1$ will not change the proof of stability. Hence, if the LMI conditions of Theorem 4.1 hold with $\delta = 0$, then the decay rate λ of original system (4.14) can be guaranteed since $z = e^{-\lambda t}\bar{z}$.

4.3 Technical Proofs

4.3.1 Well-Posedness of the Controlled System and Avoidance of Zeno Behaviour

From Lemma 4.1, the following well-posedness result can be obtained by an induction approach.

Proposition 4.1 *Let $L > 0$, $T > 0$ and assume that $z^0 \in L^2(0, L)$. Then system (4.5) under the event triggering law (4.8) has a unique solution z satisfying $z \in C(0, T; L^2(0, L)) \cap L^2(0, T; H^1(0, L))$. Furthermore, the Zeno phenomenon is avoided.*

Proof • *Existence, uniqueness and regularity of the solution:*
We proceed by induction.
(i) *Initialization.* On the first time interval, (4.5) reads

$$\begin{cases} \partial_t z + z\partial_x z + \partial_x z + \nu\partial_{xxx} z - \lambda z = -K\sum_{j=1}^{N} \bar{z}_j^0 \mathbb{1}_{\Omega_j}(x), & \forall x \in (0, L),\ t \in [0, t_1), \\ z(0, t) = z(L, t) = 0,\ \partial_x z(L, t) = 0, & \forall\, t \ge 0, \\ z(x, 0) = z^0(x), & \forall x \in (0, L), \end{cases}$$

where $\bar{z}_j^0 = \bar{z}_j(0) = \frac{1}{|\Omega_j|}\int_{\Omega_j} z^0(x)dx$, and $K > \lambda$. This is a nonlinear KdV equation with initial data $z^0 \in L^2(0, L)$ and source term $f = -K\sum_{j=1}^{N} \bar{z}_j^0 \mathbb{1}_{\Omega_j} \in L^1(0, t_1; L^2(0, L))$. Lemma 4.1 allows to conclude that there exists a unique solution $z \in C([0, t_1]; L^2(0, L)) \cap L^2(0, t_1; H_0^1(0, L))$ to the latter system.
(ii) *Heredity.* Let us only highlight that the previously obtained solution satisfies $z(t_1) \in L^2(0, L)$ so that system (4.5) considered on the next time interval $[t_1, t_2)$ has an initial condition $z(t_1) \in L^2(0, L)$ and a source term $-K\sum_{j=1}^{N} \bar{z}_j^1 \mathbb{1}_{\Omega_j} \in L^1(t_1, t_2; L^2(0, L)$ where $\bar{z}_j^1 = \bar{z}_j(t_1)$. Therefore, the same argument using Lemma 4.1 holds again and the heredity is proved similarly at any step $k \in \mathbb{N}$.
(iii) *Conclusion.* By induction, for any $k \in \mathbb{N}$, $z \in C([t_k, t_{k+1}]; L^2(0, L)) \cap L^2(t_k, t_{k+1}; H^1(0, L))$. Therefore, from the extension by continuity at the instants

t_k, one can conclude that (4.5) has a unique solution $z \in C([0, T]; L^2(0, L)) \cap L^2(0, T; H^1(0, L))$.

(iv) *Convergence.* The solution will never blow up before T as a contrary of the Zeno behaviour (i.e. $\exists t_k > T$).

• *Avoidance of Zeno behaviour:*

We aim at showing that the event-triggering mechanism (4.8) rules out the Zeno behaviour, where an infinite number of updates may occur in a finite amount of time. It is actually sufficient to show that for a given $T > 0$, there exists $\tau^* > 0$ such that all the sampling instants $t_k \le T$ complying to (4.8) satisfy $t_{k+1} - t_k \ge \tau^*$.

Let us denote by e_k the deviation from the continuous time position: for any $x \in [0, L]$ and $t \in (0, T)$, there exists $k \in \mathbb{N}$ such that $t \in [t_k, t_{k+1})$, and we set

$$e_k(x, t) \triangleq z(x, t) - z(x, t_k). \tag{4.16}$$

Since the solution of closed-loop system (4.5) satisfies $z \in C([0, T]; L^2(0, L))$ and $[0, T]$ is a compact set, this error function e_k is uniformly continuous in time with values in $L^2(0, L)$. This means that for any $\epsilon > 0$ there exists $\tau^* > 0$ such that for all $t, s \in [0, T]$, if $|t - s| < \tau^*$ then we have $\|e_k(\cdot, t) - e_k(\cdot, s)\|_{L^2(0,L)} < \epsilon$.
Thus, the following reasoning by contraposition holds:
$\forall \epsilon > 0, \exists \tau^* > 0, \forall t, s \in [0, T]$,

$$\|e_k(\cdot, t) - e_k(\cdot, s)\|_{L^2(0,L)} \ge \epsilon \implies |t - s| \ge \tau^*. \tag{4.17}$$

Since $e_k(t_k) = 0$, we have

$$\|e_k(\cdot, t_{k+1}) - e_k(\cdot, t_k)\|_{L^2(0,L)} = \|e_k(\cdot, t_{k+1})\|_{L^2(0,L)}.$$

Next the substitution $t \to t_{k+1}$ and $s \to t_k$ into (4.17), together with the definition of t_{k+1} in (4.8), leads to

$$\begin{aligned}\|e_k(\cdot, t_{k+1})\|^2_{L^2(0,L)} &\ge \gamma E(t_{k+1}) + \gamma_0 E(0) e^{-2\theta t_{k+1}} \\ &\ge \gamma_0 \|z^0\|^2_{L^2(0,L)} e^{-2\theta T}\end{aligned}$$

implying that $|t_{k+1} - t_k| \ge \tau^*$. Indeed, given $z^0 \neq 0$, we choose

$$\epsilon = \sqrt{\gamma_0 \|z^0\|^2_{L^2(0,L)} e^{-2\theta T}}$$

so that there exists $\tau^* > 0$, depending on $z^0, \theta, \gamma_0, \gamma$ and T for which for any k such that $t_k, t_{k+1} \in [0, T]$, one has $t_{k+1} - t_k \ge \tau^*$, so that the Zeno behaviour is avoided. □

4.3.2 Regional Stability Analysis

Now we focus on the regional stability analysis of the closed-loop system and prove Theorem 4.1. Let us mention two things.

On the one hand, the event-triggering mechanism (4.8) yields that the event-triggering error function is bounded on each time sub-interval as follows: $\forall t \in [t_k, t_{k+1})$

$$\|e_k(\cdot, t)\|^2_{L^2(0,L)} \leq \gamma E(t) + \gamma_0 E(0)e^{-2\theta t}. \tag{4.18}$$

On the other hand, f defined by (4.3) can be rewritten as

$$f(x, t) = -K \sum_{j=1}^{N} \mathbb{1}_{\Omega_j}(x) \left[z(x, t) - f_j(x, t) - \rho_j(t)\right],$$

$$\forall x \in [0, L],\ \forall t \in [t_k, t_{k+1}),\ \forall k \in \mathbb{N} \tag{4.19}$$

where

$$f_j(x, t) = z(x, t) - \bar{z}_j(t) = z(x, t) - \frac{1}{|\Omega_j|} \int_{\Omega_j} z(x, t)dx,$$

$$\rho_j(t) = \bar{z}_j(t) - \bar{z}_j(t_k) = \frac{1}{|\Omega_j|} \int_{\Omega_j} e_k(x, t)dx.$$

Proof of Theorem 4.1 Writing $V_\mu(t) = \mu \int_0^L x|z(x, t)|^2 dx$ with $\mu > 0$, we define the following functional (using simplified notations [10]):

$$V(t) \triangleq E(t) + V_\mu(t) = \int_0^L |z(x, t)|^2 dx + \mu \int_0^L x|z(x, t)|^2 dx. \tag{4.20}$$

First, this Lyapunov functional candidate $V(t)$ is equivalent to the energy of the system $E(t)$ in the sense that

$$E(t) \leq V(t) \leq (1 + \mu L)E(t). \tag{4.21}$$

Then, let us estimate its time derivative. For $t \in [t_k, t_{k+1})$, substituting (4.19) into (4.5) and differentiating $V(t)$ along (4.5), one gets

$$
\begin{aligned}
\dot{V}(t) &= \dot{E}(t) + \dot{V}_\mu(t) = 2\int_0^L z(x,t)\partial_t z(x,t)dx + 2\mu\int_0^L xz(x,t)\partial_t z(x,t)dx \\
&= 2\int_0^L (1+\mu x)z(x,t)\big[-\nu\partial_{xxx}z(x,t) - z(x,t)\partial_x z(x,t) - \partial_x z(x,t) + \lambda z(x,t) - Kz(x,t)\big]dx \\
&+ 2K\sum_{j=1}^N \int_{\Omega_j} (1+\mu x)z(x,t)\big[f_j(x,t) + \rho_j(t)\big]dx.
\end{aligned}
\tag{4.22}
$$

Hence,

$$
\begin{aligned}
\dot{V}(t) &= -\nu|\partial_x z(0,t)|^2 - 3\mu\nu\int_0^L |\partial_x z(x,t)|^2dx + \mu\int_0^L |z(x,t)|^2dx \\
&+ \frac{2}{3}\mu\int_0^L z^3(x,t)dx - 2(K-\lambda)\int_0^L (1+\mu x)|z(x,t)|^2dx \\
&+ 2K\sum_{j=1}^N \int_{\Omega_j} (1+\mu x)z(x,t)[f_j(x,t) + \rho_j(t)]dx.
\end{aligned}
\tag{4.23}
$$

Using (4.18), for any $\lambda_0 > 0$ we can deduce that

$$
\begin{aligned}
\dot{V}(t) &\le \dot{V}(t) + \lambda_0\left[\gamma E(t) + \gamma_0 E(0)e^{-2\theta t} - \|e_k(\cdot,t)\|^2_{L^2(0,L)}\right] \\
&\le -3\mu\nu\int_0^L |\partial_x z(x,t)|^2dx - (2K - \mu - \lambda_0\gamma - 2\lambda)\int_0^L |z(x,t)|^2dx \\
&+ \frac{2}{3}\mu\int_0^L z^3(x,t)dx - (2K-2\lambda)\mu\int_0^L x|z(x,t)|^2dx \\
&+ 2K\sum_{j=1}^N \int_{\Omega_j} (1+\mu x)z(x,t)[f_j(x,t) + \rho_j(t)]dx \\
&+ \lambda_0\gamma_0 E(0)e^{-2\theta t} - \lambda_0\|e_k(\cdot,t)\|^2_{L^2(0,L)}.
\end{aligned}
\tag{4.24}
$$

Several estimates can now be obtained to deal with each of these terms and bring this into a quadratic form. First, Cauchy–Schwarz inequality and Sobolev's inequality (see Lemma 1.7) leads to

$$
\begin{aligned}
\int_0^L z^3(x,t)dx &\le \|z(\cdot,t)\|^2_{L^\infty(0,L)}\int_0^L |z(x,t)|dx \\
&\le L\sqrt{L}\|\partial_x z(\cdot,t)\|^2_{L^2(0,L)}\|z(\cdot,t)\|_{L^2(0,L)}.
\end{aligned}
\tag{4.25}
$$

Then from Lemma 1.2, Wirtinger's inequality yields

$$
\lambda_1\left[\|\partial_x z(\cdot,t)\|^2_{L^2(0,L)} - \frac{\pi^2}{L^2}\|z(\cdot,t)\|^2_{L^2(0,L)}\right] \ge 0. \tag{4.26}
$$

for any $\lambda_1 > 0$.

Moreover, since $\int_{\Omega_j} f_j(x,t)dx = 0$, from Lemma 1.1, Poincaré's inequality rewrites

$$\|f_j(\cdot,t)\|^2_{L^2(\Omega_j)} \le \frac{\Delta^2}{\pi^2}\|\partial_x z(\cdot,t)\|^2_{L^2(\Omega_j)},$$

bringing for any $\lambda_2 > 0$

$$\lambda_2 \sum_{j=1}^{N}\left[\|\partial_x z(\cdot,t)\|^2_{L^2(\Omega_j)} - \frac{\pi^2}{\Delta^2}\|f_j(\cdot,t)\|^2_{L^2(\Omega_j)}\right] \ge 0. \tag{4.27}$$

Applying the Cauchy–Schwarz inequality, we obtain

$$\begin{aligned}
&\sum_{j=1}^{N}\int_{\Omega_j}\rho_j^2(t) = \sum_{j=1}^{N}\rho_j^2(t)|\Omega_j| \le \sum_{j=1}^{N}\frac{1}{|\Omega_j|}\left(\int_{\Omega_j} e_k(x,t)dx\right)^2 \\
&\le \sum_{j=1}^{N}\int_{\Omega_j} e_k^2(x,t)dx = \int_0^L e_k^2(x,t)dx.
\end{aligned} \tag{4.28}$$

Hence,

$$\lambda_0\left[\|e_k(\cdot,t)\|^2_{L^2(0,L)} - \sum_{j=1}^{N}\int_{\Omega_j}\rho_j^2(t)\right] \ge 0. \tag{4.29}$$

Set $\eta(x,t) = \text{col}\{z(x,t), f_j(x,t), \rho_j(t)\}$. Substituting (4.25) and (4.28) into (4.24), and adding (4.26), (4.27) and (4.29) to $\dot V(t)$, we obtain

$$\begin{aligned}
&\dot V(t) + 2\delta V(t) \le \sum_{j=1}^{N}\int_{\Omega_j}\eta(x,t)^\top\Phi(x)\eta(x,t) \\
&-\left(3\mu\nu - \lambda_1 - \lambda_2 - \frac{2\mu}{3}L\sqrt{L}\|z(\cdot,t)\|_{L^2(0,L)}\right)\|\partial_x z(\cdot,t)\|^2_{L^2(0,L)} \\
&-2(K-\lambda-\delta)\mu\int_0^L x|z(x,t)|^2dx + \lambda_0\gamma_0 E(0)e^{-2\theta t},
\end{aligned} \tag{4.30}$$

where

$$\Phi(x) = \begin{bmatrix} \phi_{11} & K(1+\mu x) & K(1+\mu x) \\ * & -\lambda_2\frac{\pi^2}{\Delta^2} & 0 \\ * & * & -\lambda_0 \end{bmatrix}$$

and $\phi_{11} = -2K + 2\lambda + \mu + \lambda_0\gamma - \lambda_1\frac{\pi^2}{L^2} + 2\delta$ as in (4.12).

Applying Schur complement theorem [11], one gets that $\Phi(x) \prec 0$ is equivalent to

$$\phi_{11} + K^2(1+\mu x)^2 \left(\frac{\Delta^2}{\lambda_2 \pi^2} + \lambda_0^{-1} \right) < 0,$$

that also writes

$$-2K + 2\lambda + \mu + \lambda_0 \gamma - \lambda_1 \frac{\pi^2}{L^2} + 2\delta + K^2(1+\mu x)^2 \left(\frac{\Delta^2}{\lambda_2 \pi^2} + \lambda_0^{-1} \right) < 0. \tag{4.31}$$

Since we need that property for all $x \in [0, L]$, and since we have $1 \leq (1+\mu x)^2 \leq (1+\mu L)^2$, then it proves

$$\Phi(L) \prec 0 \implies \Phi(x) \prec 0, \ \forall x \in [0, L].$$

Hence, denoting $\Phi = \Phi(L)$ so that (4.10) holds, we have proved that

$$\sum_{j=1}^{N} \int_{\Omega_j} \eta(x,t)^\top \Phi(x) \eta(x,t) \leq 0. \tag{4.32}$$

A final step as to be performed to handle the non-quadratic estimate (4.25). Let us first assume that

$$\|z(\cdot, t)\|_{L^2(0,L)} < R, \ \forall t \geq 0. \tag{4.33}$$

Under assumptions (4.9)–(4.10) and (4.33), from (4.30) and (4.32) and choosing $K > \lambda + \delta$, we obtain

$$\dot{V}(t) + 2\delta V(t) \leq \lambda_0 \gamma_0 E(0) e^{-2\theta t} \leq \lambda_0 \gamma_0 V(0) e^{-2\theta t}, \ \forall t \geq 0.$$

Now let $\theta > \delta$. Then, for all $t \geq 0$ we can write

$$\begin{aligned} V(t) &\leq e^{-2\delta t} V(0) + \lambda_0 \gamma_0 e^{-2\delta t} V(0) \int_0^t e^{-2(\theta-\delta)s} ds \\ &\leq e^{-2\delta t} V(0) + \frac{\lambda_0 \gamma_0 V(0)}{2(\theta-\delta)} \left[e^{-2\delta t} - e^{-2\theta t} \right] \\ &\leq \left(1 + \frac{\lambda_0 \gamma_0}{2(\theta-\delta)} \right) e^{-2\delta t} V(0) - \frac{\lambda_0 \gamma_0}{2(\theta-\delta)} V(0) e^{-2\theta t}. \end{aligned}$$

From (4.21) it follows that

$$E(t) \leq \left(1 + \frac{\lambda_0 \gamma_0}{2(\theta-\delta)} \right) (1+\mu L) e^{-2\delta t} E(0) - \frac{\lambda_0 \gamma_0}{2(\theta-\delta)} E(0) e^{-2\theta t}, \tag{4.34}$$

which implies (4.13).

In order to end the proof of Theorem 4.1, we need to prove that (4.33) holds. On the one hand, for $t = 0$, inequality (4.33) holds by hypothesis in Theorem 4.1, so that $E(0) < \frac{1}{\Gamma}$. On the other hand, let (4.33) be false for some $t > 0$ and let t^* be the smallest instant such that $E(t^*) \geq R^2$. Since E is continuous in time, we have $E(t^*) = R^2$ and $E(t) < R^2$ for $t \in [0, t^*)$. Therefore, the feasibility of inequality (4.9) and LMI (4.10) guarantee that (4.34) is true for all $t \in [0, t^*)$. Hence, by continuity,

$$\begin{aligned} E(t) &\leq \left(1 + \frac{\lambda_0 \gamma_0}{2\theta - 2\delta}\right)(1 + \mu L)e^{-2\delta t} E(0) \\ &\leq \left(1 + \frac{\lambda_0 \gamma_0}{2\theta - 2\delta}\right)(1 + \mu L)\frac{1}{\Gamma}, \quad \forall t \in [0, t^*]. \end{aligned}$$

The above inequality, together with the assumption (4.11), implies

$$E(t) \leq \left(1 + \frac{\lambda_0 \gamma_0}{2\theta - 2\delta}\right)(1 + \mu L)\frac{1}{\Gamma} < R^2$$

for all $t \in [0, t^*]$, which contradicts the definition of t^*. Therefore, (4.33) holds.

Note that the feasibility of the strict LMI (4.10) with $\delta = 0$ implies its feasibility with a slightly larger $\delta_0 > 0$. Therefore, if the strict LMI (4.10) holds for $\delta = 0$, then the closed-loop system is exponentially stable with a small decay rate.

Remark 4.4 It must be stressed that the present Lyapunov function cannot work for the case of a simple sampled-data control under averaged measurement and that the event-triggered law is critical in the proof of stability.

Remark 4.5 Given $K > \lambda + \delta$, the LMI conditions of Theorem 4.1 are always feasible for small enough γ, γ_0, Δ and large enough λ_0 such that $\lambda_0 \gamma$ is small. By Schur complement, $\Phi \prec 0$ is equivalent to (4.31) with $x = L$. The latter holds for $\mu = \lambda_1 = \gamma = \Delta = 0$, and large enough λ_0. Thus, LMIs hold for small enough μ, λ_1, γ, Δ, γ_0, R with appropriate (large enough) λ_0.

Remark 4.6 Let us explain here what prevents us from obtaining such results under point measurements. For the case of averaged measurements, in the proof of Theorem 4.1, we need the Lyapunov functional to be continuous in L^2-norm. For the case of point measurements, for a matter of continuity in the space variable, we need to guarantee that the Lyapunov functional is continuous in H^1-norm. But this requires that the solution is in $C([0, T]; H^1(0, L))$, therefore requiring more regular initial and boundary data than it is the case here.

4.4 Extension to the Controller Distributed on Subdomains

In this subsection, we are concerned with the case that the actuation does not cover the whole domain Ω and the averaged measurements are measured over the parts of the subdomains. As in [12], let

$$0 \le \tilde{x}_1 < \tilde{x}_2 \le \tilde{x}_3 < \tilde{x}_4 \le \cdots \le \tilde{x}_{2N-1} < \tilde{x}_{2N} \le L,$$

$$[\tilde{x}_{2j-1}, \tilde{x}_{2j}] \subset [x_{j-1}, x_j], \; j = 1, 2, \cdots, N.$$

Denote $\widetilde{\Omega}_j \triangleq [\tilde{x}_{2j-1}, \tilde{x}_{2j}]$. Now we study the system (4.1) under the event-triggered controller

$$f(x,t) = -K \sum_{j=1}^{N} \tilde{z}_j(t_k) \mathbb{1}_{\widetilde{\Omega}_j}(x), \quad K > \lambda, \tag{4.35}$$

where

$$\tilde{z}_j(t_k) = \frac{1}{|\widetilde{\Omega}_j|} \int_{\widetilde{\Omega}_j} z(x, t_k) dx, \; |\widetilde{\Omega}_j| = \tilde{x}_{2j} - \tilde{x}_{2j-1}. \tag{4.36}$$

By applying the first mean value theorem, since $z \in C([0,T], L^2(0,L))$ we obtain that there exists a point $\bar{x}_t^j \in \widetilde{\Omega}_j$ such that

$$\frac{1}{|\widetilde{\Omega}_j|} \int_{\widetilde{\Omega}_j} z(x,t) dx = z(\bar{x}_t^j, t). \tag{4.37}$$

Then the controller (4.35) can be rewritten as

$$f(x,t) = -K \sum_{j=1}^{N} [z(\bar{x}_t^j, t) - \tilde{\rho}_j(t)] \mathbb{1}_{\widetilde{\Omega}_j}(x), \tag{4.38}$$

where $\tilde{\rho}_j(t) = \dfrac{1}{|\widetilde{\Omega}_j|} \int_{\widetilde{\Omega}_j} e_k(x,t) dx.$
This leads to the closed-loop system

$$\begin{cases} \partial_t z + z \partial_x z + \partial_x z + \nu \partial_{xxx} z - \lambda z = -K \sum_{j=1}^{N} [z(\bar{x}_t^j, t) - \tilde{\rho}_j(t)] \mathbb{1}_{\widetilde{\Omega}_j}(x), & \\ \qquad\qquad\qquad\qquad\qquad\qquad\qquad \text{in} (0, L) \times [t_k, t_{k+1}), \; k \in \mathbb{N}, & \\ z(0,t) = z(L,t) = 0, \; \partial_x z(L,t) = 0, & \forall t \ge 0, \\ z(x,0) = z^0(x), & \forall x \in (0,L). \end{cases} \tag{4.39}$$

Denote

$$l_j \triangleq \max\{\tilde{x}_{2j} - x_{j-1}, x_j - \tilde{x}_{2j-1}\}. \tag{4.40}$$

Then we have the following result (Fig. 4.1):

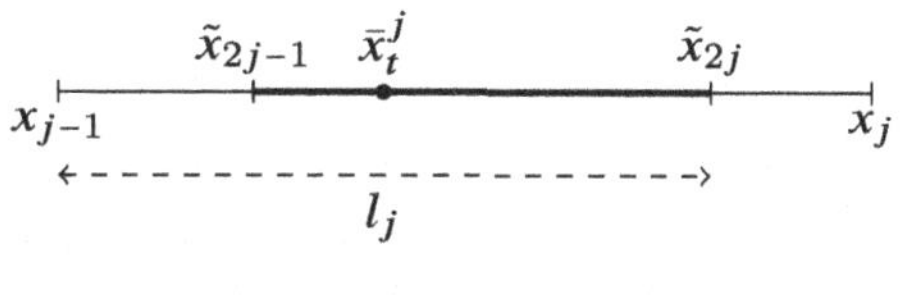

Fig. 4.1 Subdomain $\widetilde{\Omega}_j = [\tilde{x}_{2j-1}, \tilde{x}_{2j}]$, point $\bar{x}_t^j$ and l_j

Proposition 4.2 *Consider the closed-loop system (4.39). Let $L > 0, T > 0$. Denote*

$$l \triangleq \max_j l_j, \quad \bar{\Delta} \triangleq \min_j \frac{|\widetilde{\Omega}_j|}{|\Omega_j|}. \tag{4.41}$$

Given a desired decay rate $\delta > 0$, a control gain $K > \lambda + \delta$, length bounds $l > 0$, $\bar{\Delta} > 0$, and positive tuning parameters λ_0, R, $\theta > \delta$, $\gamma_0 > 0$, assume that there exist positive scalars μ, λ_1, β_i $(i = 1, 2)$, γ, and Γ that solve the following optimization problem:

$$\min \Gamma \quad \text{subject to}$$

$$-3\mu\nu + \lambda_1 + \beta_2 + \frac{2}{3}\mu RL\sqrt{L} < 0, \tag{4.42}$$

$$\Phi = \begin{bmatrix} \phi_{11} & \beta_2 \frac{\pi^2}{4l^2} & 0 & K(1+\mu L) \\ * & -2K\bar{\Delta} - \beta_2 \frac{\pi^2}{4l^2} & -K\mu L & 0 \\ * & * & -\beta_1 & 0 \\ * & * & * & -\lambda_0 \end{bmatrix} \prec 0, \tag{4.43}$$

$$(1+\mu L)\left(1 + \frac{\lambda_0 \gamma_0}{2(\theta - \delta)}\right) < R^2 \Gamma, \tag{4.44}$$

where

$$\phi_{11} = \mu + \lambda_0 \gamma + (2\lambda + 2\delta)(1 + \mu L) + \beta_1 - \beta_2 \frac{\pi^2}{4l^2} - \lambda_1 \frac{\pi^2}{L^2}. \tag{4.45}$$

Then for any initial function $z^0 \in L^2(0, L)$ satisfying $\|z^0\|_{L^2(0,L)} < \frac{1}{\sqrt{\Gamma}}$, the closed-loop system (4.39) under the event-triggering mechanism (4.8) is exponentially stable in the sense that (4.13) holds. Moreover, if the above LMIs hold with $\delta = 0$, then the closed-loop system is exponentially stable with a small enough decay rate.

Proof Consider $V(t)$ given by (4.20). Differentiating $V(t)$ along (4.39), for any $\lambda_0 > 0$ one gets

$$
\begin{aligned}
\dot{V}(t) &\le \dot{V}(t) + \lambda_0\Big[\gamma E(t) + \gamma_0 E(0)e^{-2\theta t} - \|e_k(\cdot,t)\|^2_{L^2(0,L)}\Big] \\
&\le -\nu|\partial_x z(0,t)|^2 - 3\mu\nu\int_0^L |\partial_x z(x,t)|^2 dx + (\mu+\lambda_0\gamma)\int_0^L |z(x,t)|^2 dx \\
&+\frac{2}{3}\mu\textstyle\int_0^L z^3(x,t)dx + 2\lambda\displaystyle\int_0^L (1+\mu L)|z(x,t)|^2 dx - 2K\sum_{j=1}^N z^2(\bar{x}_t^j,t)|\widetilde{\Omega}_j| \\
&-2K\mu\sum_{j=1}^N\int_{\Omega_j} x\mathbb{1}_{\widetilde{\Omega}_j}(x)z(x,t)z(\bar{x}_t^j,t)dx + 2K\sum_{j=1}^N\int_{\Omega_j}\mathbb{1}_{\widetilde{\Omega}_j}(x)(1+\mu x)z(x,t)\tilde{\rho}_j(t)dx \\
&+\lambda_0\gamma_0 E(0)e^{-2\theta t} - \lambda_0\|e_k(\cdot,t)\|^2_{L^2(0,L)}.
\end{aligned}
\tag{4.46}
$$

Cauchy–Schwarz's inequality yields

$$
\begin{aligned}
&\int_{\Omega_j}[\mathbb{1}_{\widetilde{\Omega}_j}(x)\tilde{\rho}_j(t)]^2 dx = |\widetilde{\Omega}_j|\tilde{\rho}_j^2(t) = \frac{1}{|\widetilde{\Omega}_j|}\left[\int_{\widetilde{\Omega}_j} e_k(x,t)dx\right]^2 \\
&\le \int_{\widetilde{\Omega}_j}|e_k(x,t)|^2 dx \le \int_{\Omega_j}|e_k(x,t)|^2 dx,
\end{aligned}
$$

so that

$$
\lambda_0\sum_{j=1}^N\int_{\Omega_j}\Big[|e_k(x,t)|^2 - [\mathbb{1}_{\widetilde{\Omega}_j}(x)\tilde{\rho}_j(t)]^2\Big]dx \ge 0.
$$

From $\widetilde{\Omega}_j \subset \Omega_j$, one has $\beta_1 > 0$ such that

$$
\beta_1\sum_{j=1}^N\int_{\Omega_j}\Big[|z(x,t)|^2 - [\mathbb{1}_{\widetilde{\Omega}_j}(x)z(x,t)]^2\Big]dx \ge 0.
$$

Wirtinger's inequality leads to (4.26) and

$$
\begin{aligned}
&\int_{\Omega_j}[z(x,t)-z(\bar{x}_t^j,t)]^2 dx = \int_{x_{j-1}}^{\bar{x}_t^j}[z(x,t)-z(\bar{x}_t^j,t)]^2 dx + \int_{\bar{x}_t^j}^{x_j}[z(x,t)-z(\bar{x}_t^j,t)]^2 dx \\
&\le \frac{4(\bar{x}_t^j - x_{j-1})^2}{\pi^2}\int_{x_{j-1}}^{\bar{x}_t^j}|\partial_x z(x,t)|^2 dx + \frac{4(x_j-\bar{x}_t^j)^2}{\pi^2}\int_{\bar{x}_t^j}^{x_j}|\partial_x z(x,t)|^2 dx.
\end{aligned}
\tag{4.47}
$$

From (4.41) and (4.47), it follows that

$$
\int_{\Omega_j}[z(x,t)-z(\bar{x}_t^j,t)]^2 dx \le \frac{4l^2}{\pi^2}\int_{\Omega_j}|\partial_x z(x,t)|^2 dx,
$$

which implies

$$
\beta_2\sum_{j=1}^N\int_{\Omega_j}\left[|\partial_x z(x,t)|^2 - \frac{\pi^2}{4l^2}[z(x,t)-z(\bar{x}_t^j,t)]^2\right]dx \ge 0
$$

for some constant $\beta_2 > 0$.

Set $\tilde{\eta}(x,t) = \{z(x,t), z(\bar{x}_t^j, t), \mathbb{1}_{\tilde{\Omega}_j}(x)z(x,t), \mathbb{1}_{\tilde{\Omega}_j}(x)\rho_j(t)\}$. Using (4.25), (4.26), (4.41), (4.46) and applying S-procedure, we have

$$
\begin{aligned}
&\dot{V}(t) + 2\delta V(t) \leq \dot{V}(t) + 2\delta V(t) + \lambda_0 \left[\gamma E(t) + \gamma_0 E(0)e^{-2\theta t} - \|e_k(\cdot,t)\|^2_{L^2(0,L)}\right] \\
&+\lambda_1 \left[\|\partial_x z(\cdot,t)\|^2_{L^2(0,L)} - \frac{\pi^2}{L^2}\|z(\cdot,t)\|^2_{L^2(0,L)}\right] \\
&+\lambda_0 \sum_{j=1}^{N} \int_{\Omega_j} \left[|e_k(x,t)|^2 - [\mathbb{1}_{\tilde{\Omega}_j}(x)\rho_j(t)]^2\right] dx \\
&+\beta_1 \sum_{j=1}^{N} \int_{\Omega_j} \left[|z(x,t)|^2 - [\mathbb{1}_{\tilde{\Omega}_j}(x)z(x,t)]^2\right] dx \\
&+\beta_2 \sum_{j=1}^{N} \int_{\Omega_j} \left[|\partial_x z(x,t)|^2 - \frac{\pi^2}{4l^2}[z(x,t) - z(\bar{x}_t^j,t)]^2\right] dx \\
&\leq \sum_{j=1}^{N} \int_{\Omega_j} \tilde{\eta}(x,t)^\top \tilde{\Phi}(x)\tilde{\eta}(x,t) + \lambda_0\gamma_0 E(0)e^{-2\theta t} \\
&- \left(3\mu\nu - \lambda_1 - \beta_2 - \frac{2\mu}{3}L\sqrt{L}\|z(\cdot,t)\|_{L^2(0,L)}\right)\|\partial_x z(\cdot,t)\|^2_{L^2(0,L)},
\end{aligned}
$$

where

$$
\tilde{\Phi}(x) = \begin{bmatrix} \phi_{11} & \beta_2\frac{\pi^2}{4l^2} & 0 & K(1+\mu x) \\ * & -2K\bar{\Delta} - \beta_2\frac{\pi^2}{4l^2} & -K\mu x & 0 \\ * & * & -\beta_1 & 0 \\ * & * & * & -\lambda_0 \end{bmatrix},
$$

with ϕ_{11} as in (4.45). Thus, by Schur complement, the LMIs (4.42), (4.43) yield (4.13). □

Remark 4.7 Given $\bar{\Delta} < 1$ and $K > \frac{\lambda+\delta}{\bar{\Delta}}$, the LMI conditions of Proposition 4.2 are always feasible for small enough γ, γ_0, $R < \frac{9\nu}{2L\sqrt{L}}$, and large enough λ_0. By Schur complement,

$$
\tilde{\Phi} < 0 \Longleftrightarrow \phi_{11} + \lambda_0^{-1}K^2(1+\mu L)^2 - (\beta_2\frac{\pi^2}{4l^2})^2[-2K\bar{\Delta} - \beta_2\frac{\pi^2}{4l^2} + \beta_1^{-1}K^2\mu^2L^2]^{-1} < 0.
$$

Choose $\beta_2 = 2K\bar{\Delta}\frac{4l^2}{\pi^2}$. The latter holds for $\mu = \lambda_1 = \gamma = 0$, small enough β_1 and large enough λ_0. Thus, LMIs hold for small enough μ, λ_1, β_1, γ, γ_0, $R < \frac{9\nu}{2L\sqrt{L}}$ with appropriate (large enough) λ_0, (small enough) l such that $\lambda_0\gamma$ and $2K\bar{\Delta}\frac{4l^2}{\pi^2}$ are small.

4.5 Numerical Examples

Consider the KdV system:

$$\begin{cases} \partial_t z + z\partial_x z + \partial_x z + \nu\partial_{xxx} z - \lambda z = f(x,t), \ \forall 0 < x < L, \ t \geq 0, \\ z(0,t) = z(L,t) = \partial_x z(L,t) = 0, \\ z(x,0) = z^0(x) = 0.32\left(1 - \cos(\frac{2\pi x}{L})\right), \ x \in [0,L], \end{cases}$$

where $\nu > 0$ will be chosen below.

We will give simulation for the following cases:

- Open-loop system without input (i.e. $f(x,t) = 0$)
- Closed-loop system under continuous-time controller $f(x,t) = -Kz(x,t)$
- Closed-loop system under event-triggered controller distributed on the whole domain with averaging $f(x,t) = -K\sum_{j=1}^{N} \bar{z}_j(x,t_k)\mathbb{1}_{\Omega_j}(x)$
- Closed-loop system under event-triggered controller distributed on subdomains with averaging $f(x,t) = -K\sum_{j=1}^{N} \tilde{z}_j(x,t_k)\mathbb{1}_{\tilde{\Omega}_j}(x)$

where $K > \lambda$ is a controller gain.

Example 1: For the event-triggered control law (4.3) under averaged measurements, we verify LMI conditions of Theorem 4.1 with $K = L = 1$, $\lambda = 0.5$, $\nu = 0.3$, $\delta = 0.4$, $\Delta = 0.1$, $R = 0.5$. We find that the closed-loop system under event-triggering mechanism (4.8) with $\theta = 2$, $\gamma = 0.00029$ and $\gamma_0 = 0.02$ is exponentially stable for $\mu = 0.5401$ and for any initial values satisfying $\|z^0\|_{L^2(0,1)} < \dfrac{1}{\sqrt{6.1997}} \approx 0.4$.

A finite difference method is used to illustrate the effect of the proposed event-triggered control law. The steps of space and time are chosen as 0.05 and 0.0001, respectively. Figure 4.2 illustrates the evolution of the state of the open-loop KdV system. It is seen that the open-loop system is unstable. Figure 4.3 illustrates the evolution of the state of the closed-loop KdV system under the event-triggering mechanism

$$t_{k+1} = \inf\left\{t \geq t_k \,\middle|\, \|e_k\|^2_{L^2(0,L)} \geq 0.00029E(t) + 0.02E(0)e^{-4t}\right\}.$$

with the control law (4.3) where $\bar{z}_j(t_k) = 10\int_{\Omega_j} z(\zeta, t_k)d\zeta$, $t \in [t_k, t_{k+1})$ subject to $x_j - x_{j-1} = |\Omega_j| = \Delta = 0.1$. It shows that the state of closed-loop KdV system under the event-triggered controller converges exponentially to zero. Figure 4.4 shows that the release time and release interval by event-triggering for $t \in [0, 0.2]$. Figure 4.5 demonstrates the time evolution of $\ln(E(t))$ for the open-loop system, the closed-loop system under continuous-time controller, and the closed-loop system under the event-triggered controller. The simulations show that the event-triggered controller improves the performance.

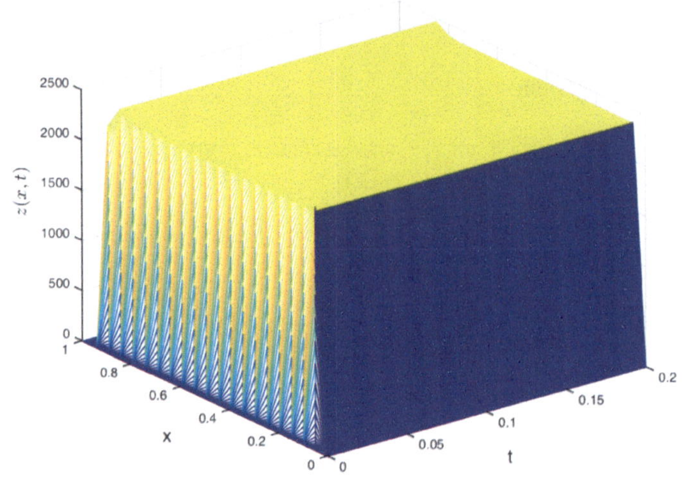

Fig. 4.2 State of the open-loop system with $\lambda = 0.5$

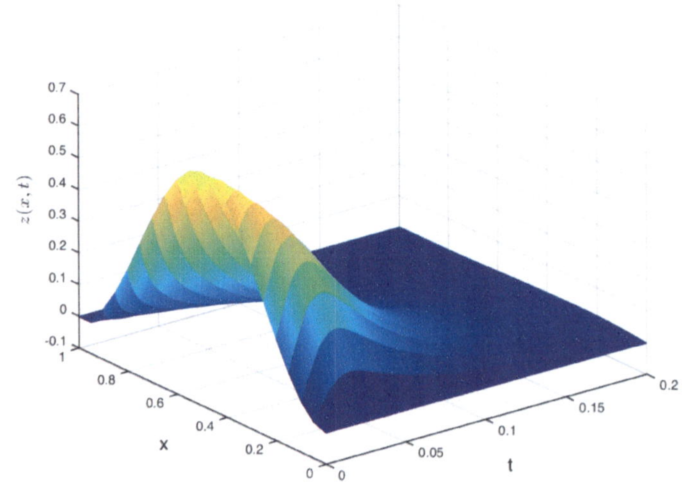

Fig. 4.3 State of the closed-loop system with the event-triggered control law (4.3) distributed over all domain

Example 2: For the event-triggered control law (4.35) under averaged and localized measurements, we verify LMI conditions of Proposition 4.2 with $K = L = 1$, $\lambda = 0.5$, $\lambda_0 = 1$, $\nu = 0.3$, $\delta = 0.4$, $R = 0.5$, $l = 0.2$, $\bar{\Delta} = 1/3$. We find that the closed-loop system under event-triggering mechanism (4.8) with $\theta = 2$, $\gamma = 0.0013$ and $\gamma_0 = 0.02$ is exponentially stable for $\mu = 0.0235$ and for any initial values satisfying $\|z^0\|_{L^2(0,1)} < \dfrac{1}{\sqrt{4.5603}} \approx 0.46$. We proceed further with numerical simulations of the closed-loop KdV system under the event-triggering mechanism

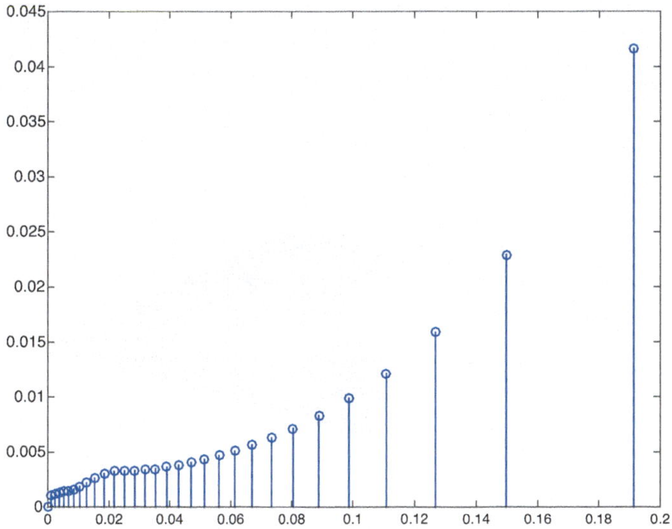

Fig. 4.4 Release instants and release interval by event-triggering

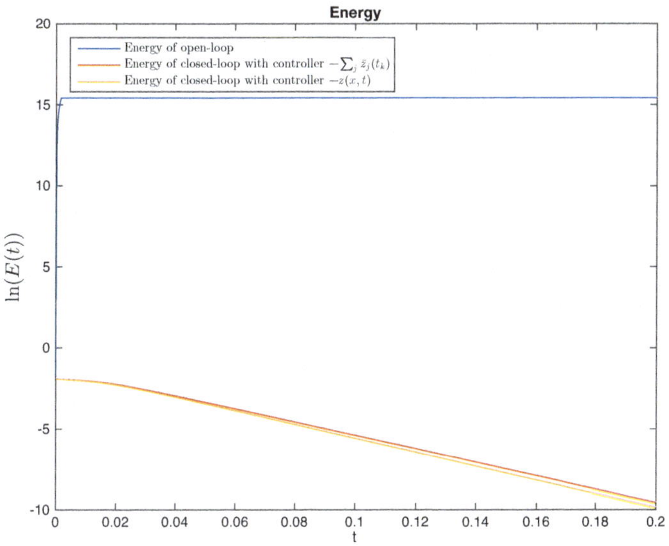

Fig. 4.5 $\ln(E(t))$ of the open-loop system, closed-loop system under continuous-time/event-triggered controller

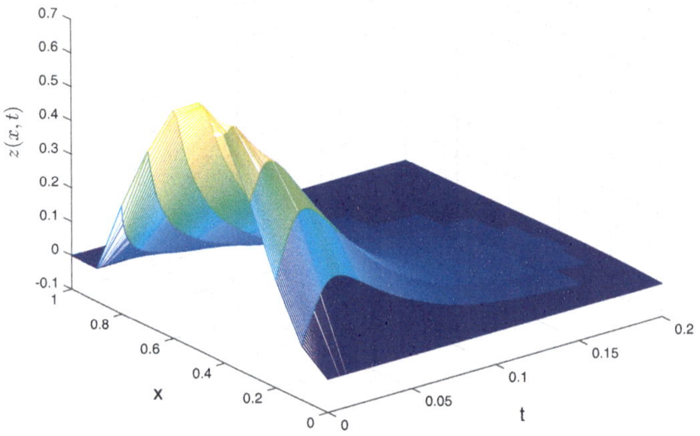

Fig. 4.6 State of the closed-loop system with the event-triggered control law (4.35) distributed over subdomains

$$t_{k+1} = \inf\left\{ t \geq t_k \,\middle|\, \|e_k\|^2_{L^2(0,L)} \geq 0.0013E(t) + 0.02E(0)e^{-4t} \right\}.$$

Let $x_0 = 0$, $x_1 = 0.3$, $x_2 = 0.6$, $x_3 = 0.9$ and $x_4 = 1$. Set $\tilde{x}_1 = 0.1$, $\tilde{x}_2 = 0.2$, $\tilde{x}_3 = 0.4$, $\tilde{x}_4 = 0.5$, $\tilde{x}_5 = 0.7$, $\tilde{x}_6 = 0.8$, $\tilde{x}_7 = 0.9$, $\tilde{x}_8 = 1$. The simulations show that the state of closed-loop KdV system converges to zero (see Fig. 4.6).

4.6 Notes

This chapter discusses event-triggered control of nonlinear PDE systems. The presented method gives efficient tools for various event-triggered control design and observer design problems for nonlinear PDEs. Recent advances for event-triggered control were extended to complex PDE systems, such as cascaded PDE-ODE [13] and PDE-PDE[14] systems.

References

1. Tang, S.X., Krstic, M.: Stabilization of linearized Korteweg-de Vries Systems with anti-diffusion. In: American Control Conference, USA, pp. 3302–3307 (2013)
2. Cerpa, E.: Control of a Korteweg-de Vries equation: a tutorial. Math. Control Relat. Fields **4**, 45–99 (2014)
3. Chapouly, M.: Global controllability of a nonlinear Korteweg-de Vries equation. Commun. Contemp. Math. **11**, 495–521 (2009)
4. Pazoto, A.F.: Unique continuation and decay for the Korteweg-de Vries equation with localized damping, ESAIM Control Optim. Calc. Var. **11**, 473–486 (2005)
5. Perla Menzala, G., Vasconcellos, C.F., Zuazua, E.: Stabilization of the Korteweg-de Vries equation with localized damping. Quart. Appl. Math. **60**, 111–129 (2002)
6. Azouani, A., Titi, E.S.: Feedback control of nonlinear dissipative systems by finite determining parameters – A reaction-diffusion Paradigm. Evol. Equ. Control Theory **3**, 579–594 (2014)
7. Fridman, E., Blighovsky, A.: Robust sampled-data control of a class of semilinear parabolic systems. Automatica **48**, 826–836 (2012)
8. Fridman, E., Bar Am, N.: Sampled-data distributed H_∞ control of transport reaction systems. SIAM J. Control Opt. **51**, 1500–1527 (2013)
9. Lunasin, E., Titi, E.S.: Finite determining parameters feedback control for distributed nonlinear dissipative systems - a computational study. Evol. Equ. Control Theory **6**, 535–557 (2017)
10. Baudouin, L., Crépeau, E., Valein, J.: Two approaches for the stabilization of nonlinear KdV equation with boundary time-delay feedback. IEEE Trans. Autom. Control **64**, 1403–1414 (2019)
11. Fridman, E.: Introduction to Time-delay Systems: Analysis and Control. Birkhäuser, Basel (2014)
12. Wang, J.W., Wu, H.N.: Lyapunov-based design of locally collocated controllers for semi-linear parabolic PDE systems. J. Frankl. Inst. **351**, 429–441 (2014)
13. Wang, J., Krstic, M.: Event-triggered adaptive control of a parabolic PDE-ODE cascade with piecewise-constant inputs and identification. IEEE Trans. Autom. Control **68**(9), 5493–5508 (2022)
14. Wang, J., Krstic, M.: Event-triggered adaptive control of coupled hyperbolic PDEs with piecewise-constant inputs and identification. IEEE Trans. Autom. Control **68**(3), 1568–1583 (2022)

Chapter 5
Disturbance Rejection of PDE Systems

After exploring event-triggered control strategies for the KdV equation in Chap. 4, we shift our focus in Chapter 5 to the stabilization of the KdVB equation, particularly in the presence of external disturbances. Two approaches, active disturbance rejection control (ADRC) and disturbance observer-based control (DOBC), are introduced. The challenge lies in designing an effective extended state observer (ESO) for ADRC and a disturbance observer (DO) for DOBC. Both continuous-time and event-triggered anti-disturbance controllers are presented, incorporating event-triggering mechanisms to reduce control updates. The Lyapunov functional method is used to establish sufficient stability conditions for the closed-loop system.

5.1 Problem Statement

Consider the one-dimensional KdVB equation as follows:

$$\begin{cases} z_t(x,t) = \epsilon z_{xx}(x,t) - \nu z_{xxx}(x,t) - z(x,t)z_x(x,t) + \sum\limits_{j=1}^{N_s} b_j(x)\left[U_j(t) + d(t)\right], \\ \hfill 0 < x < 1,\ t > 0, \\ z(0,t) = z(1,t),\ z_x(0,t) = z_x(1,t), \\ z_{xx}(0,t) = z_{xx}(1,t),\ z(x,0) = z_0(x), \end{cases} \tag{5.1}$$

where $\epsilon > 0$, $\nu > 0$, $z(x,t)$ is the state of KdVB system, $d(t)$ represents an external unknown disturbance, and $z_0(x)$ is the initial condition.

Let the points $0 = x_0 < x_1 < \cdots < x_{N_s} = 1$ divide $[0, 1]$ into N_s subdomains. Denote each subdomain by $\Omega_j \triangleq [x_{j-1}, x_j)$. The width of each subdomain is upper bounded: $0 < x_j - x_{j-1} = \Delta_j \le \Delta,\quad j = 1, 2, \ldots, N_s,$ where Δ represents the corresponding upper bound. $U_j(t)$ are the control inputs that enter (5.1) through the

W. Kang and E. Fridman, *Delay-Robust Control of Distributed Parameter Systems*, Advances in Delays and Dynamics 13, https://doi.org/10.1007/978-981-96-6344-6_5

shape functions $b_j(x)$ such that $b_j(x) = 1$ if $x \in \Omega_j$ or $b_j(x) = 0$ if $x \notin \Omega_j$. Suppose that the sensors provide averaged measurements as follows:

$$y_j(t) = \frac{1}{\Delta_j}\int_{x_{j-1}}^{x_j} z(\zeta, t)d\zeta,\ j = 1, \ldots, N_s. \tag{5.2}$$

Our objective is to investigate the continuous-time distributed control law and the event-triggered distributed control law of KdVB model (5.1) in the presence of an external unknown disturbance occurring in the control input channel. Here ADRC and DOBC strategies are employed to reject the disturbance actively.

5.2 Active Disturbance Rejection Control (ADRC): Design and Main Results

In this section, the disturbance is attenuated via ESO with constant/ time-varying high gains. Moreover, continuous-time and event-triggered ADRCs will be proposed for the stabilization of (5.1).

5.2.1 Time-Varying High Gain Estimator

Based on (5.2), we define the following variable:

$$y(t) = \sum_{j=1}^{N_s} \Delta_j y_j(t). \tag{5.3}$$

Integrating by parts along the solution of (5.1), we obtain

$$\begin{aligned} \dot{y}(t) &= \sum_{j=1}^{N_s} \int_{x_{j-1}}^{x_j} [\epsilon z_{\zeta\zeta}(\zeta, t) - \nu z_{\zeta\zeta\zeta}(\zeta, t) - z(\zeta, t)z_{\zeta}(\zeta, t) + U_j(t) + d(t)]d\zeta \\ &= \sum_{j=1}^{N_s} \Delta_j U_j(t) + d(t). \end{aligned} \tag{5.4}$$

Then the following ESO with a time-varying high gain is proposed:

$$\begin{cases} \dot{\hat{y}}_r(t) = \sum_{j=1}^{N_s} \Delta_j U_j(t) + \dot{\hat{d}}_1(t) - r(t)\left[\hat{y}_r(t) - y(t)\right], \\ \dot{\hat{d}}_1(t) = -r^2(t)\left[\hat{y}_r(t) - y(t)\right], \end{cases} \tag{5.5}$$

where $\hat{d}_1(t)$ is the approximation of $d(t)$, $\hat{y}_r(t)$ is the approximation of $y(t)$ and $r(t)$ is a time-vary high gain satisfying

$$\begin{cases} r(t) \in C^1[0, +\infty), r(t) > 0, \dot{r}(t) > 0, \lim\limits_{t\to\infty} r(t) = +\infty, \\ \dfrac{\dot{r}(t)}{r(t)} \le M \text{ for } t \ge 0 \text{ and some } M > 0. \end{cases} \tag{5.6}$$

Assumption 5.2.1 Suppose that the disturbance $d(t) \in \mathcal{H}^1_{loc}(0, \infty)$ satisfies

$$\lim_{t\to\infty} \frac{|\dot{d}(t)|}{r(t)} = 0.$$

Remark 5.1 In the recent papers [1, 2], it is usually supposed that $|\dot{d}(t)|$ is bounded, which is a precondition. If $|\dot{d}(t)|$ is bounded, then from $\lim\limits_{t\to\infty} r(t) = +\infty$, it follows that $\lim\limits_{t\to\infty} \dfrac{|\dot{d}(t)|}{r(t)} = 0$. To release the restriction, here we just assume that $\lim\limits_{t\to\infty} \dfrac{|\dot{d}(t)|}{r(t)} = 0$, which means that the gain $r(t)$ needs to have a higher order of growth than $|\dot{d}(t)|$. Such an appropriate gain $r(t)$ can be obtained by a trial and error approach. Then the proposed ESO (5.5) is effective.

Set $\tilde{y}_r(t) = r(t)[\hat{y}_r(t) - y(t)]$ and $\tilde{d}_1(t) = \hat{d}_1(t) - d(t)$. Then the estimation errors satisfy

$$\begin{cases} \dot{\tilde{y}}_r(t) = -r(t)\left[\tilde{y}_r(t) - \tilde{d}_1(t)\right] + \dfrac{\dot{r}(t)}{r(t)}\tilde{y}_r(t), \\ \dot{\tilde{d}}_1(t) = -r(t)\tilde{y}_r(t) - \dot{d}(t). \end{cases} \tag{5.7}$$

Define a Lyapunov–Krasovskii functional

$$V_0(\tilde{y}_r(t), \tilde{d}_1(t)) = \tilde{y}_r^2(t) + 1.5\tilde{d}_1^2(t) - \tilde{y}_r(t)\tilde{d}_1(t). \tag{5.8}$$

The following inequality

$$0.5\left[\tilde{y}_r^2(t) + \tilde{d}_1^2(t)\right] \le V_0(\tilde{y}_r(t), \tilde{d}_1(t)) \le 2\left[\tilde{y}_r^2(t) + \tilde{d}_1^2(t)\right] \tag{5.9}$$

means that $V_0(\tilde{y}_r(t), \tilde{d}_1(t)) \ge 0$. Then one has

$$\begin{aligned} &\dot{V}_0(\tilde{y}_r(t), \tilde{d}_1(t)) = -r(t)\left[\tilde{y}_r^2(t) + \tilde{d}_1^2(t)\right] \\ &+\frac{2\dot{r}(t)}{r(t)}\tilde{y}_r^2(t) - \frac{\dot{r}(t)}{r(t)}\tilde{y}_r(t)\tilde{d}_1(t) - \dot{d}(t)\left[3\tilde{d}_1(t) - \tilde{y}_r(t)\right] \\ &\le \left[-r(t) + \frac{5\dot{r}(t)}{2r(t)}\right] \times \left[\tilde{y}_r^2(t) + \tilde{d}_1^2(t)\right] + 3|\dot{d}(t)|\left(|\tilde{d}_1(t)| + |\tilde{y}_r(t)|\right). \end{aligned} \tag{5.10}$$

According to (5.6) and (5.9), we have

$$\begin{aligned}
&\dot{V}_0(\tilde{y}_r(t), \tilde{d}_1(t)) \\
&\leq [-r(t) + 2.5M] \times \left[\tilde{y}_r^2(t) + \tilde{d}_1^2(t)\right] + 3\sqrt{2}|\dot{d}(t)|\sqrt{\tilde{y}_r^2(t) + \tilde{d}_1^2(t)} \\
&\leq [-2r(t) + 5M]\, V_0(\tilde{y}_r(t), \tilde{d}_1(t)) + 6|\dot{d}(t)|\sqrt{V_0(\tilde{y}_r(t), \tilde{d}_1(t))}
\end{aligned}$$

for $t \geq 0$, which means that

$$\frac{d\sqrt{V_0(\tilde{y}_r(t), \tilde{d}_1(t))}}{dt} \leq [-r(t) + 2.5M]\sqrt{V_0(\tilde{y}_r(t), \tilde{d}_1(t))} + 3|\dot{d}(t)|.$$

Therefore, for $t \geq 0$, one has

$$\begin{aligned}
&\frac{d\sqrt{V_0(\tilde{y}_r(t), \tilde{d}_1(t))}}{dt} e^{\int_0^t [r(\rho)-2.5M]d\rho} \\
&\leq \left\{ [-r(t) + 2.5M]\sqrt{V_0(\tilde{y}_r(t), \tilde{d}_1(t))} + 3|\dot{d}(t)| \right\} e^{\int_0^t [r(\rho)-2.5M]d\rho}.
\end{aligned}$$

Hence,

$$\frac{d}{dt}\left(\sqrt{V_0(\tilde{y}_r(t), \tilde{d}_1(t))} e^{\int_0^t [r(\rho)-2.5M]d\rho}\right) \leq 3|\dot{d}(t)| e^{\int_0^t [r(\rho)-2.5M]d\rho},$$

which implies

$$\begin{aligned}
&\sqrt{V_0(\tilde{y}_r(t), \tilde{d}_1(t))} e^{\int_0^t [r(\rho)-2.5M]d\rho} \\
&\leq \sqrt{V_0(\tilde{y}_r(0), \tilde{d}_1(0))} + 3\int_0^t |\dot{d}(s)| e^{\int_0^s [r(\rho)-2.5M]d\rho} ds.
\end{aligned}$$

Thus,

$$\begin{aligned}
\sqrt{V_0(\tilde{y}_r(t), \tilde{d}_1(t))} &\leq \sqrt{V_0(\tilde{y}_r(0), \tilde{d}_1(0))} e^{-\int_0^t [r(s)-2.5M]ds} \\
&+ \frac{3\int_0^t |\dot{d}(s)| e^{\int_0^s [r(\rho)-2.5M]d\rho} ds}{e^{\int_0^t [r(s)-2.5M]ds}}.
\end{aligned} \tag{5.11}$$

The first term on the right side of the equation converges to 0 as $t \to \infty$. Then applying L'Hospital rule to the second term on the right side leads to

$$\begin{aligned}
&\lim_{t\to\infty} \frac{3\int_0^t |\dot{d}(s)| e^{\int_0^s [r(\rho)-2.5M]d\rho} ds}{e^{\int_0^t [r(s)-2.5M]ds}} \\
&= \lim_{t\to\infty} \frac{3|\dot{d}(t)| e^{\int_0^t [r(s)-2.5M]ds}}{e^{\int_0^t [r(s)-2.5M]ds}\,[r(t) - 2.5M]} = \lim_{t\to\infty} \frac{3|\dot{d}(t)|}{r(t) - 2.5M} = 0,
\end{aligned} \tag{5.12}$$

if the conditions in (5.6) are satisfied. Therefore, one has

$$\lim_{t\to\infty} \sqrt{V_0(\tilde{y}_r(t), \tilde{d}_1(t))} = 0,$$

which implies that

$$(\tilde{y}_r(t), \tilde{d}_1(t)) \to 0 \text{ as } t \to \infty. \tag{5.13}$$

Remark 5.2 As for the KdVB model (5.1), if the periodic boundary conditions of (5.1) are changed into the following boundary conditions:

$$z(0,t) = z(1,t),\, z_{xx}(0,t) = \frac{\epsilon}{\nu} z_x(0,t),\, z_{xx}(1,t) = \frac{\epsilon}{\nu} z_x(1,t),$$

then (5.4) also holds. By the same procedure, (5.13) can also be also obtained.

5.2.1.1 Continuous-Time Control

For the stabilization of KdVB system (5.1), the following continuous-time ADRC is presented:

$$U_j(t) = -\frac{K}{\Delta_j} \int_{x_{j-1}}^{x_j} z(\zeta, t) d\zeta - \hat{d}_1(t),\, j = 1, \ldots, N_s. \tag{5.14}$$

Define

$$f_j(x,t) = \frac{\int_{x_{j-1}}^{x_j} [z(x,t) - z(\zeta,t)]\, d\zeta}{\Delta_j}. \tag{5.15}$$

Then the closed-loop system becomes

$$\begin{cases} z_t(x,t) = \epsilon z_{xx}(x,t) - \nu z_{xxx}(x,t) - z(x,t) z_x(x,t) \\ -\sum\limits_{j=1}^{N_s} b_j(x) \left[K z(x,t) - K f_j(x,t) + \tilde{d}_1(t) \right], \\ z(0,t) = z(1,t),\, z_x(0,t) = z_x(1,t),\, z_{xx}(0,t) = z_{xx}(1,t), \\ \dot{\tilde{y}}_r(t) = -r(t) \left[\tilde{y}_r(t) - \tilde{d}_1(t) \right] + \frac{\dot{r}(t)}{r(t)} \tilde{y}_r(t), \\ \dot{\tilde{d}}_1(t) = -r(t) \tilde{y}_r(t) - \dot{d}(t). \end{cases} \tag{5.16}$$

Define the operator $\mathcal{A} : \mathcal{D}(\mathcal{A}) \to \mathcal{L}^2(0,1)$ as follows:

$$\begin{cases} \mathcal{A} f = -\nu \frac{\partial^3 f}{\partial x^3}, \\ D(\mathcal{A}) = \mathcal{H}^3_{per}(0,1) = \{\mathcal{H}^3(0,1) | f(0) = f(1),\, f'(0) = f'(1),\;\, f''(0) = f''(1)\}. \end{cases}$$

Hence, $-i\mathcal{A}$ is a positive operator in $\mathcal{L}^2(0,1)$ with the real eigenvalues and the corresponding eigenfunctions

$$\lambda_n = 8\pi^3 n^3 \nu,\ \phi_n(x) = e^{2\pi n i x},\ n \in \mathbb{Z}^+.$$

Then $\{\phi_n(x)\}_{n=1}^{\infty}$ forms an orthonormal basis for $\mathcal{L}^2(0,1)$. $\{\lambda_n^{-\frac{1}{2}}\phi_n(x)\}_{n=1}^{\infty}$ forms an orthonormal basis for $\mathcal{H}_{per}^{\frac{3}{2}}(0,1) = \{f \in \mathcal{H}^{\frac{3}{2}}(0,1) | f(0) = f(1), f'(0) = f'(1)\}$, which is a closed subspace of $\mathcal{H}^{\frac{3}{2}}(0,1)$. $\phi_n(x)$ is uniformly bounded such that $\max_{0\le x\le 1} |\phi_n(x)| \le R$ for some constant $R > 0$.

Note that $D((-i\mathcal{A})^{\frac{1}{2}}) = \mathcal{H}_{per}^{\frac{3}{2}}(0,1)$ and $D((-i\mathcal{A})^{\frac{1}{2}})$ is a canonical isomorphism from $\mathcal{H}_{per}^{\frac{3}{2}}(0,1)$ onto $\mathcal{L}^2(0,1)$. Consider $\mathcal{L}^2(0,1)$ as the pivot space. Let $[D((-i\mathcal{A})^{\frac{1}{2}})]' = \mathcal{H}_{per}^{*}(0,1)$ be the dual space of $\mathcal{H}_{per}^{\frac{3}{2}}(0,1)$ with respect to the pivot space $\mathcal{L}^2(0,1)$. Then the following Gelfand triple compact inclusions are valid:

$$\begin{aligned}&\mathcal{H}_{per}^{\frac{3}{2}}(0,1) = D((-i\mathcal{A})^{\frac{1}{2}}) \hookrightarrow \mathcal{L}^2(0,1) = [\mathcal{L}^2(0,1)]' \\ &\hookrightarrow [D((-i\mathcal{A})^{\frac{1}{2}})]' = \mathcal{H}_{per}^{*}(0,1).\end{aligned}$$

The operator $\mathcal{A}$ has an extension which is still denoted by

$$\mathcal{A} \in \mathcal{L}(\mathcal{H}_{per}^{\frac{3}{2}}(0,1), \mathcal{H}_{per}^{*}(0,1))$$

defined by

$$\langle \mathcal{A} f, \psi\rangle_{\mathcal{H}_{per}^{*}, \mathcal{H}_{per}^{\frac{3}{2}}} = i\langle(-i\mathcal{A})^{\frac{1}{2}} f, (-i\mathcal{A})^{\frac{1}{2}}\psi\rangle_{\mathcal{L}^2}$$

for all $f, \psi \in \mathcal{H}_{per}^{\frac{3}{2}}(0,1)$. Similarly, $\mathcal{A}$ also admits an extension

$$\mathcal{A} \in \mathcal{L}(\mathcal{L}^2(0,1), [D(\mathcal{A})]'),$$

where $[D(\mathcal{A})]'$ is the dual space of $D(\mathcal{A})$ with the pivot space $\mathcal{L}^2(0,1)$.

Before discussing the well-posedness of (5.16), we introduce the following definition first.

Definition 5.1 The function $z \in \mathcal{L}^2(0,T;\mathcal{H}_{per}^{\frac{3}{2}}(0,1))$ with $z_t \in \mathcal{L}^2(0,T;\mathcal{H}_{per}^{*}(0,1))$ is said to be a weak solution of the PDE subsystem of (5.16) if

$$\begin{aligned}\langle z_t, \psi\rangle &= -\nu\langle z_{xxx}, \psi\rangle + \epsilon\langle z_{xx}, \psi\rangle - \langle z z_x, \psi\rangle \\ &\quad -K\sum_{j=1}^{N_s}\langle \frac{1}{\Delta_j}\int_{\Omega_j} z(x,t)dx, \psi\rangle - \langle \tilde{d}_1(t), \psi\rangle\end{aligned} \tag{5.17}$$

on $[0,T]$ for every $\psi \in \mathcal{H}_{per}^{\frac{3}{2}}(0,1)$.

The well-posedness and stability results of (5.16) are given as follows:

Lemma 5.1 *Assume that the time-varying gain* $r(t) \in C^1[0, +\infty)$ *satisfies (5.6) and the disturbance* $d(t) \in \mathcal{H}^1_{loc}(0, \infty)$ *satisfies* $\lim_{t\to\infty} \frac{|\dot{d}(t)|}{r(t)} = 0$. *Then for any initial condition* $(z_0(x), \tilde{y}_r(0), \tilde{d}_1(0)) \in \left(\mathcal{L}^2(0, 1) \times \mathbb{R} \times \mathbb{R}\right)$, *the closed-loop system (5.16) has a unique weak solution for all* $T > 0$ *such that* $z \in C\left([0, T], \mathcal{L}^2(0, 1)\right)$, $\tilde{y}_r \in \mathcal{H}^1([0, T], \mathbb{R})$ *and* $\tilde{d}_1 \in \mathcal{H}^1([0, T], \mathbb{R})$.

Proof Before discussing the well-posedness of (5.16), we introduce the following lemma first.

Lemma 5.2 *Suppose that* $z \in \mathcal{L}^2(0, T; \mathcal{H}^{\frac{3}{2}}_{per}(0, 1))$ *and* $z_t \in \mathcal{L}^2(0, T; \mathcal{H}^*_{per}(0, 1))$. *Then* $z \in C([0, T]; \mathcal{L}^2(0, 1))$. □

The proof of Lemma 5.2 follows from [3], [4] (Lemma 1.2) and [5] (Theorem 3, pp. 287) by changing the associate spaces.

Now we are in a position to prove Lemma 5.1.

As for the ODE subsystem of (5.16), for any initial condition $(\tilde{y}_r(0), \tilde{d}_1(0)) \in \mathbb{R} \times \mathbb{R}$, there exists a unique solution $(\tilde{y}_r, \tilde{d}_1) \in \mathcal{H}^1([0, T], \mathbb{R} \times \mathbb{R})$ for all $T > 0$.

As for the PDE subsystem of (5.16), the proof is based on the Galerkin approximation method. Inspired by [6], the well-posedness is established by the following three steps.

Step 1. Define $V_N = \text{span}\{\phi_1, \phi_2, \ldots, \phi_N\}$ for any $N \in \mathbb{Z}^+$. A Galerkin approximation solution to the PDE subsystem is constructed as follows:

$$z^N(x, t) = \sum_{n=1}^{N} g_n(t)\phi_n(x),$$

where $g_n(t)$ satisfies

$$\begin{cases} \dot{g}_n(t) = \epsilon\langle g_n(t)\phi_n, \phi_n''\rangle + \nu\langle g_n(t)\phi_n, \phi_n'''\rangle \\ -\langle X^T(\{\phi_p\phi_q'\}_{p,q=1}^N)X, \phi_n\rangle - K\sum_{j=1}^{N_s}\langle\frac{1}{\Delta_j}\int_{\Omega_j} z^N(x, t)dx, \phi_n\rangle \\ -\langle\tilde{d}_1(t), \phi_n\rangle, \ \forall\phi_n \in V_N, \\ g_n(0) = \langle z_0, \phi_n\rangle, \ n = 1, 2, \ldots, N, \end{cases} \tag{5.18}$$

where $X(t) = (g_1(t), \ldots, g_N(t))^T$. Due to $\int_0^1 \overline{\phi_n(x)}dx = 0$, it follows that

$$\dot{X}(t) = AX(t) + F(X(t)), \tag{5.19}$$

where

$$\begin{aligned} A &= \epsilon(\{\langle\phi_p, \phi_q''\rangle\}_{p,q=1}^N)^T + \nu(\{\langle\phi_p, \phi_q'''\rangle\}_{p,q=1}^N)^T \\ &= \text{diag}(a_1, a_2, \ldots, a_N), \ a_n = -4\pi^2n^2\epsilon + i\lambda_n, \\ F(X(t)) &= -(f_1(X(t)), f_2(X(t)), \ldots, f_N(X(t)))^T, \\ f_n(X(t)) &= \langle X^T(\{\phi_p\phi_q'\}_{p,q=1}^N)X, \phi_n\rangle, \ n = 1, 2, \ldots, N. \end{aligned}$$

Note that the nonlinearity $F(X(t))$ is locally Lipschitz in X. The equation (5.19) possesses a unique solution

$$X(t) = e^{At}X(0) + \int_0^t e^{A(t-\tau)}F(X(\tau))d\tau. \tag{5.20}$$

Hence, the existence of $g_n(t),\ n = 1, 2, \ldots, N$ is ensured on some interval $[0, t_N)$. To extend these functions to any $t_N = T < \infty$ and to pass to the limit as $N \to \infty$, we need a-priori estimates.

Step 2. We show that $\sup\limits_N \|z^N(\cdot, t)\|_{\mathcal{L}^2(0,T;\mathcal{H}^{\frac{3}{2}}_{per}(0,1))} < \infty$. From (5.20), one has

$$\begin{aligned} z^N(x,t) &= \sum_{n=1}^N g_n(t)\phi_n(x) \\ &= \sum_{n=1}^N \lambda_n^{\frac{1}{2}}[e^{a_n t}g_n(0) + \int_0^t e^{a_n(t-\tau)}f_n(X(\tau))d\tau][\lambda_n^{-\frac{1}{2}}\phi_n(x)]. \end{aligned}$$

Denote $R_{1n} \triangleq \|f_n\|_{\mathcal{L}^\infty(0,T)}$. Note that $\{\lambda_n^{-\frac{1}{2}}\phi_n(x)\}_{n=1}^\infty$ forms an orthonormal basis for $\mathcal{H}^{\frac{3}{2}}_{per}(0,1)$, which is a closed subspace of $\mathcal{H}^{\frac{3}{2}}(0,1)$. Hence,

$$\begin{aligned} &\frac{1}{2}\left\|z^N(x,t)\right\|^2_{\mathcal{H}^{\frac{3}{2}}_{per}} \\ &\le \sum_{n=1}^N |\lambda_n^{\frac{1}{2}}e^{a_n t}g_n(0)|^2 + \sum_{n=1}^N |\lambda_n^{\frac{1}{2}}\textstyle\int_0^t e^{a_n(t-\tau)}f_n(X(\tau))d\tau|^2 \\ &\le \sum_{n=1}^N \lambda_n e^{-8\pi^2n^2\epsilon t}|g_n(0)|^2 + \sum_{n=1}^N \lambda_n R_{1n}^2 |\textstyle\int_0^t e^{a_n(t-\tau)}d\tau|^2 \\ &\le \sum_{n=1}^N \lambda_n e^{-8\pi^2n^2\epsilon t}|g_n(0)|^2 + \sum_{n=1}^N \lambda_n R_{1n}^2 \frac{(1+e^{-4\pi^2n^2\epsilon t})^2}{16\pi^4n^4\epsilon^2 + \lambda_n^2}, \end{aligned}$$

which yields $\left\|z^N(x,t)\right\|^2_{\mathcal{H}^{\frac{3}{2}}_{per}} < \infty$. Therefore, we have $\int_0^T \left\|z^N(x,t)\right\|^2_{\mathcal{H}^{\frac{3}{2}}_{per}} dt < \infty$, which means that $\sup\limits_N \|z^N(\cdot,t)\|_{\mathcal{L}^2(0,T;\mathcal{H}^{\frac{3}{2}}_{per}(0,1))} < \infty$.

Step 3. We show that $\sup\limits_N \|z_t^N(\cdot,t)\|_{\mathcal{L}^2(0,T;\mathcal{H}^*_{per}(0,1))} < \infty$, where $\mathcal{H}^*_{per}(0,1)$ is the dual space of $\mathcal{H}^{\frac{3}{2}}_{per}(0,1)$ with respect to the pivot space $\mathcal{L}^2(0,1)$. For any $\varphi \in \mathcal{H}^{\frac{3}{2}}_{per}(0,1)$ with $\|\varphi\|_{\mathcal{H}^{\frac{3}{2}}_{per}} \le 1$, it follows from (5.18) that for a.e. $0 \le t \le T$, we have

$$\begin{aligned} &\langle z_t^N(\cdot,t), \varphi\rangle_{\mathcal{H}^*_{per},\mathcal{H}^{\frac{3}{2}}_{per}} = \langle z_t^N(\cdot,t), \varphi\rangle_{\mathcal{L}^2} \\ &= -\epsilon\langle z_x^N(\cdot,t), \varphi'\rangle_{\mathcal{L}^2} + \nu\langle z_{xx}^N(\cdot,t), \varphi'\rangle_{\mathcal{L}^2} - \langle z^N(\cdot,t)z_x^N(\cdot,t), \varphi\rangle_{\mathcal{L}^2} \\ &\quad -K\sum_{j=1}^{N_s}\langle \tfrac{1}{\Delta_j}\textstyle\int_{\Omega_j} z^N(x,t)dx, \varphi\rangle_{\mathcal{L}^2} - \langle \tilde{d}_1(t), \varphi\rangle_{\mathcal{L}^2}. \end{aligned}$$

Therefore, for a positive constant R_2, we get

$$\begin{aligned}&R_2^{-1}\int_0^T ||z_t^N(x,t)||^2_{\mathcal{H}^*_{per}}dt \le \int_0^T ||z_x^N(\cdot,t)||^2_{\mathcal{L}^2}dt\\&+\int_0^T ||z_{xx}^N(\cdot,t)||^2_{\mathcal{L}^2}dt+\int_0^T ||z^N(\cdot,t)z_x^N(\cdot,t)||^2_{\mathcal{L}^2}dt\\&+\int_0^T\sum_{j=1}^{N_s}||\frac{1}{\Delta_j}\int_{\Omega_j}z^N(x,t)dx||^2_{\mathcal{L}^2}dt+\int_0^T|\tilde{d}_1(t)|^2dt.\end{aligned}$$

Note that

$$\begin{aligned}||z_x^N(\cdot,t)||_{\mathcal{L}^2}&=\left\|\sum_{n=1}^N[e^{a_nt}g_n(0)+\int_0^t e^{a_n(t-\tau)}f_n(X(\tau))d\tau]\phi_n'(x)\right\|_{\mathcal{L}^2}\\&\le\sum_{n=1}^N 2\pi n\left|e^{a_nt}g_n(0)+\int_0^t e^{a_n(t-\tau)}f_n(X(\tau))d\tau\right|\\&\le\sum_{n=1}^N 2\pi n\left[2e^{-8\pi^2n^2\epsilon t}|g_n(0)|^2+2R_{1n}^2\frac{(1+e^{-4\pi^2n^2\epsilon t})^2}{16\pi^4n^4\epsilon^2+\lambda_n^2}\right]^{\frac{1}{2}}<\infty.\end{aligned}$$

Following the similar computations, we have

$$||z_{xx}^N(\cdot,t)||_{\mathcal{L}^2}<\infty,\ ||z^N(\cdot,t)z_x^N(\cdot,t)||_{\mathcal{L}^2}<\infty.$$

From (5.11) and Assumption 5.2.1, $|\tilde{d}_1(t)|$ is bounded. Hence,

$$R_2^{-1}\int_0^T ||z_t^N(x,t)||^2_{\mathcal{H}^*_{per}}dt<\infty,$$

which yields $\sup_N ||z_t^N(\cdot,t)||_{\mathcal{L}^2(0,T;\mathcal{H}^*_{per}(0,1))}<\infty$.

From Steps 2–3, we have $z\in\mathcal{L}^2(0,T;\mathcal{H}^{\frac{3}{2}}_{per}(0,1))$ and $z_t\in\mathcal{L}^2(0,T;\mathcal{H}^*_{per}(0,1))$. Then by Lemma 5.2, we have $z\in C([0,T];\mathcal{L}^2(0,1))$. □

Theorem 5.1 *Consider the time-varying high gain estimator (5.5) and the perturbed system (5.1) under continuous-time ADRC (5.14). For parameters* $K>0$ *and* $\sigma_i>0 (i=1,2)$ *satisfying the following conditions:*

$$-K+\frac{K\sigma_1}{2}+\frac{\sigma_2}{2}<0,\ -\epsilon+\frac{K\Delta^2}{2\sigma_1\pi^2}<0, \tag{5.21}$$

it follows that the closed-loop system (5.16) is asymptotically stable in the sense that

$$\lim_{t\to\infty}\left[\|z(\cdot,t)\|^2_{\mathcal{L}^2(0,1)}+\tilde{y}_r(t)+\tilde{d}_1(t)\right]=0. \tag{5.22}$$

Proof The convergence of $(\tilde{y}_r(t),\tilde{d}_1(t))$ has been proven in (5.13). Then we need only discuss the convergence of $u(x,t)$. Consider the Lyapunov–Krasovskii functional

$$V_1(t) = \frac{1}{2}\int_0^1 |z(x,t)|^2 dx. \tag{5.23}$$

Differentiating (5.23) along (5.16) and integrating by parts, where (5.17) is used with $\psi = u$, we have

$$\begin{aligned}\dot{V}_1(t) \leq &-\epsilon\int_0^1 |z_x(x,t)|^2 dx - K\int_0^1 |z(x,t)|^2 dx \\ &+K\sum_{j=1}^{N_s}\int_{x_{j-1}}^{x_j} |z(x,t)||f_j(x,t)|dx - \int_0^1 |z(x,t)||\tilde{d}_1(t)|dx.\end{aligned} \tag{5.24}$$

Due to $\int_{x_{j-1}}^{x_j} |f_j(x,t)|dx = 0$, Poincaré's inequality yields

$$\int_{x_{j-1}}^{x_j} |f_j(x,t)|^2 dx \leq \frac{\Delta^2}{\pi^2}\int_{x_{j-1}}^{x_j} |z_x(x,t)|^2 dx,$$

which implies

$$\sum_{j=1}^{N_s}\int_{x_{j-1}}^{x_j} |f_j(x,t)|^2 dx \leq \frac{\Delta^2}{\pi^2}\int_0^1 |z_x(x,t)|^2 dx. \tag{5.25}$$

Using Young's inequality, we have

$$\begin{aligned}\dot{V}_1(t) \leq &-\left(K - \frac{K\sigma_1}{2} - \frac{\sigma_2}{2}\right)\int_0^1 |z(x,t)|^2 dx \\ &-(\epsilon - \frac{K\Delta^2}{2\sigma_1\pi^2})\int_0^1 |z_x(x,t)|^2 dx + \frac{1}{2\sigma_2}|\tilde{d}_1(t)|^2\end{aligned}$$

for positive constants σ_1, σ_2. Denote $\delta_1 \triangleq 2K - K\sigma_1 - \sigma_2$. If the conditions (5.21) are satisfied, we have

$$V_1(t) \leq e^{-\delta_1 t}V_1(0) + \frac{1}{2\sigma_2}\int_0^t e^{-\delta_1(t-s)}|\tilde{d}_1(s)|^2 ds, t \geq 0.$$

Note that $\lim_{t\to\infty}\tilde{d}_1(t) = 0$. From (5.11) the above inequality, it follows that $\lim_{t\to\infty} V_1(t) = 0$. Then the closed-loop system (5.16) is asymptotically stable in the sense that (5.22) holds. □

Remark 5.3 If we assume a particular decay rate $\bar{r}$ such that $\frac{|\dot{d}(t)|}{r(t)} = e^{-\bar{r}t}$, the equation (5.12) becomes

$$\lim_{t\to\infty} \frac{3\int_0^t |\dot{d}(s)| e^{\int_0^s [r(\rho)-2.5M]d\rho} ds}{e^{\int_0^t [r(s)-2.5M]ds}} = \lim_{t\to\infty} \frac{3|\dot{d}(t)|}{r(t)-2.5M}$$
$$= \lim_{t\to\infty} 3e^{-\bar{r}t} = 0.$$

Together with (5.11), one has

$$\lim_{t\to\infty} \sqrt{V_0(\tilde{y}_r(t), \tilde{d}_1(t))} \le \lim_{t\to\infty} \sqrt{V_0(\tilde{y}_r(0), \tilde{d}_1(0)) e^{-\int_0^t [r(s)-2.5M]ds}},$$

which means that $\lim_{t\to\infty} \sqrt{V_0(\tilde{y}_r(t), \tilde{d}_1(t))} = 0$. Hence, a better exponential stability result than Theorem 5.1 can be established. Even if $d(t)$ is unknown and we only have bounds of it, we could choose $r(t)$ such that $\frac{|\dot{d}(t)|}{r(t)}$ exponentially decays with a desired rate $\bar{r}$.

5.2.1.2 Event-Triggered Control

To significantly reduce the workloads of the network, we further propose an event-triggered ADRC on the basis of (5.14) as follows:

$$U_j(t) = -\frac{K}{\Delta_j} \int_{x_{j-1}}^{x_j} z(\zeta, t_k) d\zeta - \hat{d}_1(t), \ t \in [t_k, t_{k+1}), \ k = 0, 1, 2, \ldots, \tag{5.26}$$

where the update instants satisfy $0 = t_0 < t_1 < \cdots < t_k < t_{k+1}$ and $\lim_{k\to\infty} t_k = \infty$. The event-triggering mechanism is defined as

$$t_{k+1} = \inf\left\{ t \ge t_k \middle| \, \|z(\cdot, t) - z(\cdot, t_k)\|^2_{\mathcal{L}^2(0,1)} \ge \alpha \|z(\cdot, t)\|^2_{\mathcal{L}^2(0,1)} + \alpha_0 \|z_0\|^2_{\mathcal{L}^2(0,1)} \right\} \tag{5.27}$$

where α and α_0 are positive constants to be determined. The sampling cannot be implemented if it requires to transmit an infinite number of signals in finite time, which is called the Zeno behaviour. Therefore, $\inf\{t_{k+1} - t_k\} > 0$ must be ensured (see Theorem 5.2 below). Under the control law (5.26), the closed-loop system is governed by

$$\begin{cases} z_t(x,t) = \epsilon z_{xx}(x,t) - \nu z_{xxx}(x,t) - z(x,t) z_x(x,t) \\ - \sum_{j=1}^{N_s} b_j(x) \left[K z(x,t) - K f_j(x,t) - K \rho_j(t) + \tilde{d}_1(t) \right], \\ z(0,t) = z(1,t), \, z_x(0,t) = z_x(1,t), \, z_{xx}(0,t) = z_{xx}(1,t), \\ \dot{\tilde{y}}_r(t) = -r(t) \left[\tilde{y}_r(t) - \tilde{d}_1(t) \right] + \frac{\dot{r}(t)}{r(t)} \tilde{y}_r(t), \\ \dot{\tilde{d}}_1(t) = -r(t) \tilde{y}_r(t) - \dot{d}(t), \end{cases} \tag{5.28}$$

where

$$\rho_j(t) = \frac{\int_{x_{j-1}}^{x_j} [z(\zeta, t) - z(\zeta, t_k)]\, d\zeta}{\Delta_j} \tag{5.29}$$

and $f_j(x, t)$ is given by (5.15).

Lemma 5.3 *Assume that the time-varying gain $r(t) \in C^1[0, +\infty)$ satisfies (5.6) and the disturbance $d(t) \in \mathcal{H}^1_{loc}(0, \infty)$ satisfies $\lim_{t\to\infty} \frac{|\dot{d}(t)|}{r(t)} = 0$. Then for any initial condition $(z_0(x), \tilde{y}_r(0), \tilde{d}_1(0)) \in \left(\mathcal{L}^2(0, 1) \times \mathbb{R} \times \mathbb{R}\right)$, the closed-loop system (5.28) has a unique weak solution for all $T > 0$ such that $z \in C\left([0, T], \mathcal{L}^2(0, 1)\right)$, $\tilde{y}_r \in \mathcal{H}^1([0, T], \mathbb{R})$ and $\tilde{d}_1 \in \mathcal{H}^1([0, T], \mathbb{R})$.*

Remark 5.4 By the arguments of Lemma 5.1, the well-posedness proof of (5.28) can also be established by replacing $-\frac{K}{\Delta_j}\int_{x_{j-1}}^{x_j} z(\zeta, t) d\zeta$ with $-\frac{K}{\Delta_j}\int_{x_{j-1}}^{x_j} z(\zeta, t_k) d\zeta$ for $t \in [t_0, t_1]$. Then we apply the same arguments of Lemma 1 step-by-step for $[t_1, t_2], \ldots, [t_k, t_{k+1}], \ldots$ Following this procedure, we obtain that there exists a weak solution of the system (5.28) for all $t \in [0, \infty)$.

The stability result of the actuated system (5.41) and the avoidance of Zeno behaviour are shown as follows:

Theorem 5.2 *Consider the time-varying high gain estimator (5.5) and the perturbed system (5.1) under event-triggered ADRC (5.26) subject to the ETM (5.27). For parameters $K > 0$, $\sigma_i > 0 (i = 1, 2, 3)$, $\alpha > 0$ and $\alpha_0 > 0$ satisfying the following conditions:*

$$-K + \frac{K\sigma_1}{2} + \frac{\sigma_2}{2} + \frac{K\sigma_3}{2} + \frac{K\alpha}{2\sigma_3} < 0, \quad -\epsilon + \frac{K\Delta^2}{2\sigma_1\pi^2} < 0, \tag{5.30}$$

it follows that
(i) the closed-loop system (5.28) is uniformly ultimate bounded in the sense that

$$\lim_{t\to\infty} \|z(\cdot, t)\|^2_{\mathcal{L}^2(0,1)} \le \frac{K\alpha_0}{\delta_2\sigma_3} \|z_0\|^2_{\mathcal{L}^2(0,1)}, \quad \lim_{t\to\infty} \left[\tilde{y}_r(t) + \tilde{d}_1(t)\right] = 0, \tag{5.31}$$

where

$$\delta_2 = 2K - K\sigma_1 - \sigma_2 - K\sigma_3 - \tfrac{K\alpha}{\sigma_3}; \tag{5.32}$$

(ii) the ETM (5.27) rules out the Zeno behaviour.

Proof (i) Cauchy–Schwarz's inequality and the ETM (5.27) yield

$$\begin{aligned}
\sum_{j=1}^{N_s} \int_{x_{j-1}}^{x_j} |\rho_j(t)|^2 dx &= \sum_{j=1}^{N_s} \frac{1}{\Delta_j} \left| \int_{x_{j-1}}^{x_j} e_k(x, t) dx \right|^2 \\
&\le \sum_{j=1}^{N_s} \int_{x_{j-1}}^{x_j} |e_k(x, t)|^2 dx \le \alpha \|z(\cdot, t)\|^2_{\mathcal{L}^2(0,1)} + \alpha_0 \|z_0\|^2_{\mathcal{L}^2(0,1)}.
\end{aligned}$$

Consider the Lyapunov–Krasovskii functional (5.23). By the arguments of Theorem 5.1, we have

$$\dot{V}_1(t) \le -\left(K - \frac{K\sigma_1}{2} - \frac{\sigma_2}{2} - \frac{K\sigma_3}{2} - \frac{K\alpha}{2\sigma_3}\right)\|z(x,t)\|^2_{\mathcal{L}^2(0,1)}$$
$$-(\epsilon - \frac{K\Delta^2}{2\sigma_1\pi^2})\|z_x(x,t)\|^2_{\mathcal{L}^2(0,1)} + \frac{1}{2\sigma_2}|\tilde{d}_1(t)|^2 + \frac{K\alpha_0}{2\sigma_3}\|z_0\|^2_{\mathcal{L}^2(0,1)},$$

for $\sigma_i > 0 (i = 1, 2, 3)$ and δ_2 given by (5.32). If the condition (5.30) is satisfied, then

$$\dot{V}_1(t) \le -\delta_2 V_1(t) + \frac{1}{2\sigma_2}|\tilde{d}_1(t)|^2 + \frac{K\alpha_0}{2\sigma_3}\|z_0\|^2_{\mathcal{L}^2(0,1)}.$$

From (5.13), one has

$$V_1(t) \le e^{-\delta_2 t} V_1(0) + \frac{K\alpha_0}{2\delta_2\sigma_3}\|z_0\|^2_{\mathcal{L}^2(0,1)}.$$

Therefore, the closed-loop system (5.28) is uniformly ultimate bounded in the sense that (5.31) holds.

(ii) Denote $e_k(x,t) \triangleq z(x,t) - z(x,t_k)$. Since $z \in C\left([0,T], \mathcal{L}^2(0,1)\right)$, $e_k(x,t)$ is uniformly continuous in time. Therefore, for any $\mu > 0$ there exists a constant $\tau > 0$ such that if $|t-s| < \tau$ for all $t, s \in [0,T]$, one has $||e_k(\cdot,t) - e_k(\cdot,s)||_{\mathcal{L}^2(0,1)} < \mu$. It follows that

$$\begin{aligned}&\forall \mu > 0,\ \exists \tau > 0,\ \forall t, s \in [0,T],\\ &||e_k(\cdot,t) - e_k(\cdot,s)||_{\mathcal{L}^2(0,1)} \ge \mu \Rightarrow |t-s| \ge \tau.\end{aligned} \tag{5.33}$$

Together with the definition (5.27), substituting $t = t_{k+1}$ and $s = t_k$ into (5.33), we get

$$\begin{aligned}&||e_k(\cdot,t_{k+1}) - e_k(\cdot,t_k)||_{\mathcal{L}^2(0,1)} = ||e_k(\cdot,t_{k+1})||_{\mathcal{L}^2(0,1)}\\ &\ge \alpha\|z(\cdot,t_{k+1})\|^2_{\mathcal{L}^2(0,1)} + \alpha_0\|z_0\|^2_{\mathcal{L}^2(0,1)} \ge \alpha_0\|z_0\|^2_{\mathcal{L}^2(0,1)}.\end{aligned}$$

Indeed, by choosing $\mu = \sqrt{\alpha_0\|z_0\|^2_{\mathcal{L}^2(0,1)}}$ with $z_0 \ne 0$, one arrives at $t_{k+1} - t_k \ge \tau$, $\forall t_k, t_{k+1} \in [0,T]$, $k = 0, 1, 2, \ldots$. Hence, the Zeno behaviour is avoided. □

5.2.2 *Constant High Gain Estimator*

To design the ESO with a constant high gain, the following assumption is necessary:

Assumption 5.2.2 Suppose that the derivative of disturbance $\dot{d}(t)$ is bounded such that $|\dot{d}(t)| \le c$ for a constant $c > 0$. □

Based on (5.3) and (5.4), the following constant high-gain estimator for the disturbance $d(t)$ is employed:

$$\begin{cases} \dot{\hat{y}}_c(t) = \sum_{j=1}^{N_s} \Delta_j U_j(t) + \hat{d}_2(t) - \frac{1}{b}\left[\hat{y}_c(t) - y(t)\right], \\ \dot{\hat{d}}_2(t) = -\frac{1}{b^2}\left[\hat{y}(t) - y(t)\right], \end{cases} \tag{5.34}$$

where $\hat{d}_2(t)$ is the approximation of $d(t)$, $\hat{y}_c(t)$ is the approximation of $y(t)$, and $b > 0$ is a small tuning parameter. Let $\tilde{y}_c(t) = \hat{y}_c(t) - y(t)$ and $\tilde{d}_2(t) = \hat{d}_2(t) - d(t)$. The estimation errors satisfy

$$\begin{bmatrix} \dot{\tilde{y}}_c(t) \\ \dot{\tilde{d}}_2(t) \end{bmatrix} = A \begin{bmatrix} \tilde{y}_c(t) \\ \tilde{d}_2(t) \end{bmatrix} + B\dot{d}(t), \tag{5.35}$$

where $A = \begin{bmatrix} -\frac{1}{b} & 1 \\ -\frac{1}{b^2} & 0 \end{bmatrix}$ and $B = \begin{bmatrix} 0 \\ -1 \end{bmatrix}$. The eigenvalues of A are $\lambda_1 = -\frac{1}{2b} + \frac{\sqrt{3}}{2b}i$, $\lambda_2 = -\frac{1}{2b} - \frac{\sqrt{3}}{2b}i$. Then the solution of (5.35) is

$$\begin{bmatrix} \tilde{y}_c(t) \\ \tilde{d}_2(t) \end{bmatrix} = e^{At} \begin{bmatrix} \tilde{y}_c(0) \\ \tilde{d}_2(0) \end{bmatrix} + \int_0^t e^{A(t-s)} B\dot{d}(s)ds, \tag{5.36}$$

where $e^{At}B = \begin{bmatrix} \frac{b^2\lambda_1\lambda_2\left(e^{\lambda_1 t} - e^{\lambda_2 t}\right)}{\lambda_1 - \lambda_2} & \frac{\lambda_2 e^{\lambda_1 t} - \lambda_1 e^{\lambda_2 t}}{\lambda_1 - \lambda_2} \end{bmatrix}^T$.

As $t \to \infty$, the first term in (5.36) can be arbitrarily small due to e^{At}. As $t \to \infty$ and $b \to 0$, the second term in (5.36) can also be arbitrarily small due to $e^{At}B$ and $|\dot{d}(t)| \le c$. Therefore, one has

$$(\tilde{y}_c(t), \tilde{d}_2(t)) \to 0 \text{ as } t \to \infty, b \to 0. \tag{5.37}$$

Under the continuous-time ADRC

$$U_j(t) = -\frac{K}{\Delta_j} \int_{x_{j-1}}^{x_j} z(\zeta, t)d\zeta - \hat{d}_2(t),\ j = 1, \ldots, N_s, \tag{5.38}$$

or event-triggered ADRC

$$U_j(t) = -\frac{K}{\Delta_j} \int_{x_{j-1}}^{x_j} z(\zeta, t_k)d\zeta - \hat{d}_2(t),\ t \in [t_k, t_{k+1}),\ k = 0, 1, 2, \ldots \tag{5.39}$$

subject to (5.27), the closed-loop system becomes

$$\begin{cases} z_t(x,t) = \epsilon z_{xx}(x,t) - \nu z_{xxx}(x,t) - z(x,t)z_x(x,t) \\ -\sum_{j=1}^{N_s} b_j(x)\left[Kz(x,t) - Kf_j(x,t) + \tilde{d}_2(t)\right], \\ z(0,t) = z(1,t), z_x(0,t) = z_x(1,t), z_{xx}(0,t) = z_{xx}(1,t), \\ \dot{\tilde{y}}_c(t) = -\frac{1}{b}\tilde{y}_c(t) + \tilde{d}_2(t), \dot{\tilde{d}}_2(t) = -\frac{1}{b^2}\tilde{y}_c(t) - \dot{d}(t), \end{cases} \tag{5.40}$$

or

$$\begin{cases} z_t(x,t) = \epsilon z_{xx}(x,t) - \nu z_{xxx}(x,t) - z(x,t)z_x(x,t) \\ -\sum_{j=1}^{N_s} b_j(x)\left[Kz(x,t) - Kf_j(x,t) - K\rho_j(t) + \tilde{d}_2(t)\right], \\ z(0,t) = z(1,t), z_x(0,t) = z_x(1,t), z_{xx}(0,t) = z_{xx}(1,t), \\ \dot{\tilde{y}}_c(t) = -\frac{1}{b}\tilde{y}_c(t) + \tilde{d}_2(t), \dot{\tilde{d}}_2(t) = -\frac{1}{b^2}\tilde{y}_c(t) - \dot{d}(t), \end{cases} \tag{5.41}$$

where $f_j(x,t)$ and $\rho_j(t)$ are given by (5.15) and (5.29). The convergence of $(\tilde{y}_c(t), \tilde{d}_2(t))$ has been proven in (5.13). Since stability conditions of (5.40) and (5.41) are the same with Theorems 5.1 and 5.2, we omit the proofs of the following Theorems 5.3 and 5.4. Stability results of (5.40) and (5.41) are directly given as follows:

Theorem 5.3 *Consider the constant high gain estimator (5.34) and the perturbed system (5.1) under continuous-time ADRC (5.38). For parameters* $K > 0$ *and* $\sigma_i > 0 (i = 1, 2)$ *satisfying the conditions (5.21), it follows that the closed-loop system (5.40) is asymptotically stable in the sense that*

$$\lim_{t\to\infty, b\to 0}\left[\|z(\cdot,t)\|^2_{\mathcal{L}^2(0,1)} + \tilde{y}_c(t) + \tilde{d}_2(t)\right] = 0.$$

Theorem 5.4 *Consider the constant high gain estimator (5.34) and the perturbed system (5.1) under event-triggered ADRC (5.39) subject to the ETM (5.27). For parameters* $K > 0$, $\sigma_i > 0 (i = 1, 2, 3)$, $\alpha > 0$ *and* $\alpha_0 > 0$ *satisfying the conditions (5.30), it follows that the closed-loop system (5.41) is uniformly ultimate bounded in the sense that*

$$\begin{aligned} &\lim_{t\to\infty}\|z(\cdot,t)\|^2_{\mathcal{L}^2(0,1)} \le \frac{K\alpha_0}{\delta_2\sigma_3}\|z_0\|^2_{\mathcal{L}^2(0,1)}, \\ &\lim_{t\to\infty}\left[\tilde{y}_c(t) + \tilde{d}_2(t)\right] = 0, \end{aligned} \tag{5.42}$$

where δ_2 *is given by (5.32).*

Remark 5.5 With the constant high gain estimator, more strict condition $|\dot{d}(t)| \le c$ as Assumption 5.2.2 is required. Whereas with the time-varying high gain estimator, the condition on $d(t)$ is relaxed.

Remark 5.6 The constant high gain estimator is more effective in filtering the high frequency noise. The time-varying high gain estimator has the advantage of reducing the peaking value. In practice, a comprehensive strategy is recommended that the time-varying gain is adopted firstly to reduce the peaking value to a reasonable level and then the constant gain is applied. This will be explained numerically in Sect. 5.4.

5.3 Disturbance Observer-Based Control (DOBC): Design and Main Results

In this section, the continuous-time and event-triggered DOBCs are presented for KdVB system (5.1).

5.3.1 Disturbance Observer (DO) Design

The following assumptions are necessary :

Assumption 5.3.1 The disturbance $d(t)$ in (5.1) is formulated by the following exogenous system

$$\dot{w}(t) = Cz(t),\ \ d(t) = Dw(t), \tag{5.43}$$

where $C \in \mathbb{R}^{r\times r}$ and $D \in \mathbb{R}^{1\times r}$ are known matrices, $w(t) \in \mathbb{R}^r$ is the state of the exogenous system, and the initial condition $z(0)$ is unknown. □

Assumption 5.3.2 The pair (C, D) is observable. □

Remark 5.7 Since initial condition $w(0)$ is unknown, the above exogenous system (5.43) can describe various kinds of disturbances that are not exactly known. For instance, harmonic signals with any initial phases and step functions with arbitrary magnitude can be modeled by (5.43).

Based on (5.3), we construct the following DO:

$$\begin{cases} \hat{d}_3(t) = D\hat{w}(t),\ \hat{w}(t) = \Gamma(t) + Ly(t), \\ \dot{\Gamma}(t) = (C - LD)\left[\Gamma(t) + Ly(t)\right] - L\sum\limits_{j=1}^{N_s} \Delta_j U_j(t), \end{cases} \tag{5.44}$$

where $L \in \mathbb{R}^r$ is an observer gain such that $C - LD$ is Hurwitz, $y(t)$ is given by (5.3), and $\Gamma(t)$ is an auxiliary vector. $\hat{d}_3(t) \in \mathbb{R}$ represents the estimate of $d(t)$, and $\hat{w}(t) \in \mathbb{R}^r$ represents the estimate of $w(t)$.

5.3.2 Continuous-Time Control

For stabilization of KdVB system, the following continuous-time DOBC is proposed:

$$U_j(t) = -\frac{K}{\Delta_j}\int_{x_{j-1}}^{x_j} z(\zeta,t)d\zeta - \hat{d}_3(t),\ j = 1,\ldots,N_s. \tag{5.45}$$

Define the estimation errors as $\tilde{z}(t) = \hat{w}(t) - w(t)$ and $\tilde{d}_3(t) = \hat{d}_3(t) - d(t)$. From (5.1) and (5.44), it follows that

$$\begin{cases} z_t(x,t) = \epsilon z_{xx}(x,t) - \nu z_{xxx}(x,t) - z(x,t)z_x(x,t) \\ -\sum\limits_{j=1}^{N_s} b_j(x)\left[Kz(x,t) - Kf_j(x,t) + D\tilde{w}(t)\right], \\ z(0,t) = z(1,t),\ z_x(0,t) = z_x(1,t),\ z_{xx}(0,t) = z_{xx}(1,t), \\ \dot{\tilde{w}}(t) = (C - LD)\,\tilde{w}(t),\ \tilde{d}_3(t) = D\tilde{w}(t), \end{cases} \tag{5.46}$$

where $f_j(x,t)$ is given by (5.15). The stability result of the closed-loop system (5.46) is presented as follows:

Theorem 5.5 *Under Assumptions 5.3.1 and 5.3.2, consider the perturbed system (5.1) with continuous-time DOBC (5.45). For parameters $K > 0$, $\gamma_i > 0 (i = 1, 2)$ and $L \in \mathbb{R}^r$ satisfying the following conditions:*

$$-K + \frac{K\gamma_1}{2} + \frac{\gamma_2}{2} < 0,\ -\epsilon + \frac{K\Delta^2}{2\sigma_1\pi^2} < 0, \tag{5.47}$$

$$\lambda_{\max}\left(C - LD + \frac{1}{2\gamma_2}D^TD\right) < 0, \tag{5.48}$$

it follows that the closed-loop system (5.46) is exponentially stable in the sense that $E(t) \le e^{-\delta_3 t}E(0),\ \forall t \ge 0$, where

$$\delta_3 = \min\left\{2K - K\gamma_1 - \gamma_2, 2\lambda_{\min}\left(-C + LD - \tfrac{1}{2\gamma_2}D^TD\right)\right\}. \tag{5.49}$$

Proof Consider the Lyapunov–Krasovskii functional

$$E(t) = \frac{1}{2}\int_0^1 |z(x,t)|^2dx + \frac{1}{2}|\tilde{w}(t)|^2. \tag{5.50}$$

By the arguments of Theorem 5.1, differentiating (5.50) along (5.46), we have

$$\begin{aligned} \dot{E}(t) &\le -\left(K - \frac{K\gamma_1}{2} - \frac{\gamma_2}{2}\right)\int_0^1 |z(x,t)|^2dx \\ &-(\epsilon - \frac{K\Delta^2}{2\gamma_1\pi^2})\int_0^1 |z_x(x,t)|^2dx + \lambda_{\max}(C - LD + \frac{1}{2\gamma_2}D^TD)|\tilde{w}(t)|^2 \end{aligned}$$

for $\gamma_i > 0 (i = 1, 2)$. Choose positive parameters K, γ_1, γ_2, and $L \in \mathbb{R}^r$ such that the conditions (5.47) and (5.48) are satisfied. Then $\dot{E}(t) \le -\delta_3 E(t),\ \forall t \ge 0$, where δ_3 is given by (5.49). Thus, $E(t) \le e^{-\delta_3 t}E(0),\ \forall t \ge 0$. □

5.3.3 Event-Triggered Control

In this subsection, we further suggest an event-triggered DOBC based on (5.45) to reduce the communication cost. Consider the following event-triggered DOBC:

$$U_j(t) = -\frac{K}{\Delta_j}\int_{x_{j-1}}^{x_j} z(\zeta, t_k)d\zeta - \hat{d}_3(t), \quad t \in [t_k, t_{k+1}), \ k = 0, 1, 2, \ldots \tag{5.51}$$

subject to the ETM

$$\begin{aligned} t_{k+1} = \inf\Big\{t \geq t_k \Big| \, \|z(\cdot, t) - z(\cdot, t_k)\|^2_{\mathcal{L}^2(0,1)} \\ \geq \alpha \|z(\cdot, t)\|^2_{\mathcal{L}^2(0,1)} + \alpha_0 \|z_0\|^2_{\mathcal{L}^2(0,1)} e^{-\omega t}\Big\}, \end{aligned} \tag{5.52}$$

where α, α_0 and ω are positive parameters to be determined. Under the event-triggered DOBC (5.51), the closed-loop system becomes

$$\begin{cases} z_t(x,t) = \epsilon z_{xx}(x,t) - \nu z_{xxx}(x,t) - z(x,t)z_x(x,t) \\ -\sum\limits_{j=1}^{N_s} b_j(x)\left[Kz(x,t) - Kf_j(x,t) - K\rho_j(t) + D\tilde{w}(t)\right], \\ z(0,t) = z(1,t), z_x(0,t) = z_x(1,t), z_{xx}(0,t) = z_{xx}(1,t), \\ \dot{\tilde{w}}(t) = (C - LD)\,\tilde{w}(t), \tilde{d}_3(t) = D\tilde{w}(t), \end{cases} \tag{5.53}$$

where $f_j(x,t)$ and $\rho_j(t)$ are given by (5.15) and (5.29). The following stability result and the avoidance of Zeno behaviour can be obtained:

Theorem 5.6 *Under Assumptions 5.3.1 and 5.3.2, consider the perturbed system (5.1) with event-triggered DOBC (5.51) subject to the ETM (5.52). For parameters $K > 0$, $\gamma_i > 0 (i = 1, 2, 3)$, $L \in \mathbb{R}^r$, $\alpha > 0$, $\alpha_0 > 0$ and $\omega > \delta_4$ satisfying the conditions (5.48) and*

$$-K + \frac{K\gamma_1}{2} + \frac{\gamma_2}{2} + \frac{K\gamma_3}{2} + \frac{K\alpha}{2\gamma_3} < 0, \ -\epsilon + \frac{K\Delta^2}{2\sigma_1\pi^2} < 0, \tag{5.54}$$

it follows that
(i) the closed-loop system (5.53) is exponentially stable in the sense that

$$E(t) \leq e^{-\delta_4 t}E(0)\left[1 + \frac{K\alpha_0}{2\delta_4\gamma_3(\omega - \delta_4)}\right], \ \forall t \geq 0,$$

where

$$\delta_4 = \min\left\{2K - K\gamma_1 - \gamma_2 - K\gamma_3 - \tfrac{K\alpha}{\gamma_3}, \ 2\lambda_{\min}\left(-C + LD - \frac{1}{2\gamma_2}D^T D\right)\right\}; \tag{5.55}$$

(ii) the ETM (5.52) rules out the Zeno behaviour.

Proof (i) Consider the Lyapunov–Krasovskii functional (5.50). By the arguments of Theorems 5.1, 5.2 and 5.5, differentiating (5.50) along (5.53), we get

$$\dot{E}(t) \leq -\delta_4 E(t) + \frac{K\alpha_0}{2\gamma_3} \|z_0\|^2_{\mathcal{L}^2(0,1)} e^{-\omega t}$$

for $\gamma_i > 0 (i = 1, 2, 3)$ and δ_4 given by (5.55), if the conditions (5.48) and (5.54) are satisfied.

Choose a constant $\omega > \delta_4$. Then one gets

$$\begin{aligned} E(t) &\leq e^{-\delta_4 t} E(0) + \frac{K\alpha_0}{2\delta_4\gamma_3} e^{-\delta_4 t} E(0) \int_0^t e^{-(\omega-\delta_4)s} ds \\ &\leq e^{-\delta_4 t} E(0) + \frac{K\alpha_0}{2\delta_4\gamma_3(\omega-\delta_4)} E(0)(e^{-\delta_4 t} - e^{-\omega t}) \\ &\leq e^{-\delta_4 t} E(0) \left[1 + \frac{K\alpha_0}{2\delta_4\gamma_3(\omega-\delta_4)}\right], \quad \forall t \geq 0. \end{aligned}$$

(ii) By the arguments of Theorem 5.2, substituting $t = t_{k+1}$ and $s = t_k$ into (5.33), we have $||e_k(\cdot, t_{k+1}) - e_k(\cdot, t_k)||_{\mathcal{L}^2(0,1)} \geq \alpha_0 \|z_0\|^2_{\mathcal{L}^2(0,1)} e^{-\omega T}$. By choosing $\mu = \sqrt{\alpha_0 \|z_0\|^2_{\mathcal{L}^2(0,1)} e^{-\omega T}}$ with $z_0 \neq 0$, one arrives at $t_{k+1} - t_k \geq \tau$, $\forall t_k, t_{k+1} \in [0, T]$, $k = 0, 1, 2, \ldots$. Hence, the Zeno behaviour is avoided. □

Remark 5.8 Compared with (5.27), an improved event-triggering scheme (5.52) is utilized by introducing an exponential term $e^{-\omega t}$. Without the term $e^{-\omega t}$, we have $\dot{E}(t) \leq -\delta_4 E(t) + \frac{K\alpha_0}{2\gamma_3} \|z_0\|^2_{\mathcal{L}^2(0,1)}$, which leads to the uniform ultimate bounded result. With $e^{-\omega t}$, a better exponential stability result has been established in Theorem 5.6.

5.4 Numerical Simulation

For KdVB system (5.1) with $\epsilon = 0.02$, $\nu = 1$ and $z_0(x) = 0.5\sin(2\pi x)$, the finite difference method is utilized with the steps of space and time $dx = 10^{-1}$ and $dt = 10^{-5}$. The open-loop system is unstable (see Fig. 5.1a).

5.4.1 Simulation of ADRC Approach

Suppose that $d(t) = \sin(8\pi t)$. Choose $\Delta = 0.1$, $K = 12$ and $b = 0.01$. Consider the ETM (5.27) with $\alpha = 0.05$ and $\alpha_0 = 0.0001$. Here two types of ESOs are discussed:

Case 1: ESO (5.5) with the time-varying high gain

$$r(t) = \begin{cases} 50 + 100t, & 0 \le t \le 0.5, \\ 100, & t > 0.5. \end{cases} \tag{5.56}$$

Closed-loop responses are described as Fig. 5.1. Figure 5.1b and d show that the closed-loop system is stable under continuous-time or event-triggered ADRC. The proposed ESO (5.5) is valid as Fig. 5.1c and e. Figure 5.1f describes the event-triggered release times of $U_1(t)$.

Case 2: ESO (5.34) with the constant high gain $b = 0.01$. Figure 5.2 is the closed-loop responses. Figure 5.2a and c illustrate that both continuous-time ADRC and event-triggered ADRC can stabilize the KdVB system. Figure 5.2b and d indicate that the proposed ESO (5.34) is effective to estimate the disturbance. Table 5.1 shows that the peaking values of $\hat{d}_1(t)$ and $\hat{d}_2(t)$ can be reduced by introducing the time-varying gain.

Figure 5.3 depicts the $\mathcal{L}^2$ norm of uncontrolled system subject to disturbance and its controlled version by ADRC. Even with a fast decayed disturbance $d(t) = e^{-2t}$, a control is needed to stabilize the system (see Fig. 5.3b).

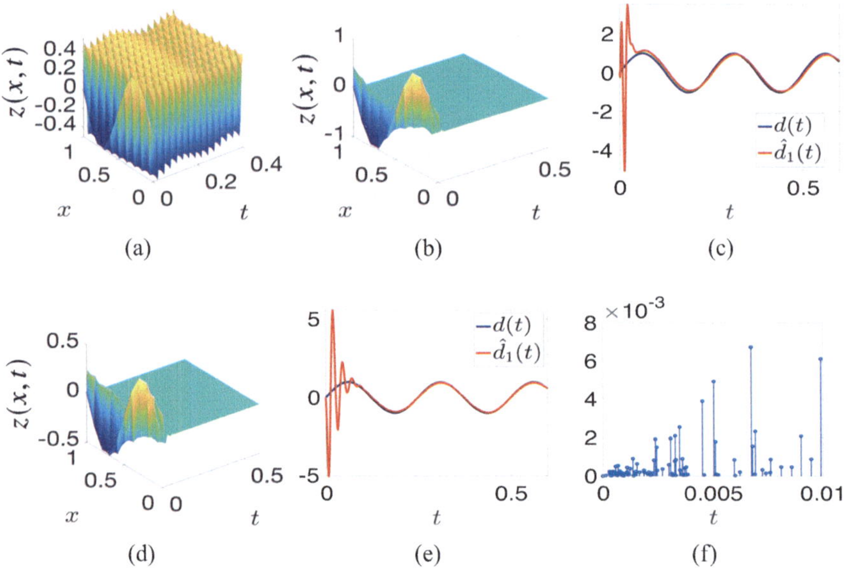

Fig. 5.1 Open-loop and closed-loop responses via ADRC with the time-varying gain (5.56). **a** State of the open-loop system; **b** State $z(x, t)$ under continuous-time ADRC (5.14); **c** $d(t)$ and $\hat{d}_1(t)$ under continuous-time ADRC (5.14); **d** State $z(x, t)$ under event-triggered ADRC (5.26); **e** $d(t)$ and $\hat{d}_1(t)$ under event-triggered ADRC (5.26); **f** Event-triggered release times of $U_1(t)$

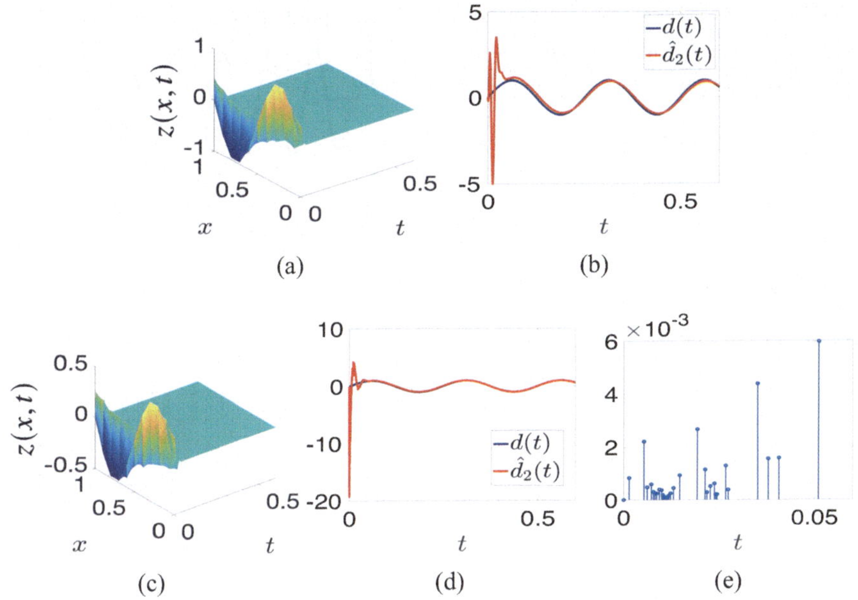

Fig. 5.2 Closed-loop responses via ADRC with the constant gain $b = 0.01$. **a** State $z(x, t)$ under continuous-time ADRC (5.38); **b** $d(t)$ and $\hat{d}_2(t)$ under continuous-time ADRC (5.38); **c** State $z(x, t)$ under event-triggered ADRC (5.39); **d** $d(t)$ and $\hat{d}_2(t)$ under event-triggered ADRC (5.39); **e** Event-triggered release times of $U_1(t)$

Table 5.1 Peaking values of $\hat{d}_1(t)$ and $\hat{d}_2(t)$

	Constant gain	Time-varying gain
Continuous-time	8.514	4.469
Event-triggered	23.506	10.619

Consider the time-varying high gain $r(t) = e^t$ and the noise corrupted measurements

$$y_j(t) = \frac{1}{\Delta_j} \int_{x_{j-1}}^{x_j} z(\zeta, t) d\zeta + \varepsilon(t), \ \ j = 1, \ldots, N_s$$

where $\varepsilon(t)$ represents the noise. Figure 5.4 describes the performance of ESO (5.5) with $\varepsilon(t) = e^{-2t}$ and $\varepsilon(t) = 0.1 \sin(10t)$ respectively. As expected, the simulation results indicate that the ESO is effective if the derivative of $\varepsilon(t)$ converges to zero(see Fig. 5.4a), otherwise the ESO is inapplicable (see Fig. 5.4b).

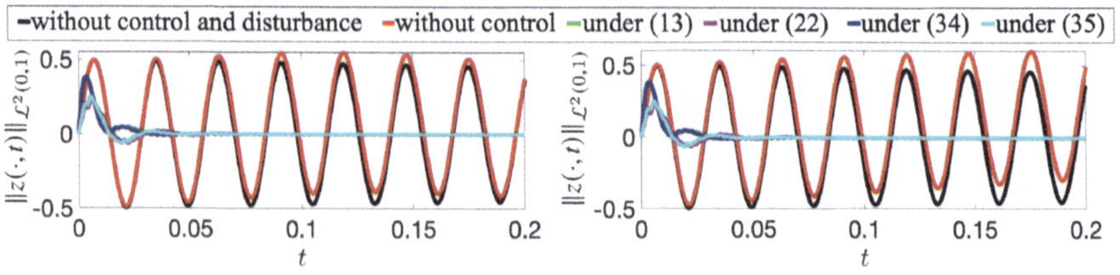

(a) $d(t)=\sin(8\pi t)$ (b) $d(t)=e^{-2t}$

Fig. 5.3 $\|z(\cdot,t)\|_{\mathcal{L}^2(0,1)}$by ADRC approach

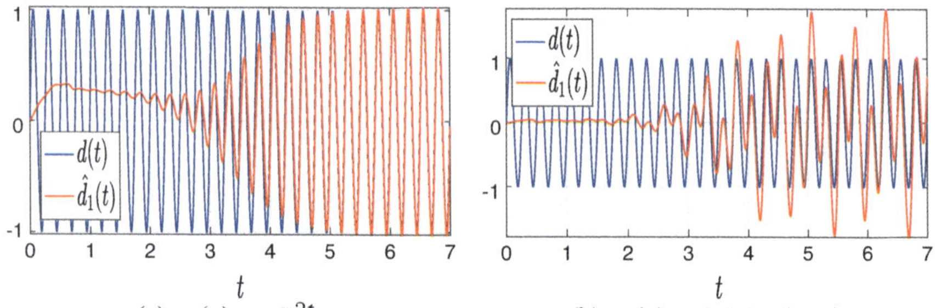

(a) $\varepsilon(t)=e^{-2t}$ (b) $\varepsilon(t)=0.1\sin(10t)$

Fig. 5.4 $d(t)$ and $\hat{d}_1(t)$

5.4.2 Simulation of DOBC Approach

Suppose that the disturbance $d(t)$ is generated by the exogenous system (5.43) with the matrices $C=\begin{bmatrix}3 & 67\\ -15 & 2\end{bmatrix}$, $D=[2\ \ 0.1]$ and the initial condition $w(0)=[0.6\ \ 0.8]^T$. Choose the gain matrix as $L=[19\ \ 6]^T$. We design the DO (5.44). Consider the improved ETM (5.52) with $\alpha=0.05$, $\alpha_0=0.0001$ and $\omega=8$. Closed-loop responses are given in Fig. 5.5. Under the continuous-time DOBC (5.45)/event-triggered DOBC (5.51) subject to (5.52), closed-loop system is stable shown in Fig. 5.5a and c. Figure 5.5b and d demonstrate that the proposed DO (5.34) is effective for disturbance estimation. Figure 5.5e shows the event-triggered release times of $U_1(t)$. Figure 5.6 describes the $\mathcal{L}^2$ norm of the uncontrolled system subject to disturbance and its controlled version by DOBC approach.

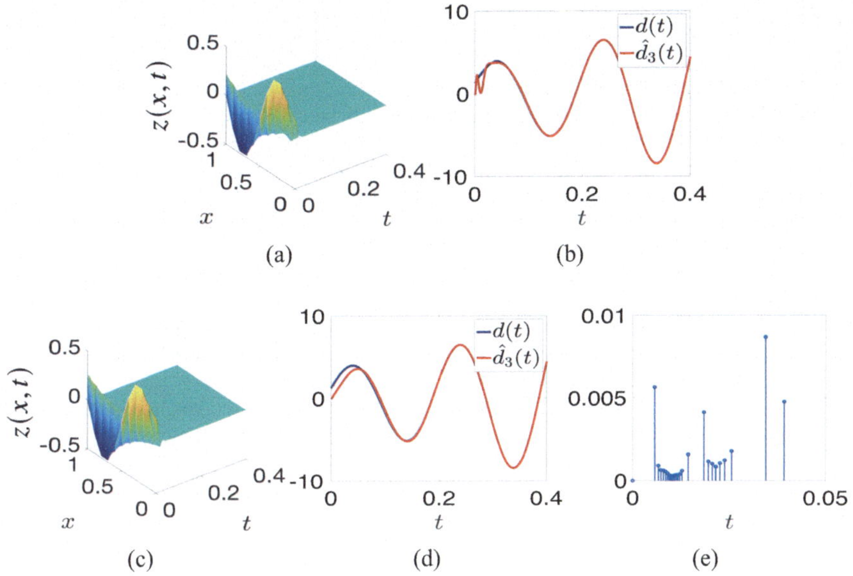

Fig. 5.5 Closed-loop responses via DOBC. **a** State $z(x, t)$ under continuous-time DOBC (5.45); **b** $d(t)$ and $\hat{d}_3(t)$ under continuous-time DOBC (5.45); **c** State $z(x, t)$ under event-triggered DOBC (5.51); **d** $d(t)$ and $\hat{d}_3(t)$ under event-triggered DOBC (5.51); **e** Event-triggered release times of $U_1(t)$

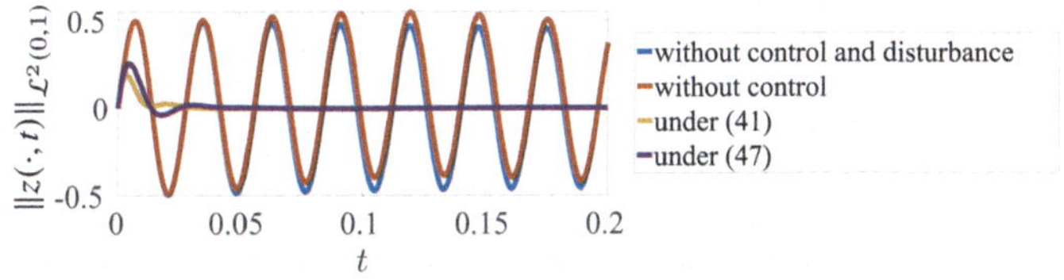

Fig. 5.6 $\|z(\cdot, t)\|_{\mathcal{L}^2(0,1)}$ by DOBC approach

5.5 Notes

This chapter investigates the anti-disturbance control design of KdVB model in the presence of an external unknown disturbance occurring in the control input channel. The stabilization of KdVB equation in presence of a space-dependent disturbance has been considered in [7], which inspires us to investigate the robustness of ADRC/DOBC with respect to the space-dependent disturbances for PDE systems, such as KdVB, Navier–Stokes–Voight and wave equations.

References

1. Guo, B.Z., Jin, F.F.: The active disturbance rejection and sliding mode control approach to the stabilization of the Euler-Bernoulli beam equation with boundary input disturbance. Automatica **49**(9), 2911–2918 (2013)
2. Zhou, H.C., Guo, B.Z.: Stabilization of ODE with hyperbolic equation actuator subject to boundary control matched disturbance. Int. J. Control **92**(1), 12–26 (2019)
3. Lions, J.L., Magenes, E.: Non-homogeneous Boundary Value Problems and Applications. Springer, Berlin (1972)
4. Temam, R.: On the theory and numerical analysis of the Navier-Stokes equations. Lecture Note # 9 (1973)
5. Evans, L.C.: Partial Differential Equations, vol. 19. American Mathematical Society (2022)
6. Larkin, N.A.: Korteweg-de Vries and Kuramoto-Sivashinsky equations in bounded domains. J. Math. Anal. Appl. **297**, 169–185 (2004)
7. Kalantarov, V.K., Titi, E.S.: Global stabilization of the Navier-Stokes-Voight and the damped nonlinear wave equations by finite number of feedback controllers. Discrete Contin. Dyn. Syst.-B **23**(3), 1325–1345 (2018)

The manufacturer's authorised representative in the EU is Springer Nature Customer Service Centre GmbH, Europaplatz 3, 69115 Heidelberg, Germany. If you have any concerns regarding our products, please contact ProductSafety@springernature.com

Printed and bound by CPI Group (UK) Ltd, Croydon, CR0 4YY

01/10/2026

02225922-0005